Constructing Regional Security

Constructing Regional Security

The Role of Arms Transfers, Arms Control, and Reassurance

William J. Durch

A Century Foundation Book

palgrave

First published 2000 by
PALGRAVE™
175 Fifth Avenus, New York, N.Y. 10010 and
Houndmills, Basingstoke, Hampshire, England RG21 6XS.
Companies and representatives throughout the world.

PALGRAVE™ is the new global publishing imprint of St. Martin's Press LLC Scholarly
and Reference Division and Palgrave Publishers Ltd (formerly Macmillan Press Ltd).

ISBN 978-1-349-63107-0   ISBN 978-1-137-08052-3 (eBook)
DOI 10.1007/978-1-137-08052-3

Library of Congress Cataloging-in-Publication Data
Durch, William J.
Constructing regional security : the role of arms transfers, arms control, and reassurance
/ by William J. Durch.
     p.   cm.
   Includes bibliographical references and index.

   1. National security.   2. Arms transfers.   3. Arms control.   4. Military assistance.
5. United States—Military policy.   6. World politics—1989-   I. Title.
UA10.5.D87   2000
355'.032—dc21

                                                            00–041726

A catalogue record for this book is available from the British Library.

Design by Letra Libre, Inc.

First edition: December 2000
10   9   8   7   6   5   4   3   2   1

The Century Foundation, formerly the Twentieth Century Fund, sponsors and supervises timely analyses of economic policy, foreign affairs, and domestic political issues. Not-for-profit and nonpartisan, it was founded in 1919 and endowed by Edward A. Filene.

Contents

List of Figures

List of Tables

Foreword

Richard C. Leone

There have been two great eruptions of arms production in the United States. The first was essentially coterminous with the Civil War; the second began with the outbreak of World War II and, despite the end of the Cold War, continues to this day. For more than a half century now, America has been committed to global arms superiority. In effect, we have become a nation in arms. Entire generations of citizens have grown up thinking this military commitment is normal, perhaps unaware of how extraordinary it is in historical terms. It is, moreover, one of the few areas of public activity—and it is a very expensive one at that—that is not a political battleground.

The reasons for this commitment to maintaining our military strength can be found in our determination to continue to fill our international role as the great twenty-first-century power with far-flung interests and responsibilities, a role that justifies our continuing to arm others even though the direct Cold War threats to our nation have been significantly reduced. And the lack of political debate over the expense involved has much to do with the contemporary insistence that, when it comes to armaments, our boys (and now girls) in uniform must have the best.

The result is that U.S. spending on arms research, development, and procurement far exceeds that of any other nation. Indeed, it exceeds the spending of all other nations combined. This focus results in more and better weapons, weapons that are so good that they are in demand. It also underpins the scale of our transfers of arms to others. Arms sales help keep down the cost of defense in part by amortizing some of the costs involved in creating these weapons in the first place, and weapons sales are "good" for the American companies in this business, good as well for their shareholders and workers. If the products were not mostly lethal, we might describe the whole process as just another virtuous cycle of capitalism.

During the Cold War, there was a well understood strategy behind the arms transfers from the center of the two major blocs to allies and to the nations in the middle. One helped—that is to say, armed—one's friends so that they could successfully compete with those who were being armed by the other side. It was not impossible that some of the arms would wind up being used against your side—weapons are obviously subject to capture, illicit sale, and a change of government. But such risks were acceptable given the fact that the "survival of the Free World" was at stake. Among the many questions about arms sales and transfers today is the basic issue of whether a balanced analysis of the risks and rewards still favors continuation of the practice.

In this book William Durch, senior associate of the Henry L. Stimson Center, addresses those questions with specific emphasis on U.S. policy. There are good reasons for this focus; the United States is, by far, the dominant player in the arms trade since World War II. Durch contends that regulating the commerce in conventional arms—the tanks, fighter aircraft, munitions, and artillery that fuel wars around the world—has always been a subordinate priority for policymakers. This is a paradox. For it is these weapons—not the weapons of mass destruction so often discussed in current policy debates—that have been used in hundreds of regional conflicts in the past few decades and which account for the millions of casualties that have resulted. Nevertheless, the regulation of the arms trade remains an orphan of international diplomacy.

Durch also makes a compelling case that as the few efforts to address this security challenge have failed, those failures, in turn, encourage fatalism among governments about their ability to make a difference. What accounts for this failure? A central thesis of this book is that it was the inevitable product of the neo-colonial, mechanistic approaches adopted by industrial countries to control arms shipments to smaller states. "Recent efforts to contain conventional arms proliferation have shared key characteristics with Prohibition," Durch argues. "A top-down, supply-oriented policy" has ignored the root causes and security dilemmas that account for the continued demand for arms around the world.

This book makes a unique contribution to international security literature by focusing on the need to understand and redefine the security interests of both buyers and sellers of arms as a prerequisite of effective policy. It also provides a comprehensive analysis of the history of the arms trade phenomenon, not only capturing the problems and challenges but also offering practical recommendations for future action.

The Century Foundation has, for more than eight decades, developed projects, conducted research, and published reports on American foreign policy issues. Durch's book is part of a continuing series of publications relating to foreign and security policy after the Cold War, including Richard Ullman's *Securing Europe,* Tony Smith's *America's Mission,* Michael Mandelbaum's *The Dawn of Peace in Europe,* Jonathan Dean's *Ending Europe's Wars,* John Ruggie's *Winning the Peace,* as well as ongoing work by Walter Russell Mead, Henry Nau, and Gregory Treverton.

The Century Foundation is grateful to William Durch for his careful analysis of this difficult and increasingly troublesome aspect of our foreign policy.

Richard C. Leone, President
The Century Foundation
June 2000

Preface and Acknowledgments

Two foundations and many individuals made this book possible. In 1992, the then-Twentieth Century Fund (now The Century Foundation; TCF) accepted the proposal to reexamine the source of demand for conventional armaments in the developing world that resulted in this book. The Gulf War was a very recent memory that had sparked arms buyers' interest in the "revolution in military affairs." The end of the Cold War and the conventional arms control agreements in Europe that accompanied it caused a cascade of heavy armaments from Europe's central plains to the North Atlantic Treaty Organization's Mediterranean tier, as well as sales of surplus weapons to developing states at a few cents on the dollar. At the same time, those governments able to pay for cutting-edge weapons could obtain higher technology and greater lethality than at any time in history. Plunging defense budgets in the developed industrial democracies and the transitional economies of the former Soviet bloc sent defense firms scrambling for overseas sales. It seemed a good time to take another look at the arms market, its impacts on regional security, and ways to mitigate those impacts that went beyond the usual politics of supply-side arms control. Richard C. Leone, president of TCF, agreed that it was time to revisit the sources of demand for conventional armaments, and TCF generously supported the project.

In addition, to complement a synoptic examination of the arms trade, the Henry L. Stimson Center, my intellectual home, where I am a senior associate, commissioned short studies of key countries, with the welcome support of Sally Lilienthal and the Ploughshares Fund, the second foundation whose support made this work happen. The Ploughshares grant funded three Stimson Center occasional papers, the findings from which are woven into chapter 3 of this volume: Duygu Bazoglu Sezer examined the political and security interests and policies of Turkey, which has long played a pivotal role at the crossroads of Europe and Asia. Ahmed Salah Hashim evaluated the national security policies of postrevolutionary Iran, while Shekhar Gupta and Chris Smith analyzed the nuclear and conventional dimensions of security, respectively, in India and Pakistan. I would like to thank them once again for their gracious assistance and their keen insights.

Several researchers at the Stimson Center helped to pull together the voluminous data that supports the analyses in this book, including Jonathan Henick, Erica Warner, Jill Junnolla, and Marya Zamindar. I would also like to thank the leadership of the Stimson Center, Michael Krepon and Barry Blechman, for their support of this project and, above all, for their patience. Barry Blechman and Kevin O'Prey were also good enough to read early manuscript drafts and offer valuable advice.

I would like to thank the people at TCF who at one time or another were instrumental in moving this project toward completion, including Harry Ozeroff, Greg Anrig, David Callahan, and Morton Halperin. I would especially like to thank the final team of TCF editors, Betsy Feist and Jason Renker, for their invariably thought-ful and incise critiques, which on more than one occasion provoked serious re-structuring that greatly improved the final product. I would also like to thank Karen Wolny and the editorial team at Palgrave for taking the whole project over the fin-ish line. As always, of course, any remaining errors or omissions are the sole respon-sibility of the author.

Long-running writing projects can test the patience of Job, not to mention Job's spouse. So, finally, I would like to thank my wife, Jane, for her great patience and sup-port of this particular effort, and I would like to dedicate this book to her.

WJD
Arlington, Virginia
March 2000

Introduction

For nearly a half-century, the Cold War's major contestants and their principal allies poured conventional arms into the developing countries of Africa, Asia, and Latin America in a competition to reinforce friends, outflank enemies, and make a few francs, pounds, rubles, or dollars while doing so. These are the regions in which the great, growing, and largely unhappy majority of humanity lives, regions in which grievance is rife and governance often unstable, and in which most of the regional wars of the last half-century have been fought. Conventional arms—fighting vehicles, warships, combat aircraft, and missiles—are the tools of regional conflict. The international trade in conventional arms has declined from its Cold War peak but remains a $30 billion business, most of it conducted with developing states.

After every recent regional war there has been increased interest, both official and academic, in the role that arms transfers have played in making conflict possible and sustaining it once it has started. But during war, attention more often has focused on how arms can be used to help our guy win.[1] The eight-year (1980–1988) war between Iran and Iraq, for example, attracted arms from all points of the compass, supplier regrets being registered after the fact, if at all. While the war was underway, *realpolitik* and economics ruled decision-making. Efforts to constrain arms flows to the combatants consisted largely of Washington's Operation Staunch, directed exclusively at Iran, which Washington itself violated in attempting to free U.S. hostages in Lebanon. Iraq, though the initial aggressor, attracted far more international support because it served as the West's de facto proxy in containing Iran's Islamic revolution. Even the near-sinking of the USS *Stark*—with heavy casualties—by an errant Iraqi missile failed to bring American wrath down upon Baghdad. Nor were Washington's relations with Paris soured by the fact that French-supplied missiles burned the *Stark*'s crew and melted its superstructure. Politics dictated that all such emotions be directed toward Tehran as the ultimate cause of all grief because its fervid clerics refused to stop fighting.

When Saddam Hussein used his armies to invade Kuwait in 1990, both the international reaction and the outcome were of course much different. Pointed in an inappropriate direction, Iraq's arms became a major threat to Western interests and Saddam Hussein became the new fount of all evil, which he remained as the decade closed, at least to American policymakers. Because Saddam's attack on Kuwait appeared to threaten Western oil interests directly, the UN Security Council's five permanent members, who are also the globe's top arms suppliers, unanimously backed an arms embargo that prevented Iraq from further building up its forces while the U.S.-led international coalition arrayed against it was preparing for battle.

Capitalizing on this unaccustomed unanimity, the "Perm Five" gathered briefly after the war at the behest of President Bush to talk about restraining future weapon sales to the Middle East and Persian Gulf. The discussions lasted less than a year, however, stalling when the Chinese walked out over U.S. sales of modern combat jets to Taiwan. But even while the Perm Five met to discuss restraint, the value of American arms contracts in the Persian Gulf soared. Every oil-fired state in the region wanted to own some of the late war's winningest weapons, and then-recession-strapped America and its cutback-stressed defense industry could not say no.

Earlier supply-side restraint efforts during the Carter administration failed for similar reasons: good intentions were overwhelmed by "contracts in the pipeline" on the one hand, and by regional and Cold War politics on the other. I experienced that particular policy failure firsthand as an analyst with the American delegation to the U.S.-Soviet Conventional Arms Transfer talks. It struck me then, and again after the Gulf War, that the supply-side approach to managing conventional arms proliferation and regional conflict cannot succeed unassisted. Washington would never tolerate third parties dictating the means by which to defend the United States, and other states can be expected to be no less resentful of externally imposed constraints on their own defense acquisitions.

This study arose, then, both from the failure of the Bush supply-side initiative following the Gulf War and from a prior recognition of the basic contradictions and limitations inherent in supply-side measures to control the spread of conventional arms. The work has been several years in gestation, and in that time has evolved from an assessment of the sources of demand for arms to include assessment of regional efforts to promote security and stability, with the objective of combining supply- and demand-side restraint measures in mutually reinforcing fashion.

Recent efforts to contain conventional arms proliferation have shared key characteristics with Prohibition, a top-down, supply-oriented policy that brushed aside demand as either irrelevant or too hard a problem to tackle directly.[2] After Prohibition caused more problems than it solved, government abandoned it and developed alternative schemes to manage a public policy issue that would not go away, while penalizing abuse. Conventional arms, similarly, are not going to go away. They are far too useful—even necessary—to states. Moreover, like alcohol, they have some positive traits: while injurious if abused, they are protective when consumed in moderation; while dangerously addictive to a few, they are less so to the many; while capable of causing death and destruction, they are equally capable of deterring war, unused for years at a time. Conventional arms, in other words, present an issue to be managed, not an evil to be banished.

So what does it take to manage an arms trading system in the interests of regional peace? First, it takes as many states as are closely connected by geography, rivalry, and/or recent conflict history. British scholar Barry Buzan has dubbed such groupings "regional security complexes."[3] Second, it takes the cooperation of that region's arms suppliers. Both steps require redefinition of the security interests of recipients and suppliers, and willingness on the part of the latter to place longer-term security interests ahead of short-term economic considerations—that is, willingness to forego sales.

South America provides an example of a policy ideal to work toward. Its member states have a regional arms import restraint regime in the works. Canadian scholar K. J. Holsti goes so far as to argue that something like a "no-war zone" is emerging there, accompanying widespread, if still fragile, democratic governance.[4] Such a fledgling security regime, mirroring regimes in North America and Western Europe, is eminently worthy of outside support.

Encouraging a security complex to move toward arms import restraint and resolution of its outstanding political conflicts is much more work than imposing an arms embargo. Steps toward peace in regions of tension can and usually must be partial in nature, although not all positive steps involve holding back armaments. When regional rogues threaten the peace, shipping arms to would-be victims may be one way of shoring it up—two steps forward, one step back—as may sending in outside troops and equipment. The United States and Great Britain have done that with some regularity in recent years to enforce Iraqi compliance with UN weapon inspection orders. Some combination of regional restraint and external assurances is likely to be necessary to promote regional peace or, where peace is the norm, sustain it at lower cost. This book looks for cost-effective mixes of restraint and reassurance that might promote regional stability without undermining the security of those states and peoples who adopt the policies of restraint.

In the first half of 1999, there was one major interstate war underway, between Ethiopia and Eritrea, over a patch of barren borderland on which was focused all the pride and much of the stubbornness in those two states, one only recently independent of the other. At the same time, the complex and deadly politics of central Africa involved many of its states in desultory battles over who would ultimately rule how much of what is currently called the Democratic Republic of Congo. Precisely because there is no guarantee of peace (indeed, more than a dozen internal conflicts continue to burn in developing regions), it behooves those states and institutions with the necessary political and economic resources and a stake in regional stability to bring their resources to bear *in their own interests* to prevent and to manage conflict and, where feasible, to help to resolve it.

In a rapidly changing international environment, the United States has a particularly strong and abiding interest in "peace, prosperity, and freedom" abroad.[5] Expanding that phrase a bit, the United States has an interest in an international system of legitimately governed states and strong societies, at peace internally and with their neighbors, that are able to protect their own territory but at the same time are willing to come to the aid of others. Such a system would give American (and other) businesses stable, open international markets to sell and invest in, on the one hand, and give the U.S. and other democratic governments able new partners to help keep the peace, on the other. The United States also has an interest in helping developing countries and regions—especially regions of tension—refocus public spending on the human development (public health, education, and the like) that a globalizing economy demands.

Structure of the Book

This book is divided into two parts: problems and potential solutions. The first three chapters examine the conventional arms trade of the last half-century, the security problems of developing states and how arms transfers may contribute to them, and case studies of the impact of threat perceptions, domestic politics, and state weakness on demand for arms. Chapters 4 and 5 evaluate tools to address these problems—supply-side arms control and regional confidence and security-building measures—while chapter 6 combines what we know about the problem and what we know about existing policy tools and their shortcomings to draw implications for U.S. government policy.

Chapter 1 traces the history and structure of the arms trade and describes the market of the late 1990s: who buys, who sells, what do they sell, and how do they do it? The chapter also traces the diffusion of major weapon systems—combat aircraft, missiles, armored vehicles, and naval systems—to developing regions from 1950 through the mid-1990s, charting not just numbers but also the evolution of capabilities as weapons became technologically more sophisticated.

Chapter 2 looks at the sources of developing states' security problems, the difficulty of building a state in a political goldfish bowl, and the impact of arms transfers on four elements of state and regional stability (economic, political, arms-competitive, and crisis-related). The chapter uses simple but powerful statistical techniques to examine the relations between arms transfers and regional conflict over a sequence of five-year periods, starting in 1970, when foreign military sales contracts really began to skyrocket. It finds a weak positive relationship between arms transfers and external conflict and little relation between transfers and internal conflict.

Since aggregate analysis can paint only part of the picture, chapter 3 gets down to cases, looking at the political, institutional, and regional contexts that drive demand for armaments. The chapter analyzes these contexts for key countries in some of the most conflict-prone zones of the developing world, from Turkey to North Korea. It notes the large asymmetries in size and conventional military capability that create fundamental insecurities in smaller states, which, in more instances than we like to think about have turned to weapons of mass destruction in part to even up the odds. Although security experts emphasize that nuclear deterrence produced a stable standoff between the United States and the Soviet Union, note that this standoff and the fears of military escalation that accompanied it may have led both sides to channel their competitive urges into developing regions. To a degree, the standoff made the world safe for low-intensity conflict on third-party battlefields. Similarly, some students of South Asian security politics see nuclear weapons increasing the sorts of low intensity fighting and third-party raids that consumed the spring and summer of 1999 in the contested province of Kashmir, precisely because they are believed to make conventional military reprisals too risky to contemplate.

While this book is not focused on nuclear weapons or their proliferation, the connection between regional conventional arms imbalances and incentives to acquire weapons of mass destruction is a recurring theme, not just in chapter 3 but in other chapters as well. This dynamic is at play in the Persian Gulf and elsewhere in the Middle East, where nuclear bombs are Israel's ace in the hole. It is also at play in Northeast Asia, where North Korea turned to nuclear arms once it lost its external patrons, and of course in South Asia. By a twist of logic, however, a country may well believe that its nuclear capabilities will be given greater credence if it continues to acquire and deploy conventional armaments as a visible signal of continuing determination to defend its interests. In other words, in the right conditions of asymmetric standoff, conventional arms acquisitions and operations and nuclear arms may well represent linked rather than separate paths to national security for some countries.

Chapter 4 is a critical review of supply-side arms control efforts. It examines both the power-political, utilitarian perspective on arms transfer restraint (that what you sell today you may have to fight tomorrow), and the stability-oriented, normative perspective, which argues against "excessive" and "destabilizing" transfers but has a devilishly difficult time defining either one prospectively. A norm-based approach must continually assess the goodness of recipients and make continuing tradeoffs between state security and human development in deciding what arms to buy or to sell. The chapter critically reviews the Carter restraint policy, the 1991 Perm Five talks, and, among other initiatives, the Missile Technology Control Regime, the Wassenaar Arrangement for controlling technology exports, international arms embargoes, and the proposed codes of conduct for arms sales. Most such supply-side controls historically have been hampered by divergent supplier needs and interests, but a crippling political shortcoming has been their one-way nature—that is, their fundamental lack of political-military reciprocity, a concept at the heart of successful arms control.

Chapter 5 focuses on measures that promote conflict prevention and military transparency, the net effect of which is hoped to be reduction in demand for armaments. The chapter examines the development and limitations of the UN Register of Conventional Arms, as well as the evolution in Europe and subsequent spread to other regions of various confidence and security building measures (CSBMs)—military hotlines, advance notice of military exercises, exchanges of military observers, no-fly zones, and supervised inspection flights—all designed to reduce the risk of war by misperception or miscommunication. Non-governmental organizations (NGOs) have had a surprisingly large role to play in constructing CSBM regimes, especially in eastern Asia. Successful CSBMs, even those aimed at basic conflict avoidance rather than reconciliation, require something other than a purely competitive relationship; otherwise, communications channels may fall into disuse or, worse still, have their utility destroyed by deliberate misinformation.

Chapter 6 draws together the data and analyses of the preceding chapters to lay out a series of initiatives that arms recipients and suppliers might take to support peace and stability in developing state security complexes, initiatives that are only partly conventional-arms-trade-related, because the arms trade is only part of the problem and as often a symptom as a cause: neither arms control nor CSBMs will do much for peace if such fundamental attributes of statehood as national borders are in dispute, for example. When such fundamental problems have been overcome, a security complex may turn to arms restraint as part of a package of tools to promote peace and stability, and supplier countries should be agreed in advance of such initiatives not to undercut them. Supply-side arms control efforts might also usefully focus on constraining a few dangerous weapon systems like ballistic missiles with ranges over 100 kilometers, which, the chapter argues, should be subject to a global ban if not already constrained by arms control agreement. In general, however, the main focus of supply-side restraint regimes should be on states, not weapons. The attributes and behavior of states—how they treat their neighbors and how they treat their own peoples—are better indicators of potential future violations of international norms and are more readily observed and measured than such fuzzy concepts as "destabilizing" accumulations of arms. An international restraint regime based on a code of conduct for transfers could be implemented in phases, keying on such behavior initially to determine eligibility for arms or related technology. The chapter closes with observations and implications for U.S. policy.

CHAPTER 1

Elements of the Arms Trade
and Patterns of Proliferation

In the half-century between the end of World War II and the mid-1990s, more than 150,000 major conventional weapons worth at least $750 billion flowed into the developing world in patterns of supply and demand that changed decade by decade.[1] This chapter introduces the suppliers, the recipients, and the weapon systems involved in that trade, traces its patterns from 1950 to 1997, and examines in some detail, going beyond simple bean counts, the increasing sophistication of weapons transferred to developing countries and how quickly they reached this export market. First-line weapons have been reaching politically favored or militarily threatened "early adopters" among developing countries since the 1950s, but the number of developing countries eligible for cutting-edge equipment has greatly increased over time (as, for that matter, has the number of developing countries). The overall flow of arms also increased substantially, from less than $10 billion a year in the mid-1960s, during the height of the Vietnam War, to between $50 billion and $60 billion a year in the mid-1980s, measured in constant 1995 dollars. Note that the *range of uncertainty* in the numbers for the 1980s is equal to the *entire* annual value of the arms trade in the 1960s, reflecting both limitations in the available data and the decisions that different providers of data make about what to count as part of the arms trade and how to assign value to it. These differences are a principal reason why, in this book, after these introductory remarks, I assess the arms trade mostly in terms of actual weapons transferred and the relative sophistication of those weapons, rather than in terms of dollar values. Given the limitations of the data in this field of research, readers should become accustomed to thinking of such values as ballpark figures rather than precise numbers, whatever their source. (For more on the data and dollar value problem, please see appendix B.)

Later chapters will look more closely at the sources of demand for arms in the developing world but in this chapter, as we trace the 50-year diffusion of modern weapon systems into the developing world, I will note events and motivations most likely associated with key arms transfers to round out the narrative for readers who prefer a little meat on their Terminator's bones.[2] Before getting into the details, however, I want to introduce the constituent parts of the trade and the distinct phases through which it has passed in five decades, from essentially a soup kitchen for governments to an outlet mall for manufacturers. The chapter closes with a brief recap on how the arms trade has evolved in the second half of the 1990s.

The Elements of the Arms Trade

At its most basic level, the arms trade, like any other, needs a seller, a buyer, an object or service, and a transaction. Sellers have mostly been states, or corporations acting under state license. The buyers that we track here are also states, although opposition/insurgent groups in a number of states have gained access to at least some types of heavy weapons. The objects of interest are major conventional weapon systems, new and used—combat aircraft and helicopters, armored ground vehicles, naval surface ships and submarines, and guided missiles—and the training, maintenance, and upgrade services that keep them functioning. Transactions vary from retail sales to outright gifts. Each element is treated in its own section below.

The existence of so much secondhand or otherwise discounted equipment in the arms trade compounds the problem of assigning monetary value to it. The several standard published sources of data on international arms transfers assign differing valuations. We will look at how their reporting differs and the degree to which those differences matter.

The Arms Suppliers

For most of the postwar period (that is, since World War II), the United States dominated the global arms trade. For that reason, much of the policy focus in this book and the emphasis in discussing the various phases of the arms trade will be on the United States. It has been far from the only player in that trade, however, and for one roughly 10-year period from the late 1970s to the late 1980s, it was second to the Soviet Union in the export of major armaments. The United Kingdom, Germany, France, Italy, and the Netherlands have been the leading western European arms suppliers, while China, Czechoslovakia, and Poland were the top sellers among communist states, after the USSR. Table 1.1 shows the rankings of the top twenty suppliers, according to data published by the Stockholm International Peace Research Institute (SIPRI), for the 1980s and then for the 1990s. Note how the market positions of some countries (Brazil or Austria, for instance) plummeted in the later period, while the rankings of the top eight suppliers changed very little. This change reflects in part the decrease in the 1990s of major interstate war. It also reflects the top suppliers' ability to provide the latest technology in what became, for a time, a buyers' market after the end of the Cold War. Suppliers that could not reliably provide cutting-edge technology saw the inflation-adjusted worth of their export sales drop by as much as 90 percent. However, table 1.1 shows data for transfers to all recipients (SIPRI stopped publishing individual supplier data differentiated by market in 1994), and percentage changes for sales to developing countries alone could be different.[3]

Why do countries supply arms internationally? During the Cold War, U.S. and Soviet motivation was primarily political. The United States concentrated first on building up capable allies from states that had been badly damaged in World War II, distributing among them some of the enormous stockpile of ships, tanks, trucks, artillery, mortars, and rifles left over from that war and no longer needed by now-smaller but rapidly modernizing postwar American forces. (Europe and East Asia were the two regions where the early Cold War was most intense. Soviet troops occupied central and eastern Europe, there was open warfare in Korea, an uneasy standoff across the

Table 1.1 Leading Arms Suppliers to All Countries, 1980–1989 and 1990–1997

	Values 1980–89	Values 1990–97	Rank for 1980–89	Rank for 1990–97	Change in Rank	Change in Value
USA	116,549	99,616	2	1	1	–15%
Russia/USSR	167,225	38,664	1	2	–1	–77%
UK	21,931	15,769	4	3	1	–28%
Germany	14,812	14,308	7	4	3	–30%
France	34,226	14,106	3	5	–2	–59%
China	20,909	7,756	5	6	–1	–63%
Netherlands	5,210	3,556	9	7	2	–32%
Italy	11,295	3,108	8	8	0	–72%
Spain	2,198	1,912	12	9	3	–20%
Canada	1,736	1,866	16	10	6	7%
Switzerland	1,746	1,737	15	11	4	0%
Israel	2,608	1,735	11	12	–1	–33%
Sweden	2,328	1,570	13	13	0	–33%
Czechoslovakia	17,552	1,536	6	14	–8	–91%
Ukraine	0	1,467	n/a	15	n/a	n/a
Czech Republic	0	1,149	n/a	16	n/a	n/a
North Korea	911	883	22	17	5	–3%
Poland	3,779	741	10	18	–8	–80%
Australia	226	737	36	19	17	227%
Yugoslavia	345	730	28	20	8	112%
Brazil	2,027	405	14	28	–14	–80%
Austria	1,599	111	17	42	–25	–93%
Romania	1,543	162	18	36	–18	–89%
East Germany	1,173	197	19	32	–13	–83%

Note: SIPRI Trend Indicator Values, converted to millions of 1995 U.S. dollars using U.S. Dept. of Defense deflators (1995 = 1990/.8639).
Source: SIPRI Arms Transfers Project, updated July 17, 1998. Internet: http://www.sipri.se. U.S. Department of Defense, Office of the Comptroller, *National Defense Budget Estimates for FY 1999.*

Taiwan Strait, and communist/nationalist revolt in French-held Indochina, which led eventually to America's own long military involvement in Vietnam.)

The Soviet Union first had to recover from the devastation of World War II, firm up its grip on central Europe, and match American strategic nuclear power. Before it finished that latter task in the late 1960s, its major forays as an arms supplier were relatively few: major support for the North during the Korean War; arms for President Sukarno's pugnaciously nationalistic regime in Indonesia; and, closely tied to its greatly inferior position in nuclear arms, the substantial arming of Fidel Castro's Cuba from 1961 to 1962 as preface to the Cuban Missile Crisis. In the latter half of the 1960s, as American forces poured into Southeast Asia, Soviet arms deliveries and technical assistance to North Vietnam escalated dramatically, as did assistance to Israel's Arab adversaries.[4]

Moscow's satellite allies in eastern Europe collaborated in the effort to build eastern bloc political influence in the developing world. For example, Czechoslovakia (which broke into its two constituent, eponymous parts in 1993) was a significant supplier of heavy armor to India, Syria, Iraq, and Iran (although the latter effort, to the Ayatollah Khomeini's regime, was probably much more opportunistic than political in nature).

America's European allies were always much more economically motivated. The United Kingdom (Great Britain), whose arms industries survived World War II, was a major arms merchant in the 1950s, soon joined by France, Germany, and Italy as they rebuilt their arms industries. They did not sell across the Iron Curtain, and German law forbade sales in "regions of tension," but otherwise the allies applied fewer political litmus tests than did the United States when buyers had money to put on the table. With much smaller internal markets for their arms and industries that were often government-owned, these and other arms producers had to find external markets to increase the size of their production runs and spread the costs of development over a larger number of units so that their own armed forces could afford the weapons they produced. France, in particular, used exports of its long-lived series of Mirage jet fighters to keep its military aircraft industry solvent despite a domestic market far too small to support economical production. Between 1973 and 1992, France exported 60 percent of its fighter production, while the USSR, United States, and UK exported 35 to 40 percent.[5] Consortia of European arms makers also found export markets for products like the Sepecat Jaguar and Panavia Tornado strike aircraft, and sales of Dutch and German submarines have been relatively brisk. Nonetheless, in the post–Cold War world, sales opportunities are down in a market in which flat oil prices cap the acquisitions of most oil producers; in which newly industrialized countries (NICs) have relatively small economies, modest defense shopping lists, and, most recently, serious economic troubles; and in which the larger NICs like South Korea and Taiwan have large domestic arms industries and limited appetites for complete imported weapon systems other than aircraft.

The smaller suppliers in table 1.1 who were not in the Soviet orbit also had largely economic interests in arms sales, including a notion of defense industries as stepping stones to general industrial development. Their big opportunity came in the 1980s with the Iran-Iraq conflict, a grinding war of attrition that used up large quantities of basic arms and ammunition and sought resupply wherever it could be found. When it ended, a substantial market for smaller, apolitical suppliers of such goods dried up. Three years later, the Gulf War demonstrated how readily basic arms could be chewed up by cutting-edge weapons and well-trained forces, and the smaller suppliers' market opportunities shrank still further. Rather than stepping stones to greater good fortune and high technology, their defense industries became economic liabilities.

After the end of the Cold War, in the United States and elsewhere, those who could make export sales used them to help cushion and pay for defense industry downsizing. There was never any expectation that they could substitute for domestic arms procurement. The U.S. Defense Department's annual procurement budget is about $40 billion. The annual fraction of U.S. foreign military sales attributable to weapons procurement is $6 to 7 billion. Even doubling exports could not sustain anything like the Cold War industrial base.[6]

U.S. defense industry overcapacity was cut by corporate buyouts, mergers, and slashed payrolls. In the mid-1980s, for example, there were ten U.S. producers of military aircraft, but by the mid-1990s there were just two: the merged Boeing-McDonnell Douglas and the defense conglomerate Lockheed-Martin. There is just one tank production plant. There are still two shipyards for nuclear-powered warships, but only because the Congress insists upon it.[7] The total number of U.S. firms doing defense-related business at any level fell, by one account, about 75 percent between 1986, a year of peak defense budgets, and 1995. In the same period, the aerospace industry lost nearly half of its defense workforce.[8]

Russia, trying to claw its way back to a leading role in the international system, has just three major hard-currency exports: oil, natural gas, and armaments. Its defense industry—the product of a seven-decade Soviet command economy—was the largest chunk of a Soviet military infrastructure, employing more than seven million people in the late 1980s. When the Soviet Union broke up, defense production was broken up as well, because some elements were located in what became independent states to Russia's south and west. That is why Ukraine is listed in table 1.1 among the top twenty international arms suppliers in the 1990s.

Between 1991 and 1992, Russian government orders for tanks and artillery fell to just 3 percent of their Soviet-era levels and equipment orders in other sectors such as aircraft and missiles dropped by 80 percent.[9] Russian firms compensated somewhat for those lost orders through exports, but at nothing close to levels needed to sustain the huge industrial base. Meanwhile, government restrictions on privatization kept unprofitable plants from being sold off or closed down. As the Russian economy continued to slide, idle defense plant capacity exceeded 40 percent. This excess capacity was carried on the industry's books as overhead costs of "900 percent or more," which "completely nullified Russian industry's comparative advantages, such as relatively low costs for qualified manpower, raw materials, fuel, and power."[10] Russian analysts see privatization, restructuring, and downsizing as necessary to save—indeed, to create—a modern Russian defense production sector. On the other hand, they also tend to see arms exports as a necessary source of support for that restructuring.

The Arms Recipients

Essentially every country in the world, including the United States, imports major weapon systems or components, but in this study we are looking exclusively at the impact of arms imports on the security of developing countries and regions.[11] In the early 1950s, fewer than half of present-day developing countries were politically independent. Although many incorporate very old societies and cultures, as *states* most are post–World War II in origin. Many countries in the Middle East (for example, Jordan, Syria, Lebanon, and Israel) achieved independence only in the 1940s. Few places in sub-Saharan Africa were not under European colonial rule in the 1950s. The UK began to relinquish its African colonies in 1957, France in 1958, Belgium in 1960, Portugal in 1974, and South Africa in 1989.

As a result, the number of countries covered by this study grows with each decade (see figure 1.1). By 1990, 106 developing states (excluding a number of island states that never, or only rarely, imported major weapons) are subjects of this study.

The developing world (or states, or countries, terms that will be used interchangeably here) includes all of the Western Hemisphere south of the U.S.-Mexican border, all of Africa, the Middle East, the Persian Gulf region, and all of Asia and the Pacific save China, Japan, Australia, New Zealand, and the countries spun off from the former Soviet Union. At one point, the World Bank classed some European countries (Spain, Greece, and Turkey) as developing, but I follow SIPRI's example in classifying all European countries as developed or "industrialized," partly for the sake of consistency in historical analyses. China presents something of a dilemma, being both very poor for much of the postwar period but also a nuclear power, a permanent member of the UN Security Council, and an otherwise growing and increasingly powerful presence in international affairs. Unlike SIPRI, I opted not to count China as a member of the developing world but to consider it instead as a kind of class unto itself.

Unlike the case with the top suppliers, there have been major shifts in rankings among the top arms recipients in the last two decades. Saudi Arabia, long associated in many people's minds with lavish weapon shopping lists, was indeed customer number one in the 1990s but only number four in the 1980s, when Iraq imported about twice as much as the Saudis, in the course of fighting its war with Iran (see table 1.2, bearing in mind that the SIPRI numbers presented in the table do not reflect another $1 to 2 billion a year in construction and services provided to Saudi Arabia by the United States in the 1980s).[12] The measured value of India's imports in the 1980s was also far greater than Saudi Arabia's, and the value of Syrian imports was roughly equal. Neither of the latter two states likely *paid* the amounts listed in the table, however, since both were major clients of the Soviet Union and Soviet prices to favored clients were often deeply discounted (see the discussions of data and transactions below).

Table 1.2 Leading Arms Recipients, Developing States, 1980–1989 and 1990–1997

	Values 1980–89	Values 1990–97	Rank for 1980–89	Rank for 1990–97	Change in Rank	Change in Value
Saudi Arabia	17,570	14,137	4	1	3	–20%
India	29,108	9,607	2	2	0	–67%
Taiwan (ROC)	5,263	9,333	13	3	10	77%
Egypt	12,490	9,066	5	4	1	–27%
South Korea	5,151	7,622	14	5	9	48%
Kuwait	1,909	5,202	26	6	20	172%
Thailand	3,484	4,947	22	7	15	42%
UAE	2,703	4,693	24	8	16	74%
Israel	7,606	4,632	7	9	–2	–39%
Pakistan	6,419	4,577	9	10	–1	–29%
Iran	5,985	3,950	11	11	0	–34%
Afghanistan	5,460	3,598	12	12	0	–34%
Malaysia	1,758	3,259	27	13	14	85%
Indonesia	3,649	2,141	20	14	6	–41%
Brazil	1,611	1,812	29	15	14	12%
Syria	17,880	1,791	3	16	–13	–90%
Chile	1,747	1,679	28	17	11	–4%
Algeria	4,906	1,529	16	18	–2	–69%
Singapore	2,083	1,521	25	19	6	–27%
Peru	3,908	1,012	19	20	–1	–74%
Iraq	34,285	750	1	21	–20	–98%
North Korea	6,301	741	10	22	–12	–88%
Angola	7,466	653	8	23	–15	–91%
Argentina	5,010	640	15	24	–9	–87%
Cuba	4,456	353	18	27	–9	–92%
Jordan	4,687	179	17	28	–11	–96%
Libya	12,437	7	6	29	–23	–99%

Note: SIPRI Trend Indicator Values, converted to millions of 1995 U.S. dollars using U.S. Dept. of Defense deflators (1995 = 1990/.8639).
Source: SIPRI Arms Transfers Project, updated July 17, 1998. Internet: http://www.sipri.se.

Figure 1.1   Numbers of Developing Countries, 1951–1995 by Level of Arms Imports

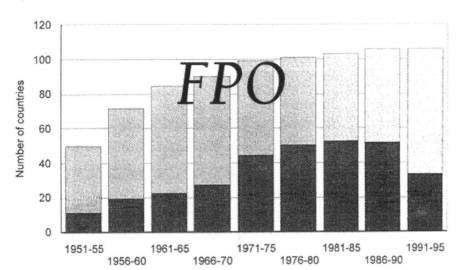

Countries importing less than $100 million in any year of a five year period.
Countries importing at least $100 million in arms in any year of a five year period.

Sources: Laurance, *International Arms Trade*, p. 107. Brzoska and Ohlson, *Arms Transfers to the Third World*, pp. 332–37. U.S. ACDA, *World Military Expenditures and Arms Transfers, 1996*, table II.

The number of developing states importing substantial amounts of armaments (that is, more than $100 million worth in 1995 dollars) grew much faster than the total number of countries, from fewer than one-third of the total to more than one-half by the mid-1980s (see fig. 1.1). By the early 1990s, the proportion of big spenders had trailed off once again to just one-third of the states in the study.

With few exceptions, the inflation-adjusted values of arms imported in the 1990s by leading recipients in the developing world were far below those of the 1980s. The exceptions include countries in East and Southeast Asia (South Korea, Taiwan, Thailand, and Malaysia) that were part of the "Asian economic miracle" that preceded the Asian economic nightmare of late 1997 and 1998. At the other end of Asia, Kuwait and the United Arab Emirates (UAE) escalated their purchases during and after the Gulf War. Iraq's major arms imports were cut off, and its overseas assets were frozen, by United Nations Security Council resolution for its role as the aggressor in that war. The embargoes left substantial unpaid bills for weapons supplied by countries like France and Russia during Iraq's previous war with Iran (1980–1988, which Iraq started as well). The steepest drop in arms imports, however, was Libya's, whose erratic leader, Moammar Qaddafi, habitually imported large amounts of weaponry in the 1970s and 1980s. Like Iraq, Libya was placed under UN arms embargo in March 1992 for its "refusal to hand over to foreign jurisdiction two of its citizens who had allegedly been involved" in the mid-air bombing of Pan Am Flight 103 over Lockerbie, Scotland, in 1988.[13]

Why do developing countries acquire arms internationally? One obvious, only somewhat facetious, answer would be that nobody local sells what they want. Another, more theoretical answer would observe that the state has been a war-making machine from its origin in European history to the present day. The military is therefore the state's most basic institution. Weapons are the tools of that institution's trade. One should therefore expect the state to acquire weapons, and if they can't be made locally they must be imported. In specific cases, the military may need the weapons to secure the state's borders or to deter threatening neighbors. They may also be needed to defend insecure governments facing threats from within. Militaries may undergo periodic force modernization either to keep pace with the neighbors, as a point of institutional pride, or as the price for not themselves posing a threat to government. Inasmuch as military coups have afflicted some developing states with some regularity, the cost of new weapons is a comparatively low price to pay for political stability. The availability of large amounts of relatively cheap weapons from both superpowers during the earlier decades of the Cold War made such strategies even cheaper to implement. Different combinations of these motives help to drive demand for conventional arms in developing countries.

Because smaller countries are relatively dependent on outside sources of arms, the urge to reduce that dependence has been substantial, as has the belief in defense production as the ticket to greater industrialization. The twin desires to reduce dependence on outside arms suppliers and to create higher-technology civilian "spinoff" industries have led a fair number of developing states to seek greater local capacity to build their own arms. The first step in the process is local assembly of weapon system kits to build basic workforce skills and familiarity with the rhythms of modern defense production. Subsequent steps include licensed production of some or all of the parts for a weapon system, starting with lower-technology components and leading in some cases to domestic design, development, and production of a growing fraction of those components in a widening variety of weapons.[14] A handful of advanced developing and newly industrialized countries (Argentina, Brazil, India, Indonesia, Israel, South Korea, Singapore, South Africa, and Taiwan) now produce the great majority of the armaments they require for national use, and most of that production is indigenous rather than licensed by supplier states. The pace of change in technology is such, however, that few such industries have the wherewithal in people or capital to compete in the global arms market, as the trends in table 1.1 indicated.

The Arms Themselves

The international arms trade involves weapons that range from aircraft carriers to handguns, and ammunition that ranges from ballistic missiles to bullets, plus services as diverse as training and technical support, base construction, and system maintenance. As noted earlier, this study looks at a slice of that trade: major conventional weapon systems. SIPRI tracks these items in part because they are more easily trackable from the public information that it uses to build its data bases.[15] They are also the "sharp end" of a military force, its principal killing power. While they will not work (for long) without training, tech support, spare parts, and good routine maintenance, nor be effective in the field without appropriate operational doctrine, competent leadership, and well-structured command and control systems, all of these things together amount to little more than piles of procedures and pointless hierarchy without the weapon systems themselves. Yet only a tiny portion of any military force—by weight or by value—is lethal, that is, actually goes "boom" or pierces targets. The vast majority of weapons technology and its supporting services makes ready, provides intelligence and communications support, and transports and guides the boom to the target. So what SIPRI assigns value to in its arms trade registers, to produce what it calls "trend indicator values," is not only a portion of the arms trade but also an even smaller portion of what it takes to operate and maintain a modern military. Other sources of data on the arms trade are more inclusive in their accounting.

Later in the chapter we will be looking at the diffusion of several categories of weapons into developing countries' armories. These include fixed-wing combat aircraft (divided into heavy and light); military helicopters (missile-armed and transport); armored combat vehicles (tanks, infantry fighting vehicles, armored personnel carriers); surface warships (ocean-going and coastal) and submarines; and guided missiles (including naval cruise missiles).

Heavy, supersonic fighter aircraft and main battle tanks are the glamour weapons of the late 20th century, the ultimate tools of industrialized warfare as it has evolved in the 130 years between the U.S. Civil War (1861–1865) and the Gulf War (1990–1991). They are, in consequence, also status symbols, as are large warships and submarines. (If it's being used right, a submarine can neither be seen nor easily detected, so its status derives less from display than from the skill level required to operate it properly, and from the advantage conferred and threat implicitly posed by any stealthy weapon.) Light attack aircraft are less sexy than larger fighters but cheaper and less complex to maintain, and therefore more useful to many developing countries. Helicopters are much harder to maintain than other aircraft but useful to countries fighting insurgencies (or up country drug production, or natural disasters). Missile-armed helicopters are primary foes of tanks, and guided missiles of all sorts are primary threats to large weapon platforms of all sorts. Surface-to-air missiles (SAMs) target aircraft; shorter range air-to-surface missiles target mobile military installations, armored vehicles, radars and SAM sites; long-range cruise missiles target fixed installations and warships.

The cost of all these items today can be substantial. Whether a buyer pays full price depends on whether the item is new or used (missiles, of course, tend to be new, and hence pricey). Cost also depends upon the nature of the transaction involved: full-price sale, discounted sale, or outright gift. The gift may, of course, require substantial refurbishing and upgrading to be usable, and that work may be done either before or after the transfer. Different countries negotiate different deals from the same supplier, not just different base prices for the weapon system but different armaments or electronics packages (a substantial portion of the value of any new system), different after-sale support and training packages, and so on. Table 1.3 lists a sampling of weapon systems whose sales agreements were inked in the early- to mid-1990s. For all of the reasons just noted, the associated prices should be used merely to get a sense of the order of magnitude costs of modern weapons, from nearly $800 million for a new guided-missile-equipped naval frigate to about $20,000 for the latest version of the TOW antitank guided missile (launcher not included).

This study does not deal with all kinds of conventional arms. It does not deal with crew-served infantry weapons (machine guns and mortars, for example). It does not address small arms (pistols, assault rifles, and the like), which have garnered increasing attention from non-governmental organizations (NGOs), governments, and the United Nations for their destabilizing effects in the aftermath of civil war and because of the ease with which they can be moved by the black market from one country to another. For example, starting in 1986, the United States gave the Afghan mujahedeen

Table 1.3 Rough Unit Costs for a Range of Recent Arms Sales

System Name	System Type	Source Country	Unit Cost (US $ mill)	Comments
Lafayette-class	Frigate	France	$783.0	Fully equipped
Lekiu-class	Frigate	UK	$300.0	
Dolphin-class	Submarine	Germany	$285.0	Diesel-powered
E-2C Hawkeye	Surveillance aircraft	USA	$175.0	
F-15S Strike Eagle	Strike fighter	USA	$125.0	With full armament, targeting pods, spare engines
Patriot	Surface-to-air missile system	USA	$79.0	Complete missile battery
SA-11	Surface-to-air missile system	Russia	$77.0	With 96 missiles
F-16C/D Fighting Falcon	Jet fighter	USA	$49.0	
KC-135 Stratotanker	Refueling aircraft	USA	$44.0	Used, re-engined
AH-64A Apache	Attack helicopter	USA	$42.0	
MiG-29 Fulcrum	Jet fighter	Russia	$33.0	
MLRS system	Multiple rocket launcher	USA	$12.2	With rockets, tech support
Leclerc	Main battle tank	France	$10.5	
Mirage F-1B/C	Jet fighter	Qatar	$10.0	Used, for Spain
MCV-80 Desert Warrior	Infantry fighting vehicle	UK	$2.9	
Mirage V	Jet fighter	Belgium	$2.2	Used, refurbished
BMP-3	Infantry fighting vehicle	Russia	$1.6	
M-60A3 Patton II	Main battle tank	USA	$1.4	Used
Mistral launcher	Surface-to-air missile system	France	$0.8	Hand-held system, with 4 missile rounds
AMRAAM	Long-range air-to-air missile	USA	$0.7	
LAV-25	Wheeled APC	Canada	$0.6	Multiple variants
AGM-88B HARM	Anti-radar missile	USA	$0.5	
YPR-765	Armored personnel carrier	Netherlands	$0.2	Used
TOW 2D	Anti-tank guided missile	USA	$0.02	Cost of missile round

Source: Stockholm International Peace Research Institute, *World Armaments and Disarmament, SIPRI Yearbook, 1996.* Arms trade registers.

the Stinger shoulder-fired, infrared-homing, surface-to-air missile. The Stingers played havoc with Soviet air power but also found their way via the black market into the Ayatollah Khomeini's Iran. Many of the several hundred thousand light weapons made available largely by the United States to the mujahedeen found their way through Pakistani bazaars into the hands of insurrectionists in Kashmir, the Indian Punjab, and Pakistan's own smoldering province of Sind.[16] Sub-Saharan Africa is similarly awash in automatic weapons after decades of anticolonial wars and follow-on struggles to answer the question, "who rules?"

Although light arms (and land mines) do contribute greatly to war and human suffering, the wars with the greatest historical potential for escalation and for drawing in outsiders have involved heavy arms and regular or quasi-regular armies. The several Arab-Israeli wars, the Iran-Iraq war, the Gulf War, the three wars between India and Pakistan, the Indochina wars, the long-running Angolan civil war, and the struggle for Bosnia all involved extensive use of heavy weapons. While interstate war receded from view in the 1990s, it is neither gone nor forgotten. As the Ethiopia-Eritrea border war of 1998–1999 should remind us, states and their leaders are still capable of fighting over symbolic issues and misunderstandings.

India and Pakistan have many symbolic issues at play in their continuing border dispute, and now they have the potential to escalate any future war to the nuclear level. Pakistan's inability to balance Indian military power by means of conventional forces alone contributed to its quest for nuclear arms (see chapter 3). This link between major conventional armaments, regional arms balances, and weapons of mass destruction (WMD), also visible in the case of Israel and its Arab neighbors, is missing in the case of light weapons. Policymakers interested in curbing arms sales have long had to confront the "doves' dilemma," the argument that failure to provide requested conventional weapons could drive jilted recipients to seek unconventional alternatives in recompense. But there is a sort of "hawks' dilemma" as well: sales of advanced conventional weapons not only may fail to convince a recipient to abandon its WMD programs, but may even give the recipient a better means by which to deliver the ordnance that its programs create. (A multirole F-16 aircraft, for example, makes a very good nuclear weapon delivery vehicle.)

Light weapons are the focus of an international consortium of 80 NGOs and a campaign to rein in black market transfers.[17] The trade in major conventional arms, like the thousands of nuclear weapons that still tip U.S. and Russian long-range missiles, has been relegated to the back benches of international security discourse. But like those nuclear arsenals, it is still active at $30 to 40 billion per year, and still dangerous.

Types of Transactions

A wide variety of arrangements govern the transfer of major conventional weapons to developing countries. Most are government-to-government arrangements. The contracts ranged from essentially retail sales, to deep discounts that help to cement political relations, and extensive lines of low-interest, government-funded or government-guaranteed credit with long repayment periods.[18] Used weapon systems, in particular, may be presented as gifts to recipients, and particularly durable goods like warships may be leased. Different suppliers establish different degrees of end-use and retransfer restrictions on the weapons they sell. An end-use restriction may limit how or where a particular weapon system may be used. A retransfer restriction may require that the original recipient obtain permission from the original supplier before reselling a ship, tank, or aircraft to a third country once the recipient further modernizes its forces and no longer has use for a weapon system.

Sometimes the end-use/retransfer restriction may apply to just part of a major weapon system (for example, the engines), but that may suffice to stymie some transactions. If engines built by General Electric or Pratt-Whitney in the United States are used in an aircraft built by, say, Sweden, then Washington effectively controls whether or not Sweden can sell that aircraft to Cuba, against which the U.S. embargoes trade, even though other countries do not recognize or support that embargo. The Swedes could sell only the aircraft without the engines unless Washington agreed to the deal.

In the early 1990s, when global arms supply capacity far exceeded global demand, interested buyers could to some extent set the terms of trade, insisting on and getting direct or indirect "offset" arrangements built into arms contracts. Direct offsets are closely related to the sale and may include, for example, local assembly or production of some weapon components. Indirect offsets may be anything else that obliges the arms seller to buy from the recipient country an amount of locally produced goods. In general, they are "compensatory, reciprocal trade agreements arranged as a condition of the export sale of military materiel and support services." The value of such offsets can exceed the value of the arms sale itself. The arms supplier subsequently tries to turn these goods into cash on local or world markets.[19]

Instead of, or sometimes in addition to, offset deals, some arms transfers involve commodities rather than cash—that is, some form of barter. Recent Russian sales have incorporated substantial elements of barter. Analysts estimate that 65 to 75 percent of the cost of contracts to supply jet fighters and submarines to China, for example, was paid for in "sneakers" and other consumer commodities. Malaysia paid for roughly 30 percent of the half-billion-dollar cost of 18 Russian MiG-29 fighters with commodities like palm kernel oil, and also required offsetting purchases by Russia worth another 40 percent of the deal.[20]

While the data in table 1.1 were grouped by country, implying that governments are the only entities that supply major weapons, a significant fraction of all major arms transfers are direct commercial sales. Over the last two decades, for example, since the U.S. Congress took the ceilings off commercial arms transactions, about 20 to 25 percent of American arms transfers to developing states have been commercial sales. International defense industrial cooperation, moreover, goes far beyond simple sales transactions, ranging from the local assembly and licensed production arrangements noted earlier, to technology transfer, codevelopment of new systems, and corporate joint ventures. The majority of these arrangements have been between companies based in developed industrial countries.[21]

While most of the companies making defense export sales have been licensed to do so by the countries in which they are incorporated, significant diversion of arms and arms-related technology into illicit markets occurs nonetheless. Important examples include the dual-use technologies (those adaptable to civilian or military use) shipped by German firms that ended up in the service of the Iraqi mass destruction weapons program, or the black-market connections used by Iran to resupply its war effort in the mid-1980s in the face of U.S. efforts to block Tehran's access to global arms markets.[22] Conversely, Washington tended to look the other way at shipments of arms reaching the Croatian government and Bosnia's Muslims in the early- to mid-1990s to fuel their fight against the Serbs. Croatia reportedly even managed to rebuild a modest air force from aircraft shipped to it in pieces despite a UN arms embargo.[23]

Operating in between the legitimate and illicit markets for arms are the so-called gray markets, which involve covert arms transfers by governments (or elements within governments) that seek deniability for their actions. For example, the United States quietly shipped hundreds of thousands of assault rifles and dozens of Stinger antiaircraft missiles to Afghan guerrilla groups fighting Soviet troops in the 1980s. Around the same

Figure 1.2   Value of U.S. Arms Transfers, All Countries, 1950–1977

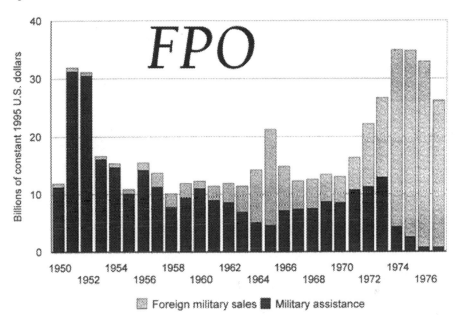

Source: U.S. Congress, House Committee on International Relations, "Review of the President's Conventional Arms Transfer Policy," Hearings before the Subcommittee on International Security and Scientific Affairs, 95th Cong., 2nd sess., February 1, 1978, p.39.

time, members of the Reagan administration's National Security Council staff contrived to use arms deliveries to Iran, contrary to declared U.S. policy, to buy freedom for Americans held hostage in Lebanon by Iran-supported militants, and to generate cash for sub rosa shipments of arms to the Nicaraguan contras, contrary to congressional wishes.[24]

As the most politically driven of arms suppliers, the United States and the Soviet Union had the greatest incentives to be creative in their provision of equipment and financing for developing state clients. All but a handful of Moscow's main arms clients during the Cold War paid for their armaments on long-term, low-interest credit or by trade in commodities. Data published by Ed Laurance indicate that the hard currency earnings from Soviet arms sales in the late 1970s did not exceed $8 billion per year (in 1995 dollars) on perhaps twice that value in total sales. Brookings scholars Clifford Gaddy and Melanie Allen cite Russian arms export czar Mikhail Maley to the effect that peak Soviet arms sales in the late 1980s netted $5 to 7 billion in "cash" per year (also adjusted to 1995 dollars) or about one-third of what it cost the USSR to manufacture the weapons. Just over half the total remaining value of late-Soviet sales were made on credit, and much of that debt has since been viewed as uncollectible.[25] For many years the USSR was, in effect, running a $15 to 20 billion per year military aid program based on rough estimates of the value of transferred equipment alone.

U.S. arms transfers from 1950 through the 1960s were provided mostly through the Military Assistance Program (MAP), that is, grants-in-aid, as figure 1.2 indicates. Foreign Military Sales (FMS) increasingly dominated U.S. arms transfer programs from

1974 onward, however. FMS involves direct, government-to-government arms transfers in which the U.S. government places orders with defense firms on behalf of foreign customers, charging those customers a 3 percent fee for its services. The U.S. government often extends credit to recipient governments under the FMS program, in the form of direct loans or government-guaranteed commercial loans.

U.S. grant aid did not disappear entirely after 1974. Following the 1979 peace treaty between Israel and Egypt, the United States heavily subsidized arms transfers to both countries, providing about $3 billion per year in direct governmental loans under the Foreign Military Financing (FMF) program. FMF money could be used to pay for FMS purchases, military construction, or direct commercial arms sales. Since 1985, repayment of these loans has been routinely waived by the Congress. From 1983 to 1989, moreover, the MAP Merger program extended grants for purchases of arms and construction services from the U.S. government. These funds were given primarily to states fighting counterinsurgencies (in 1987, for example, El Salvador, Turkey, and the Philippines received 60 percent of these funds).[26]

Global Politics and Heavy Weapons

The international trade in conventional weapons has always been a creature of its political setting and of prevailing technologies, but never more so than in the half-century examined here, when conventional arms were also major political weapons—sources of influence and cement of alliances. And just as political trends and events shaped interactions in the world at large, they also shaped the character of the postwar arms trade, which has passed through several distinct phases since good data began to be kept in 1950.

While scholars agree that there have been several phases to the arms trade since World War II, they differ in their assessments of timing and in the characteristics emphasized. Ed Laurance, at California's Monterey Institute, has divided the postwar arms trade into three periods: 1946 to 1966, 1966 to 1980, and 1980 onward. These periods were marked, he argues, by government grants of mostly surplus equipment; then by growing competition among supplier states and greater willingness to supply first-line weapons to developing states; and finally by the growing dominance of commercial interests in the arms trade. By the late 1980s, dwindling internal demand for arms in developed countries caused their arms industries to scramble for export sales, and recipient countries gained "enough military capability to deter, directly threaten, and influence the behavior of major states."[27]

Analysts David Louscher and James Sperling at the University of Akron devised a somewhat different set of periods, with a U.S.-dominant phase through the mid-1960s followed by a 10-year (roughly 1965–1975) period of peak U.S.-Soviet competition. From roughly 1975 to 1989, they argue, the arms trade became increasingly commercialized. Finally, after 1990, the market entered into a period of "diffuse economic competition."[28]

I agree with Louscher and Sperling's timing for the two later periods, but have difficulty agreeing with their "period of peak U.S.-Soviet competition," if we are talking about competition in the Third World of developing countries. There, outside East Asia, intense U.S.-Soviet competition continued in the latter half of the 1970s and the first half of the 1980s, that is, ten years later than Louscher and Sperling specify. The first Reagan administration in particular marked a period of intense, if secondhand, rivalry with the USSR in places like southern and eastern Africa (Angola, Ethiopia, Somalia), central America (Nicaragua and El Salvador), Afghanistan, and the Persian Gulf.

Since others' phasing of the earlier arms trade seemed not quite right, I set out to define a preferred set of benchmarks. These include the growing presence of the USSR in the trade after 1965 (which Laurance, and Louscher and Sperling, do emphasize); the proclamation of the Nixon Doctrine in 1969; America's shift from mostly grants-in-aid to mostly cash-and-credit sales after 1973–1974; the explosive growth in oil producers' purchasing power for more than a decade after the 1973 Middle East War; and the end of the Cold War.

In 1969, President Richard Nixon proclaimed what became known as the Nixon Doctrine, designed to help America avoid direct involvement in future regional wars like the one in Vietnam, then tying down a half-million American troops, by helping regional powers build up their own military capacities.[29] The U.S. Military Assistance Program (MAP) tapered off rapidly once the United States completed transfers to the failing government of South Vietnam in 1974–1975.

The Nixon Doctrine set the stage for substantial increases in U.S. arms transfers to developing regions. Four years later, a sharp jump in oil prices following the October 1973 Middle East War gave oil producers the revenues to pay cash for new weapons, increasing demand to match the supplies made available by the new American arms sales policy.

Petroleum prices first zoomed due to Arab oil producers' refusal to sell oil to states that supported Israel. Once the Arab oil embargo ended, oil producers endeavored to keep prices high. Imperial Iran benefited from the boom and became the first major beneficiary of the Nixon Doctrine. When the shah fell from power in late 1978, oil prices soared again as Iranian production was disrupted. Supply was hampered yet again by the lengthy Iran-Iraq War (1980–1988). Booming oil revenues made Middle Eastern oil producers both the leading buyers and financiers of arms in the developing world and helped transform the arms trade into a profitable commercial enterprise, for every tier of arms supplier.

Figure 1.3 uses SIPRI data, broken out by supplier, to show how the arms trade expanded through 1985. It is easy to see the USSR's move into the arms market, with peaks for deliveries to Cuba in 1962, and to Syria and Egypt, in particular, before and after the Middle East wars of 1967 and 1973. American arms transfers grew fairly steadily from the early 1970s to about $10 billion annually (1975–1983) before leveling off at about $6 billion a year, where they remained for several years. As noted earlier, rising oil prices and the Iran-Iraq War made a larger market for smaller suppliers, whose deliveries jumped in 1978.

Under Mikhail Gorbachev's leadership (1985–1991), the Soviet Union progressively lost interest in subsidizing Third World armies. Gorbachev's December 1988 speech before the United Nations, in which he declared his government's intent to withdraw Soviet occupation forces from Europe, signaled the beginning of the end of the Cold War. Gorbachev had announced, in effect, that the USSR would no longer defend or sustain Euro-communist governments by force of arms. People soon began to flee across the erstwhile Iron Curtain to the West by the thousands. The Berlin Wall—main symbol of the division of Europe—came down less than a year after Gorbachev's speech.

Deliveries from most arms suppliers to developing regions continued to grow until 1987, fueled in part by demand from the ongoing war in the Persian Gulf (see figure

Figure 1.3   SIPRI Trend-indicator Values for Arms Deliveries to Developing States, by Supplier Country, 1951–1985

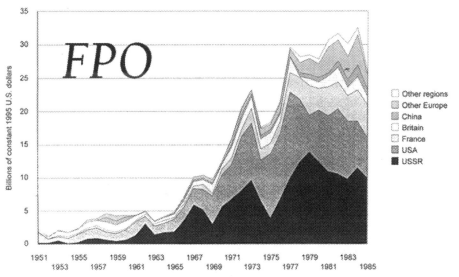

Source: Brzoska and Ohlson, *Arms Transfers to the Third World*, pp. 324–27.

1.4). However, deliveries began to tail off sharply thereafter.[30] The rapid decline in Soviet/Russian deliveries accounts for much of the total drop in the market. Between 1988 and 1991, the value of Soviet arms deliveries, as calculated by Western analysts, declined 75 percent. Russian deliveries continued to drop until 1994, to just 10 percent of their Soviet Cold War peak.

The Cold War arms trade had come to an end. In its place arose a still partly political but mostly economics-driven marketplace that we will come back to at the end of the chapter. In that marketplace, the United States resumed a dominant position. In fact, American arms sales hit their highest levels in a decade during and immediately after the Persian Gulf crisis of 1990–1991. American deliveries, based on those sales, consequently increased through the first half of the 1990s.

In sum, the four phases of the postwar arms trade appear to have played out as follows: the first phase began in 1950 with the Korean War and the sudden shifting into high gear of the military side of the Cold War. That phase consisted of U.S. grants-in-aid to allies in Europe, to Japan, and to selected countries under direct pressure from communist neighbors (especially South Korea and Taiwan), plus British sales primarily to countries within the Commonwealth, like India and South Africa. The Soviet Union and a handful of additional European arms suppliers began to play a somewhat bigger role in the trade in the 1960s, as described earlier. This phase continued until 1969 and the declaration of the Nixon Doctrine.

A relatively brief transitional phase followed, I would argue, from 1970 to 1974. U.S. transfers shifted from mostly grants to mostly sales as America's tragic Vietnam involvement came to an end and its equally tragic involvement with the shah of Iran deepened. This phase also incorporates the relatively brief period of U.S.-Soviet détente under Nixon and the watershed 1973 Middle East War. Détente essentially evap-

Figure 1.4   U.S. ACDA Values for Arms Deliveries to Developing States, by Supplier Country, 1986–1996

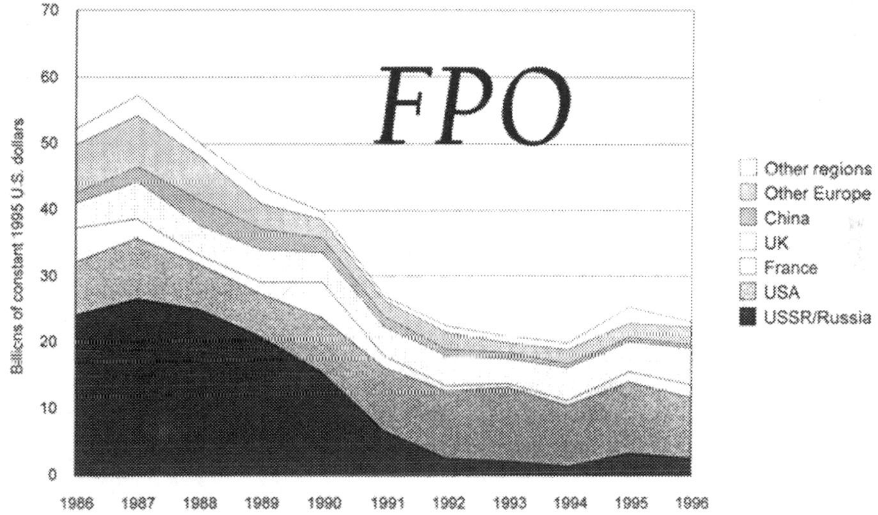

Source: U.S. ACDA, *World Military Expenditures and Arms Transfers, 1997*, table IV.

orated after 1974, a victim of U.S. domestic politics and growing Soviet activism in developing regions, and once the oil embargo ended in 1974, oil producing countries began to accumulate astounding amounts of cash very rapidly, some of which they re-cycled into arms.[31]

A long third phase of expanding arms deliveries thus began in 1975. This phase was characterized not only by a burgeoning arms trade but also by distinct worsening of U.S.-Soviet relations through the Ford, Carter, and first Reagan administrations. U.S.-Soviet rivalry became especially sharp in Africa, Central America, and Central Asia. The wars of this phase gave smaller arms suppliers expanded market share. By the end of this phase, in 1989, the Soviet Union had pulled back from most of its commit-ments in the developing world, and concessional arms sales were available to very few countries (principally Israel and Egypt).

The fourth phase, also the current one, began in 1990 with the invasion of Kuwait and the ensuing Persian Gulf crisis. Commercial competition in this period replaced the old political-ideological contest for customers. For several years, there was a buyer's market in armaments, as the arms supply capacity built up during the Cold War far exceeded combined domestic and export market demand. Offsets gained new visibil-ity as buyers sought the best economic deal they could squeeze out of suppliers. Wash-ington continued to worry about the spread of advanced arms and related technologies to "rogue" states and terrorist organizations, but by and large this latest phase of the arms trade has been driven by ability to pay. At the upper end of that abil-ity, states like Saudi Arabia and the UAE have been buying first-line, latest generation equipment. At the lower end, states have been signing commercial contracts for com-ponent upgrades to extend the life or increase the effectiveness of weapons already in their inventories.

How do American arms transfers map onto the four phases just sketched? As we know, U.S. transfers in the first phase were largely grants to all recipients but the United States delivered far more arms to developed than to developing countries in that phase (figures 1.5 and 1.6, leftmost bars), to build up militarily capable allies that could help it face down the military threats posed by the Soviet Union and communist China. The rate of transfers to developing states leaps upward in phase two. Most of the increase is in military aid, and that represents Washington's efforts to extricate its own forces from the war in Vietnam. Virtually all U.S. military aid between 1970 and 1974 went to South Vietnam and to the few Asian countries (Thailand, Philippines, and South Korea) that were helping out with the war effort either by providing troops of their own or by hosting rear bases for U.S. forces. But FMS to developing countries also increased in phase two, so that setting aside grants-in-aid, U.S. deliveries to developing states still equaled deliveries to developed states. From 1975 onward, transfers to developing states far outstripped sales to the developed world.

The patterns of U.S. arms transfers varied over regions as well as time (see table 1.4).[32] The top third of the table presents aggregate figures, worldwide and for each region, by type of transaction and for each phase, in constant 1995 dollars. Between 1950 and 1969, the United States gave away $160 billion worth of weapons and support through MAP, another $47 billion between 1970 and 1974, a little less than $8 billion from 1975 to 1989, and just over $1 billion from 1990 to 1995, for a total of $215 billion.[33] Over the same 46-year period, the United States and U.S. firms *sold* more than $357 billion worth of arms and related services to developing countries.

Figure 1.5   Annual Average Value of U.S. Arms Transfers to Developed Countries

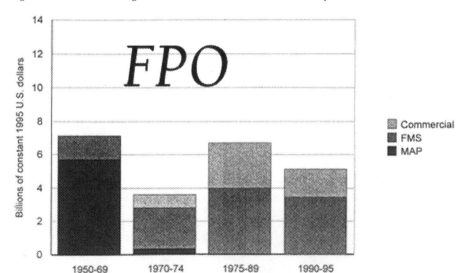

Source: U.S. Department of Defense, Defense Security Assistance Agency, *Foreign Military Sales, Foreign Military Construction Sales and Military Assistance Facts*, 1977, 1987, 1992, and 1996 editions.

Figure 1.6   Annual Average Value of U.S. Arms Transfers to Developing Countries

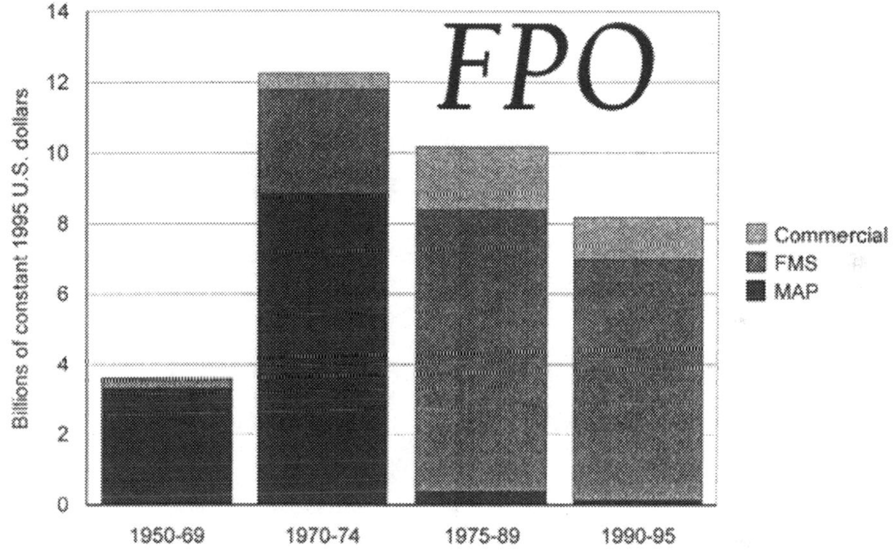

Source: U.S. Department of Defense, Defense Security Assistance Agency, *Foreign Military Sales, Foreign Military Construction Sales and Military Assistance Facts*, 1977, 1987, 1992, and 1996 editions.

The middle third of table 1.4 shows how aid and sales were distributed across regions. In phase one, more than half of U.S. arms aid went to developed states and regions, as did 84 percent of FMS transfers. In phase two, as noted earlier, East Asia absorbed most of American aid; developed states' share of FMS dropped, and the Middle East's share of deliveries jumped as the United States and Soviet Union rushed resupplies to their respective clients during the 1973 war. Reporting on U.S. commercial sales began in 1971 and, as table 1.4 indicates, the developed states have always been the majority recipients of direct commercial transfers.

The bottom third of the table shows the distribution of U.S. transfers *within* each region. Even developed countries received mostly military assistance in the first phase, and because of the Vietnam War, 97 percent of all transfers to East Asia in phase two were still MAP. By phase three, FMS was the dominant mode of transfer to all regions of the world, as it remains today, but commercial sales account for a significant fraction of deals in East Asia, in Latin America, and among developed states.

The Diffusion of Major Weapon Systems, 1950–1995[34]

For most of postwar history, transfers of complete weapon systems rather than subsystems was the norm. The following sections look at transfers of arms by major weapon category (combat aircraft, ground equipment, naval forces, and missles). The chapter introduces two measurement concepts—weapon system "generation" and transfer "delay." The former is a rough indicator of performance and helps to measure the sophistication of the technology spreading to the developing world, while

Table 1.4 Distribution of U.S. Arms Transfers, 1950–1995

	MAP 1950–69	FMS 1950–69	MAP 1970–74	FMS 1970–74	COM 1971–74	MAP 1975–89	FMS 1975–89	COM 1975–89	MAP 1990–95	FMS 1990–95	COM 1990–95
Program distributions by region (millions of constant US 1995 dollars)											
Worldwide	$160,001	$32,917	$46,775	$26,847	$5,637	$7,681	$179,340	$66,792	$1,262	$61,704	$16,855
East Asia Pacific	$55,990	$226	$43,853	$1,081	$303	$6,042	$19,390	$8,243	$7	$9,117	$3,588
Near East & So. Asia	$8,267	$3,387	$467	$12,128	$1,327	$488	$93,933	$14,944	$965	$30,291	$2,452
Europe, Canada, Japan, Australia, New Zealand	$92,203	$24,325	$2,098	$12,172	$3,686	$826	$48,945	$40,318	$11	$15,917	$9,960
Africa	$684	$29	$164	$72	$50	$147	$1,601	$436	$88	$163	$38
American Republics	$2,858	$1,642	$192	$1,393	$271	$178	$4,695	$2,852	$191	$1,323	$817
Program distributions across regions (read down columns)											
East Asia Pacific	35%	1%	94%	4%	5%	79%	11%	12%	1%	15%	21%
Near East & So. Asia	5%	10%	1%	45%	24%	6%	52%	22%	76%	49%	15%
Europe, et al.	58%	84%	4%	45%	65%	11%	33%	60%	1%	34%	59%
Africa	0.4%	0%	0.4%	0.3%	1%	2%	1%	1%	7%	0%	0%
American Republics	2%	5%	0.4%	5%	5%	2%	3%	4%	15%	2%	5%
Program distributions within regions (read across rows within time periods)											
Worldwide	83%	17%	59%	34%	7%	3%	71%	26%	2%	77%	21%
East Asia Pacific	100%	0%	97%	2%	1%	18%	58%	24%	0%	72%	28%
Near East & So. Asia	71%	29%	3%	87%	10%	0%	86%	14%	3%	90%	7%
Europe, et al.	77%	23%	12%	68%	21%	1%	59%	40%	0%	68%	32%
Africa	96%	4%	57%	25%	17%	7%	73%	20%	30%	56%	13%
American Republics	63%	37%	10%	75%	15%	2%	61%	37%	8%	57%	35%

Note: FMS = Foreign Military Sales; MAP = Military Assistance Program; COM = commercial sales.
Source: U.S. Defense Security Assistance Agency, "FMS, FMCS and Military Assistance Facts" (Washington, D.C.: DSAA, 1978, 1987, 1992, and 1996 editions).

the latter measures the time between a new system's entry into service with the producing country's armed forces and its initial delivery to a developing country. The shorter the delay, the faster technology is spreading.

Heavy Combat Aircraft

This category of aircraft includes bombers (aircraft designed to drop ordnance from internal bomb bays in level flight), ground attack aircraft with a combat payload greater than 6,000 pounds, and fighter-bombers (aircraft designed to engage in air-to-air combat but useful for ground attack as well).[35] Heavy combat aircraft hold a particular fascination for analysts of the arms trade in part because they are visible, high-prestige items that represent the leading edge of high-tech weapons proliferation and because their border-leaping capabilities—their ability to lift warfare into the third dimension—suggest an offensive, escalatory potential that is matched only by heavy armor. Certainly, some models (heavy bombers, for example, or longer-range aircraft like the Russian Su–24 Fencer or American F-111) are designed for offensive strike operations,[36] and modern fighter-bombers can carry a ground attack warload equal to many World War II heavy bombers.[37] But just as the best antitank weapon is, in many circumstances, another tank, so the best defense against offensive aerial capabilities is often an airborne defense. (I will discuss efforts to define "offensiveness" and "defensiveness" for purposes of arms control in chapter 4.)

Generations of Aircraft

Aircraft in the data set were assigned to one of six performance generations, based on the following rule-of-thumb criteria:[38]

> First generation: World War II designs, propeller-driven;
> Second generation: early jet aircraft designed in the 1940s, with weak engines and low payload;
> Third generation: the first high-subsonic jet aircraft with early air-to-air radars, early-model infrared-guided missiles, except on models designed to be long-range interceptors;
> Fourth generation: transonic jets, more capable than the third generation but rapidly overtaken by fully supersonic models;
> Fifth generation: supersonic jets with capable avionics, multibarreled Gatling guns, radar-guided missiles;
> Sixth generation: highly maneuverable airframes, high thrust-to-weight-ratio engines, with such advances as solid state avionics and fly-by-wire controls instead of standard hydraulics.

Examples of aircraft models assigned to each of these notional generations are listed in table 1.5 by period of initial deployment. The correlation of generation with year of deployment is fairly close, but the notion of generation allows aircraft of roughly comparable performance vintage to be grouped together even though their initial deployments may differ by thirty years. For example, the Soviet MiG-21, first deployed in the late 1950s, was the model for the Chinese F-7M Airguard first deployed in the 1980s. I thus rate both of them as fifth-generation aircraft.

Table 1.5 Heavy Combat Aircraft Entering Service, by Generation and Time Period, Examples

	First Generation	Second Generation	Third Generation	Fourth Generation	Fifth Generation	Sixth Generation
1940–44	DH Mosquito (UK) F-4U Corsair (US)					
1945–49	F-8F Bearcat (US) A-1 Skyraider (US)	DH-100 Vampire (UK) F-80 Shooting Star (US) Sea Fury (UK)	F-86A Saber (US) MiG-15 (USSR)			
1950–54		Meteor F.8 (UK)	F-86D Saber (US) MD-450 Ouragan (FR) Hunter F.1 (UK) MiG-17 (USSR)	F-100 Super Saber (US)		
1955–59			Hunter F.6 (UK) TU-16 Badger (USSR) A-4B/E Skyhawk (US)	MiG-19 (USSR) SU-7 Fitter-A (USSR) Super Mystere (FR)	F-104A Starfighter (US) MiG-21F (USSR)	
1960–64				F-6 (PRC) F-5A/B Freedom Ftr (US) TU-22 Blinder-A (USSR)	F-4B Phantom II (US) Mirage III (FR)	
1965–69				A-4H Skyhawk (US)	Lightning (UK) A-7 Corsair-2 (US) MiG-21M/PFM/MF (USSR)	

Table 1.5 (continued)

	First Generation	Second Generation	Third Generation	Fourth Generation	Fifth Generation	Sixth Generation
1970–74					A-4M Skyhawk II (US), MiG 23/27 (USSR), F-5E/F Tiger II (US)	F-14 Tomcat (US), Mirage F-1A (FR)
1975–79					F-7 (PRC, Mirage-50 (FR), Super Etendard (FR)	F-15A Eagle (US), AV-8B Sea Harrier (US)
1980–84					Nesher (Israel), Kfir C-7 (srael)	F-16C Fighting Falcon (US), Torrado (Panavia)
1985–89					F-7M Airguard (China)	F/A-18C Hornet (US), MiC-29C (USSR/ Russia)

The generational notion is also useful for durable designs like the American A-4 Skyhawk, which remained in production from the mid-1950s to 1979. The Skyhawk evolved through a series of models that incorporated a range of propulsion plant, avionics, and armament improvements, hence Skyhawks appear in more than one aircraft generation.

Figure 1.7 shows how many models of each generation of aircraft entered service in each half-decade between 1950 and 1995 from all suppliers. First-generation, propeller-driven aircraft models continued to enter the market in the early 1950s and were still used extensively a decade later for close support of U.S. and South Vietnamese forces in Indochina. Second-generation aircraft, on the other hand (the early jets) stopped entering service in the early 1950s and were displaced from most developed states' inventories by the mid-1950s by the more robust third generation. Nearly 80 different aircraft models comprise this generation. American Saber Jets, British Hunters, Soviet MiG-17s, and French Mystères all found their way into developing states' arsenals in substantial numbers (more than 21,000 aircraft). By comparison, the fourth generation included relatively few models and its distribution was equally limited.

The fifth generation, on the other hand, was a big one. Early entries, like the U.S. F-104 Starfighter, began to enter service in the late 1950s. American F-4 Phantoms, French Mirages, and Soviet MiG-21s and −23s appeared originally in the 1960s and new models continued to appear, in some cases, into the 1970s.

This generation was sold to developing states in great numbers starting in the 1960s and extending well into the 1990s (see figure 1.8). This generation, moreover, is currently the focus of most countries' component upgrade programs.

The sixth generation dates from roughly the mid-1970s and the deployment of high-performance aircraft like the U.S. Navy's F-14 Tomcat, the U.S. Air Force F-15 Eagle, or Russian MiG-29. Still internationally competitive as much as 25 years and a

Figure 1.7   New Combat Aircraft Models Entering Service, by Generation and Time Period

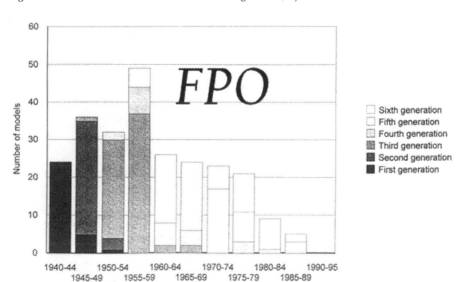

Figure 1.8 Heavy Combat Aircraft Deliveries to All Developing Regions, by Generation

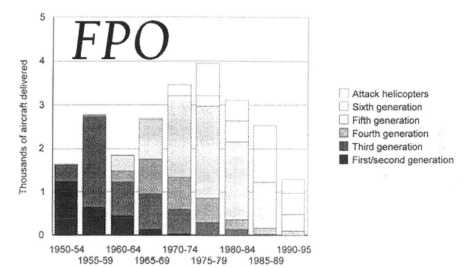

few model changes later, these aircraft are highly maneuverable designs that often can withstand greater combat stresses than can their human pilots. As a result, further performance improvements are likely to come from abilities to see and fight at greater distance, rather than from greater ability to maneuver at close range ("dogfight").

Sixth-generation aircraft have been exported to developing state markets in relatively small numbers (about 100 per year on average) since the mid-1970s. Only in the shrinking market of the 1990s did they come to be a majority of the heavy combat aircraft sold. Roughly 500 were delivered to developing states between 1990 and 1995.

Relative Delay in Distribution of Combat Aircraft

Heavy combat aircraft diffused to different developing regions at different rates. Table 1.6 measures the difference between the earliest appearance of each generation of aircraft in the developing world and its appearance in a particular region. The top half of the table shows years of first appearance (with additional data at the bottom of each column). The bottom half of the table shows the time lag in appearance across regions. For all but the sixth generation, Northeast Asia was the region of first appearance, but it lagged badly in the sixth generation. South Korea did not acquire sixth-generation aircraft until the mid-1980s, North Korea the late 1980s, and Taiwan not until the late 1990s. Following Richard Nixon's 1971 trip to China, U.S. combat aircraft deliveries to Taiwan dried up for a decade, and when they resumed under the Reagan Administration, the aircraft sent were fifth-generation F-104s, quite capable against China's own largely obsolescent air force but a design more than two decades old nonetheless. The Bush administration finally decided to sell used F-16s to Taiwan in 1992, over protests from China, and the Clinton administration began deliveries in 1997.

Several regions have been decades behind the diffusion curve. Central America lags the leaders by more than two decades from the third generation onward (1950s airframes reached countries like Honduras, for example, in the late 1960s). The

Table 1.6 Diffusion of Heavy Combat Aircraft Over Time

| | First Appearance in Region, by Aircraft Generation | | | | | |
	First	Second	Third	Fourth	Fifth	Sixth
Middle East	1950	1950	1955	1959	1962	1976
Northeast Asia	1950	1950	1950	1959	1960	1986
Persian Gulf	1956	1951	1955	1960	1963	1976
South Asia	n/a	1950	1953	1966	1962	1982
South America	1950	1950	1953	1972	1968	1977
Southeast Asia	1950	1954	1955	1960	1964	1988
Southern Africa	n/a	1952	1956	1973	1963	1975
Horn of Africa	1946	1967	1960	1966	1970	n/a
West Africa	1952	1968	1961	1969	1975	1984
Central Africa	1964	n/a	1966	1973	1974	n/a
Central America	1950	n/a	1975	1981	1987	n/a
First introduction	1946	1950	1950	1959	1960	1975
Average	1952	1955	1958	1966	1967	1982
Latest	1964	1968	1975	1981	1987	1984

| | Years Behind the Leading Region | | | | | | |
	First	Second	Third	Fourth	Fifth	Sixth	Average
Middle East	0	0	5	0	2	1	1.3
Northeast Asia	0	0	0	0	0	11	1.8
Persian Gulf	6	1	5	1	3	1	2.8
South Asia	—	0	3	7	2	7	3.8
South America	0	0	3	13	8	2	4.3
Southeast Asia	0	4	5	1	4	13	4.5
Southern Africa	—	2	6	14	3	0	5.0
Horn of Africa	0	17	10	7	10	—	8.8
West Africa	2	18	11	10	15	9	10.8
Central Africa	14	—	16	14	14	—	14.5
Central America	0		25	22	27		18.5

Caribbean would be ranked with Central America except for Soviet deliveries to Cuba, and even Cuba experienced a 20-year gap between receipt of fifth- and sixth-generation equipment (a squadron of Soviet MiG-29s, received in 1989). Segments of sub-Saharan Africa vary. Southern African numbers reflect close early ties between the white-ruled regimes of Rhodesia and South Africa and European suppliers, ties severed by United Nations arms embargoes against Rhodesia in 1965 and South Africa in 1977. Ethiopia, Somalia, and Sudan were favored by a number of aircraft suppliers in the 1960s and 1970s, especially after Soviet naval deployments into the Indian Ocean and growing Western dependence on Persian Gulf oil made the Horn of Africa something of a strategic prize for both camps.

Certain countries within regions have long been favored recipients of major power armaments. Table 1.7 sorts individual recipient countries according to the consistency with which they have received new-model combat aircraft with minimal delay (i.e., five or fewer years after initial deployment with the supplier country's armed forces). Egypt and Israel lead the list, meeting the criterion in seven of nine periods. Pakistan, Iraq, and India follow closely behind. The impact of changing U.S. policy on Taiwan,

Table 1.7 Consistency of First Deliveries of Heavy Combat Aircraft (delays of one to five years)

	1950–1954	1955–1959	1960–1964	1965–1969	1970–1974	1975–1979	1980–1984	1985–1989	1990–1995	Count
Israel	X	X	X	X	X	X	X			7
Egypt	X	X	X	X	X	X	X			7
Pakistan	X	X	X	X	X		X			6
Iraq	X	X	X		X		X	X		6
India	X	X	X		X		X	X		6
South Africa	X	X	X	X		X				5
Syria	X	X			X	X		X		5
Saudi Arabia		X		X	X	X		X		5
Taiwan	X	X	X	X						4
Venezuela	X	X			X		X			4
N. Korea	X	X				X		X		4
Thailand	X		X	X		X				4
Jordan		X	X		X	X				4
S. Korea		X		X	X	X				4
Iran		X		X	X	X				4
Peru		X		X		X		X		4
Cuba			X		X	X		X		4
Ecuador	X					X		X		3
Libya				X	X	X				3
Kuwait				X		X			X	3
Lebanon	X	X								2
Rhodesia	X		X							2
Brazil	X					X				2
Philippines		X		X						2
Colombia		X			X					2
Afghanistan		X				X				2
S. Vietnam				X	X					2
Ethiopia				X		X				2
Morocco				X		X				2
Bangladesh					X		X			2
UAE					X			X		2
Chile						X	X			2
Malaysia						X			X	2
Oman						X			X	2
Burma	X									1
Indonesia			X							1
N. Vietnam				X						1
Tanzania					X					1
Sudan						X				1
Kenya						X				1
Singapore						X				1
Nigeria						X				1
Qatar						X				1

which was a very early adopter of new aircraft through the late 1960s, is clearly visible in the table.

In general, those developing countries engaged in long-term, nose-to-nose political confrontations not only have received the greatest numbers of combat aircraft but also consistently have received the newest versions, initally as a function of suppliers' political interests rather than these states' ability to pay. A dozen countries were early adopters of heavy combat aircraft in the 1950s. More than two dozen were early adopters in the latter 1970s, and more of these transactions reflected basic ability to pay.

Figure 1.9 plots average transfer delays for combat aircraft of all generations, by region, over time. The lines show a gradual *upward* trend over time (a regression line would predict a 5.5-year delay in the early 1950s and an 11.4-year delay in the 1990s). The upward drift of the final ten years of these trend lines might be explained in part by the fact that major suppliers have not introduced major new models of aircraft recently at anything like the rate of earlier decades. The upward drift may be most consistent with the broadening of the market beyond the initial favored few political clients and former colonies to a larger market that was politically relevant but also economically interesting to suppliers. Some of these new buyers bought the latest generation available, while others were satisfied with older models or "previously owned" aircraft that nonetheless substantially upgraded their air forces.

Light Attack Aircraft

Heavy, high-performance combat aircraft are not the ticket for every security concern and, given internal security worries, may be a waste of money and effort. Aircraft able to use smaller airfields and to deposit people and ordnance on targets not so very far away from those airfields would seem to be more useful than heavier, more expensive aircraft to developing countries with greater problems of internal cohesion than external military threat. This section thus addresses briefly the distribution of light, fixed-

Figure 1.9   Average First-delivery Transfer Delay, Heavy Combat Aircraft

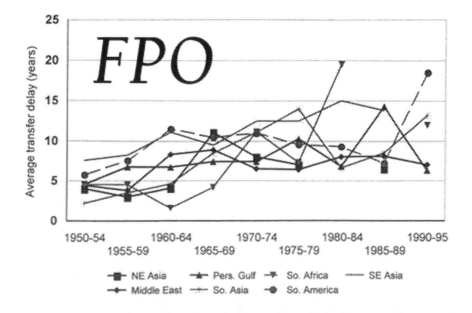

wing attack aircraft to developing countries. The category includes aircraft with a weapons payload of less than 6,000 pounds and used primarily for counterinsurgency and short-range air intercept missions. None of them have been as widely or numerously transferred as heavy combat aircraft. Were ability to pay a strong driver of transfers historically, one might have expected developing states to acquire more of this lighter, cheaper equipment than of heavier, more expensive items. But the politically driven availability of the latter, often secondhand, allowed states in each of the heavily conflictual security complexes to "trade up" readily and continually for several decades.

About half of the light attack aircraft sold to developing states have been of U.S. manufacture. About half of these went to Southeast Asia, almost all of them (nine out of ten) during the Indochina War, to support the Nixon administration's program of "Vietnamization."

Andean countries fighting multiple insurgencies have been the next-most-prominent recipients of light attack aircraft. A kind of mini-arms race ensued in the mid-1970s among Chile, Peru, Bolivia, Ecuador, and Colombia in this category of arms. The United States dominated the regional market until the 1970s, although European suppliers and Brazil, building European designs under license, made some inroads. Central America, on the other hand, remained a virtual American monopoly; about half of a small number of light attack imports went to Honduras, where the United States maintained major exercise and staging bases through the 1980s. For the distribution of light attack aircraft over time, by region, see figure 1.10.

Missile-armed Attack Helicopters

Missile-armed "attack" helicopters include models equipped with antitank guided missiles or other weaponry such as unguided rockets that enable them to damage ground

Figure 1.10   Light Attack Aircraft Deliveries, All Developing Regions

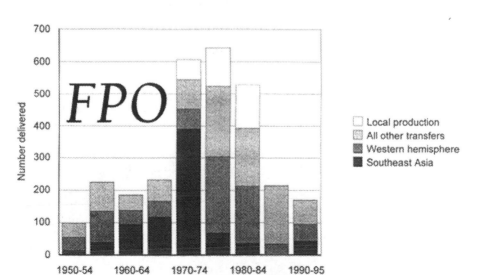

targets other than personnel from modest stand-off ranges. Soviet/Russian Mi–8/17 Hip helicopters, for example, are routinely armed with a variety of ground attack packages and thus are classed here as attack helicopters even though they can carry a squad of troops. The category also includes the American AH-1 Cobra and AH-64 Apache, the Soviet/Russian Mi–24/25/35 Hind, the French SA-342K missile-armed Gazelle, and the British Westland Lynx. (Only the largest arms suppliers manufactured attack helicopters through most of the period under analysis.)

This class of weapons emerged from the U.S. Army's need to escort troop transport helicopters in combat assaults during the Vietnam War. It rapidly evolved into a close air support platform with serious antitank capabilities when armed with wire-guided missiles. While Hips and Gazelles are essentially transport helicopters adapted to fire missiles and rockets, purpose-built attack helicopters are a leaner, tougher, and higher-technology breed and, as a result, have been sold to a relative handful of major clients (Cobras to seven countries, Apaches to four, and Hinds to a dozen). The distribution of other models has been not that much broader (for example, Gazelles to just 14 countries and Hips to 20).

Imperial Iran was the first developing country to import Cobras, in 1974, matched three years later by Iraq, the first country to import Soviet Hinds. Iraq took delivery of about 40 Hinds before going to war with revolutionary Iran in 1980. Israel acquired its first helicopter gunships in 1977 and used them to good effect against Syrian armor in a short, sharp war in and over Lebanon in 1982. One year after that conflict, Syria itself took delivery of its first Hinds. In Northeast Asia, South Korea acquired its first attack helicopters in 1976 and has built a U.S.-designed light attack helicopter, the then–Hughes 500MD. North Korea did not import attack helicopters until the mid-1980s, when it acquired several squadrons of Soviet machines, matched almost immediately by a South Korean order for the latest-model Cobras from the United States. Taiwan ordered several dozen Super Cobra gunships and armed scout helicopters from the United States as Chinese rhetoric directed at Taiwan escalated in the 1990s, suggesting that a short-range attack helicopter based on that island might actually be a threat to engage. Overall, by 1995, more than 2,200 missile-armed helicopters had been delivered to developing countries. Their distribution by region is shown in figure 1.11.

Transport Helicopters

Far more numerous and widely distributed are transport helicopters, lightly armed or unarmed. The single large delivery spike in figure 1.12 reflects yet another part of the cascade of American materiel into South Vietnam, in this case transfers of the ubiquitous UH-1 "Huey" troop transport. The number two recipient of transport helicopters was imperial Iran, which began acquiring license-built Hueys from Italy's Agusta-Bell in 1969. Iran increased its rate of acquisition in the mid-seventies to more than 100 units per year. These imports halted abruptly after the shah fell from power in late 1978.

Transport helicopters made their way in significant numbers into South America. The mountains, jungles, and large, sparsely inhabited expanses of many South American countries make military helicopters a merely logical mode of transport. But they also gave mobility to regional militaries facing or fearing guerrilla insurgencies supported by the Castro regime in Cuba, and began to appear in South American military inventories not long after Castro took power.

Figure 1.11   Missile-armed Helicopter Deliveries, by Region

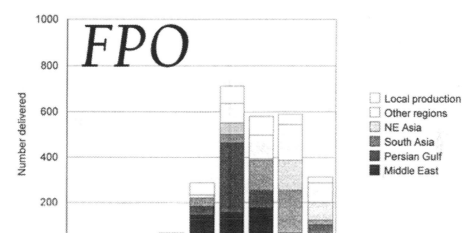

From the late 1970s, a handful of developing countries built transport helicopters under license. Brazil and India favored militarized French designs, while South Korea built American scout helicopters. But the technology is complex and the results are cranky pieces of machinery, which helps account for the fact that their production remains concentrated primarily in the developed industrial countries.

Figure 1.12   Transport Helicopter Deliveries, by Region

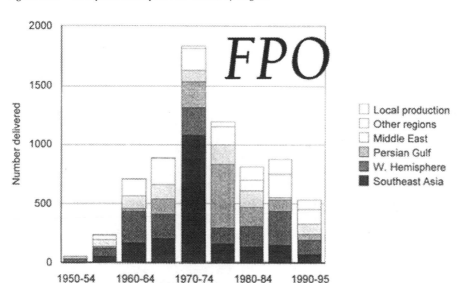

Armored Vehicles

Heavy mechanized armor became the 20th century's principal military tool for the taking and holding of other people's territory. Once heavy artillery and air strikes soften up an objective, tanks push through or destroy remaining obstacles to seize it.[39] Armored personnel carriers (APCs) ferry into battle the infantry who are, somewhat ironically, needed to protect the tanks from enemy infantry. APCs have evolved from relatively thin-skinned vehicles that protected only against light weapons and shrapnel, to fast, more heavily armored "infantry fighting vehicles" designed to keep pace with the latest, fastest tanks. Armored cars, usually wheeled vehicles with space for three or four troops, are suited for scouting or constabulary duty. Light tanks have evolved from simply scaled-down, scout versions of their heavier cousins to tracked or wheeled vehicles with lighter armor but often substantial firepower and greater road speed.

Classic armored warfare as it has evolved over the last half-century is best suited to relatively open, relatively flat, and relatively dry terrain, like the desert pans of the Middle East, North Africa, the Great Indian Desert, or the North European Plain. Political and/or religious rivals have amassed armored forces in each of these areas over the years and at various times used them. Where the terrain is suitable, the politics are hostile, or national intentions are ambiguous, many members of a region and certainly its largest members are likely to acquire not just heavy armor but also combined-arms ground forces with air components. (In the sort of terrain where armor works best it is also most vulnerable to attack from the air. Hence the proliferation of heavy armor in the past quarter-century has been accompanied by the proliferation of antitank helicopters to counter that armor and of mobile surface-to-air missile [SAM] systems to protect it.)

Light tanks and armored cars are more likely to be acquired where terrain is not so favorable to 60-ton tanks, where roads are fragile, or where the need is related to internal security and the operating arena may be urban. A number of states (Saudi Arabia, for example) find such equipment useful for their secondary military structures—the units dedicated to guarding the regime against internal threats, including possible threats from the regular military.

Defining Generations of Armor

As with heavy combat aircraft and missiles, the substantial evolution of battle tanks over the past four or five decades suggests that it may be useful to separate them into generations. Table 1.8 gives examples of each of four generations of tanks entering service between 1940 and 1995. These are potentially very long-lived vehicles. Long production time lines as well as interim upgrades for many tank types mean that generational analysis and the concept of transfer delay used in this chapter, if applied at the level of basic tank models, may understate the amount of generational evolution in transferred weapons and overstate the actual delay involved in a given transfer. This is particularly true with older Soviet models. That is, a Soviet T-55 built and transferred to India in 1969 was very likely less sophisticated than a T-55 transferred to Angola in 1987. (On the other hand, that Angolan tank may have come from Soviet operational stocks after long and hard use, following replacement by newer models.) SIPRI data are somewhat more discerning when it comes to the versions of Western armor transferred abroad.

This analysis treats "medium" and "heavy" tanks as a single category, although they are separately labeled in the data. A medium tank weighs 20 to 35 tons and carries at

Table 1.8 Main Battle Tanks Entering Service by Generation and Time Period, Examples

	First Generation	Second Generation	Third Generation	Fourth Generation
1940–50	Centurion (UK) M-4 Sherman (US) T-34 (USSR)	T-54 (USSR)		
1950–54		M-47/48 Patton (US)		
1955–59		T-55 (USSR) Type-59 (PRC)	Centurion Mk9 (UK)	
1960–64		M-60 (US)	T-62 (USSR)	
1965–69		AMX-30 (FR) Vickers Mk1 (UK)	Chieftain (UK)	
1970–74			Chieftain Mk3/5 (UK)	T-72 (USSR)
1975–79		M-48A5 (US) AMX-30S (FR)	M-60A3 (US) TAM (W. Germany) Vickers Mk3 (UK)	
1980–84			OF-40 (Italy) Type 69 (PRC)	Challenger (UK) M-1 Abrams (US) Merkava Mk2 (Israel) T-80 (USSR)
1985–89				M-1A1 (US)
1990–95				Leclerc (FR) M-1A2 (US) T-90 (Russia)

least a 76 millimeter—three inch—main gun. A World War II–era Soviet T-34 or American M-4 Sherman qualifies as a medium tank. A heavy tank weighs 36 tons and up and carries a 90 millimeter or larger main gun. A Soviet T-54 defines the weight threshold for heavy tanks.

Tank generations were defined on the basis of speed, armor, firepower, accuracy, and night fighting capacity. First-generation tanks were slow or lightly armed and armored, had guns that could be fired accurately only when the tank was at rest, and had no night vision equipment. Second-generation tanks have heavier armor, greater horsepower per ton, larger main guns (up to 105 mm), and some night vision (e.g., image intensifiers). Most third-generation tanks have gun stabilizing systems that permit accurate firing on the move, night vision (image intensifiers and/or infrared sights), and laser range finding systems. The fourth generation has these characteristics plus higher speed, advanced composite armor, and more advanced vehicle electronics.

Patterns of Diffusion for Armored Vehicles

Armored vehicles went into developing state arsenals at steadily increasing rates through the early 1980s (figure 1.13). Battle tanks account for roughly half of the transfers in each five-year period. Figure 1.14 breaks them out by generation. As transfers began to accelerate in the late 1960s, second-generation tanks dominated the trade and continued to do so through the 1970s. Third-generation tanks started to

Figure 1.13 Deliveries of Armored Vehicles, All Developing Regions

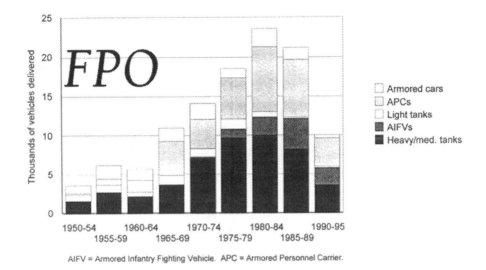

AIFV = Armored Infantry Fighting Vehicle. APC = Armored Personnel Carrier.

Figure 1.14 Deliveries of Main Battle Tanks, All Developing Regions

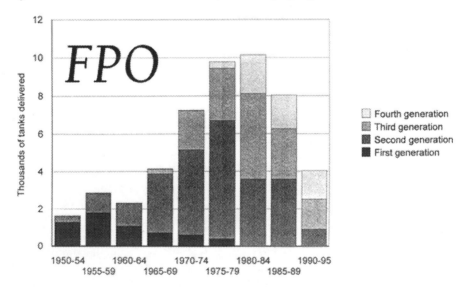

reach developing states beginning in the late 1960s. Deliveries grew to five thousand units in the 1970s and seven thousand in the 1980s. The fourth generation first appeared in quantity in the 1980s but still does not dominate the trade due to its high cost and due to the availability of older models at cut-rate or discard prices. Egypt, for example, has bought used American M-60A3 tanks for about $250,000 apiece, ammunition and supplies included, one-sixteenth the cost of buying a new M-1A2 Abrams tank (although Egypt has a license to build the Abrams as well).[40]

Most of the transferred armor went to the Middle East and Persian Gulf (figs. 1.15 and 1.16). About one-fourth of the tanks acquired by Middle East countries since 1972 are estimated to have been destroyed in wars (in 1973 and 1982) or otherwise taken out of service by 1993. The estimate for the Persian Gulf, principally the holdings of Iran and Iraq, is closer to 50 percent attrition, with the two wars there (1980–1988 and 1990–1991) taking a heavy toll.[41] Assessing the arms trade's contributions to war is an

Figure 1.15   Deliveries, Medium and Heavy Tanks, Middle East

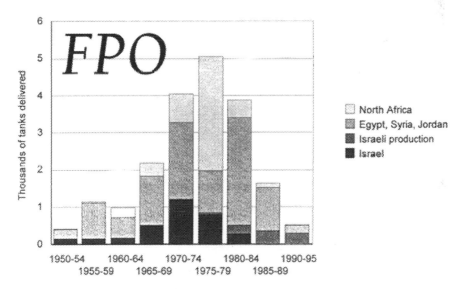

Figure 1.16   Deliveries, Medium and Heavy Tanks, Persian Gulf

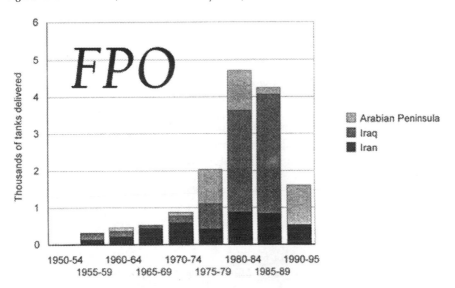

objective of chapter 2, and a detailed discussion of regional cases must wait for chapter 3, but these two regions have absorbed much of the land-grabbing armor transferred to developing states at times consistent with a buildup to war and resupply afterward: the Middle East shows the heaviest buys in the early 1970s, and the Persian Gulf in the 1980s. Even ignoring Libya's hobby of stockpiling weapons (which boosts the North Africa segments of fig. 1.15), tank acquisitions by Israel's neighbors far outnumbered those of Israel itself in all time periods, even allowing for Israeli production of its own Merkava tank in the 1980s. In the Persian Gulf (fig. 1.16), Iraq far outdid Iran in armor imports during their eight-year war, partly reflecting U.S. efforts to embargo arms to the Khomeini regime but also indicating the extent to which fierce Iranian resistance was unexpected: Iraq bought most of its armor *after* Saddam Hussein started the war, to fend off Iranian counterattacks. Gulf Arab armor buys follow the oil market: little before 1975, a lot in the first decade of oil price hikes, much less after oil prices declined again in the latter 1980s. The Gulf War, however, generated a wave of postconflict purchases, low oil prices notwithstanding.

Together, these two high-tension regions have taken in 77 percent of the medium and heavy tanks exported to developing countries since 1950, 86 percent of the infantry fighting vehicles, just over half of the APCs, and just under half of the armored cars.

Other developing regions exhibit much more modest appetites for armored vehicles. South and Northeast Asian armies have emphasized battle tanks over other types of armored equipment. Note that in the South Asian case, virtually all of the 5,000-odd tanks imported into the region went into the armories of just two feuding countries: India and Pakistan. Overgrown Southeast Asian countries have favored troop-protecting APCs over high firepower. Deliveries of armored cars and heavy tanks to sub-Saharan African states have been concentrated in just five states engulfed in long-term civil wars: Ethiopia, Angola, Somalia, Mozambique, and Sudan.

In short, a lot of armor has been shipped to developing countries over time. Most of it has gone to the most conflict-prone regions, or to some of the most conflict-ridden states within regions. Armored vehicles can last a long time, if properly maintained, and there are upgrades to be made. Indeed, such upgrade programs may represent the bulk of the market in heavy armor for the foreseeable future, because the unit cost of cutting-edge vehicles is so high. Acquiring more than a few squadrons of fourth-generation tanks would carry a price tag well over $1 billion, and the buyer would have to be concerned about the relentless march of measure and countermeasure that adds still more defensive weight to sumo-like vehicles that are increasingly complex and hence maintenance-dependent, increasingly fuel-hungry, dependent on truck transporters to move any great distance, and hell on roads when they move themselves. They are, moreover, increasingly vulnerable to smart and powerful antitank weapons. Although it is too soon to predict the demise of the market in battle tanks, cost and technology ultimately may combine to do it in.

Naval Systems

With very few exceptions, naval acquisitions by developing states have as their purpose (1) coastline and sea-based resource protection, or (2) national prestige. The former purpose generates demand for patrol craft, corvettes, and for states like Indonesia with vast archipelagic waters to patrol, longer-range frigates and destroyers, as well as submarines. But the latter, larger ship types meet prestige requirements as well. To the extent that navy budgets divert resources from domestic economic and social needs for purposes of prestige, for either the nation or the naval service itself, such spending may be unwise. However, to the extent that they divert funds from land forces spending in a fixed defense budget, they may serve a useful domestic purpose in addition to whatever national security service they provide. With rare exceptions, by dint of force structure and purpose, navies tend not to instigate seizures of power by the military. There are cases, such as Argentina's "Dirty War," in which naval units have played unsavory roles, but by and large coups are a pastime of developing state armies. All other things equal, then, naval spending might be considered at least neutral toward and perhaps favorable to the emergence and longevity of civilian and/or democratic governments in at least some developing states.

Warships can be divided into two basic categories: those that float on the surface and those that sink (deliberately) beneath the waves. We will look at warship proliferation in that order.

Surface Vessels

Of all weapon systems, naval platforms come in the widest variety of sizes and firepower, from 100 ton coastal patrol craft to 90,000 ton nuclear-powered aircraft carriers. To account for the wide variation in naval vessel size, this section discusses not only quantities of vessels transferred but also tonnage, which is a more useful distinction for warships than "generations," partly because hulls and propulsion plants last a long time (a half-century is not uncommon) and ships may go through several upgrades of communications and sensor and weapon suites along the way. Thus a ship may have capabilities at the end of its days that may have been unheard of when it was first commissioned.

In the discussion that follows, the term *patrol craft* refers to vessels of less than 500 tons full load displacement (FLD, the weight of seawater displaced by the hull of a fully loaded ship); *corvette* refers to vessels of 500 to 1,000 tons FLD; *frigate* refers to ocean-going warships displacing at least 1,000 tons; and *destroyer* refers to warships displacing at least 3,000 tons. The latter two categories have experienced substantial weight gain over the time period covered here. A World War II–vintage destroyer (e.g., the U.S. Gearing class at 3,500 tons) displaced less than a modern frigate like the U.S. Perry class (at 4,000 tons). Moreover, modern destroyers like the U.S. Burke or Spruance class are the tonnage equivalents of 1960s-vintage U.S. cruisers like the Leahy and Belknap classes (all weighing in at 8,000 tons). Although the term *cruiser* is still applied to a number of U.S. and Russian ship classes and once referred to distinct ship types with high speed, light armor, and heavy guns, contemporary cruisers and destroyers have become functionally and, to a large extent, physically indistinguishable. The United States even used the hull and propulsion system from the Spruance class as a base for its Ticonderoga-class air defense cruisers, while its newest destroyers carry an updated version of the Aegis air defense system that was the Ticonderogas' reason for being.

Through the 1960s, many naval transfers involved what the U.S. government calls "excess defense articles," or military surplus provided at marginal cost to recipients. Because ships are durable items, many enjoyed long second careers in their foster countries after a decade or more in U.S., British, Dutch, or Soviet service. Secondhand aircraft carriers and cruisers went to just two regional destinations: South Asia and

South America. Destroyer-sized and smaller ships were more evenly distributed, both to places like Taiwan that faced genuine oceangoing naval threats and to destinations like Brazil, Argentina, and Chile, where service status, opportunities to exercise with the U.S. Navy, and U.S. desires to forge and maintain military ties within the Western Hemisphere clearly outweighed any specific naval threats these countries faced as reasons for the transfers.

Total estimated tonnage of large and small vessels transferred to developing states between 1950 and 1995 is charted in figure 1.17.[42] Tonnage, as noted, is a reasonable surrogate for warship capability, as larger vessels have greater range and carry more capable sensors and a wider array of weapons.

Whether a warship is equipped with cruise missiles is a second rough measure of capability: gun-armed vessels have a weapons radius of a few miles at best; and rarely beyond the horizon (about 20 km) for the 76 to 100 mm guns that equip modern warships. Missile-armed ships, on the other hand, can deliver a 100–500 kg smart explosive over 50 to 100 km, theoretically giving them giant-killing capabilities.[43] The cruise missile gives small warships the sort of heavy punch once reserved to ships of cruiser size and larger.

In the three earliest periods depicted in figure 1.17, transfers of surplus blue-water (oceangoing) ships dominated the trade. The United States, in particular, had a large stockpile of convoy escort vessels to dispose of after World War II. These vessels made their way into the navies of a number of new allies as the United States signed mutual security pacts around the globe in the 1950s. With the explosion of new states in the 1960s, however, there arose greater demand for gun-armed patrol craft to secure coastal waters, demand that slowly dropped over time as initial requirements were met. That demand then shifted in the 1970s to larger, more seaworthy vessels as states increasingly claimed 200-mile-wide coastal economic zones and began to patrol them. The 200 mile zones were codified in the third United Nations Law of the Sea Con-

Figure 1.17   Rough Tonnage of Warships Delivered, All Developing Regions

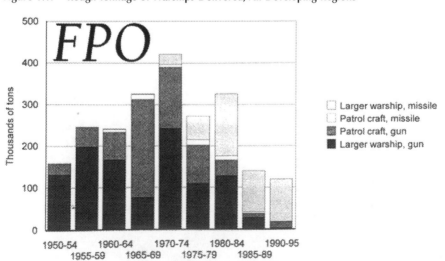

to ratify the convention, the 200 mile fisheries zone has become the de facto law of the sea, claimed by the United States and most other countries with coastlines.

Despite the dominance of missile-armed patrol craft in popular and trade reporting on developing regions' navies, the great majority of patrol craft transferred have been gun-armed, even in the 1990s. Soviet-built missile boats first appeared in the early 1960s, when Komar-class craft equipped with Styx guided missiles were sent to Indonesia, Cuba, Egypt, and Syria. An Egyptian missile boat sank the Israeli destroyer *Eilat* during the 1967 Middle East War, in the first such David-and-Goliath engagement. Israel itself soon built a fleet of fast, formidable, missile-equipped patrol craft and corvettes to police its shorelines and engage other missile boats. However, early fears that proliferating missile craft could be used to inhibit free navigation through international straits around the world, or otherwise disturb peace and commerce, proved unfounded.

Missile-armed frigates made their first appearance in the early 1970s, not as Western surplus but as new orders for the navies of Chile, Peru, South Africa, and Libya, and all came from European shipyards. Missile-armed ships increasingly dominated the market over time; few larger warships are now sold that do not sport cruise or, more rarely, surface-to-air missiles. The Iran-Iraq War stimulated other states along the Persian Gulf to upgrade their naval capabilities in the early 1980s. Together with three Andean nations (Ecuador, Peru, Chile) and three Maghreb states (Algeria, Libya, and Tunisia) they accounted for most of the surge in deliveries of missile-equipped corvettes in that time period. Colombia and Venezuela received most of the missile frigates. India, Chile, and Argentina accounted for most of the early–1980s missile destroyers. The same relative handful of countries or their neighbors (e.g., Pakistan and Bangladesh) bought most of the ships in the late 1980s and early 1990s.[44]

Several developing states that operate ocean-going warships now build their own. Argentina builds German-design MEKO 140 frigates and Brazil builds frigates of its own and of British design. India builds corvettes, frigates, and (more slowly) destroyers, mostly of British or Russian design but some indigenous. South Korean shipyards have turned out patrol craft, corvettes, frigates and, in the 1990s, a series of destroyers to replace World War II–vintage, U.S.-origin ships. Taiwan builds U.S. Perry-class frigates under license and, like Korea, soon will be building destroyers to replace its own aging fleet of ex–U.S. Navy ships.[45]

Submarines

Since 1950, SIPRI has recorded the transfer of about 160 submarines to developing-country destinations (see table 1.9). Through the 1960s, many of these were refurbished U.S. World War II and immediate postwar designs, plus Soviet boats based on late World War II designs from Germany. The U.S. Navy stopped buying diesel submarines in the mid-1950s and stopped U.S. shipyards from building them for other countries, thereby burning the bridges that might have let the Congress push the Navy back to non-nuclear submarine propulsion. But the Navy's decision also left the market for new conventional submarines to the Soviets and to various European builders (primarily the Germans, French, and British, and secondarily the Dutch and Italians). This American absence has made the diesel submarine market unique among conventional weapon systems and different from other instances of U.S. arms trade abstinence, where we owned the technology but refused to sell it.

Table 1.9 Submarine Acquisitions by Region and Time Period

	1950–1954	1955–1959	1960–1964	1965–1969	1970–1974	1975–1979	1980–1984	1985–1989	1990–1995	Country Totals
Middle East										
Egypt		7	1	6			4	2		20
Israel		1	1	3		3				8
Libya						3	3			6
Algeria							1			1
Syria								2		2
Persian Gulf										
Iran						3			2	5
South Asia										
Pakistan			1		2	3	3			9
India				4	3	1		10	1	19
Bangladesh							1			1
SE Asia										
Indonesia		2	12				3			17
NE Asia										
N. Korea				2	6	3				11
S. Korea						1			1	2
Taiwan					2			2		4
So./Cent. America										
Peru	2	2				2	4			10
Brazil		2	2		9	1		1		15
Argentina			2		3	1	1	1		8
Chile			2			2	2			6
Venezuela			1	1		4				6
Colombia					2					2
Ecuador						1				1
Cuba						3	1			4
So. Africa										
So. Africa					3	2				5
Time Period Totals	2	14	22	16	30	33	23	18	4	162

Twenty-two developing countries operate diesel-powered submarines and several (including Argentina, Brazil, India, Taiwan, and both Koreas) produce them under license. The U.S. Navy has found this proliferation to its horrified liking as it provides a helpful threat against which to gauge the navy budget in the post–Cold War era. Planners have turned from open ocean sea control and in-your-face strategies for dealing with the old Soviet navy, to support of land operations in regional conflicts. However, if shallow water, coastal operations are tricky for submariners, they also pose difficulties for antisubmarine warfare, as the acoustics of shallow waters complicate detection and tracking.[46] The British worried about Argentine submarines during the Falklands/Malvinas War, spooked by false sonar echoes and torpedo tracks that may or may not have been real.[47] Western analysts worry about Russian Kilo-class submarines that have been sold to Iran, which could make a future Persian Gulf operation riskier, even though the risk may involve unintended encounters rather than intentional belligerence.

In 1997, about 110 of the 160 boats ever transferred to developing countries remained in operation, according to the London International Institute for Strategic

Studies.[48] With the exceptions of Iran, North Korea, and Libya, the countries that operate most of the submarines in developing and newly industrialized countries are not notable enemies of the international status quo. Some submarine operators like Egypt, Turkey, or Pakistan might at some future date be ruled by Islamist regimes not friendly to Western interests, as Iran has been since 1979, but all of them are dependent to one degree or another on seaborne commerce and thus have a stake in its unhindered flow. On the Korean peninsula, submarines could be indirect contributors to the course of a future conflict, but the obsolescence of North Korea's submarine force and the low probability of its modernization make it a relatively minor threat to South Korean or Western navies.

The most widely distributed diesel-powered submarine is the German Type-209, sold to ten countries, mostly in South America, where it is de rigeur for a coastal state to have at least two submarines in its fleet. Peru is the outlier with eight boats. In Asia, India has bought or built four Type-209s, and South Korea six.

After the Type-209, the most widely distributed boat is the Soviet/Russian Kilo class, of which 13 units were in operation in developing states by 1995. Revolutionary Iran took delivery of three.

Debate about the nature and role of the submarine is as old as the technology itself. The original stealth weapons, the earliest submarines were intended to sneak up on and sink surface ships at anchor. By the second decade of the twentieth century they were capable enough to attack ships on the open ocean. Whether that basic purpose makes the submarine an offensive or defensive weapon depends on one's perspective. In the 1920s, the French wanted to keep submarines out of the Washington Naval Treaty, insisting they were ideal coastal defense weapons.[49] During the Cold War, America's NATO allies built and operated quiet diesel-electric submarines that were well suited to defending ocean "choke points" like the relatively narrow waters to either side of Iceland, through which Soviet warships and submarines based in the Arctic had to pass to reach the North Atlantic, the North Sea, or the Mediterranean. What NATO submarines would have done to Soviet naval traffic to and from the Arctic, Iranian submarines, it is feared, might do to Western petroleum traffic into and out of the Persian Gulf. Whether that would be in Iran's interests (discussed further in chapter 3) is unclear. It is clear, however, that modern submarines are becoming more capable weapons.

In addition to submerged-launch cruise missiles and advanced acoustic homing torpedoes, submarine capabilities will be advanced over the next decade by so-called air-independent-propulsion (AIP) systems, which carry an onboard oxygen supply for their engines and offer substantially improved underwater performance without nuclear power.[50] AIP-equipped boats will be able to deploy quietly up to 1,000 miles from home, should military circumstances require it (and should they have a couple weeks to spend getting there). Alternatively, such boats could spend two weeks quietly submerged nearer home, on blockade duty or laying mines (versus a few hours for current submarines reliant on batteries for performance while submerged).

Such technology will complicate twenty-first-century naval operations in most regions of the world. As naval technologies (and not just submarines) become more sophisticated and proliferate in coming decades, the U.S. Navy will face an increasingly leveled playing field. The United States has been accustomed to such conditions only in dealing with another superpower like the USSR. To face it in microcosm in regional engagements will be new, discomfiting, potentially dangerous, and likely to give decisionmakers further pause before committing U.S. forces to fight. On the other hand, submarines are not that widely distributed, and not that many modern boats are in the hands of countries against which the United States may feel impelled to act militarily.

Missiles

Of all the new weapon technologies of the Cold War arms trade, the guided missile most distinguishes the period from all those coming before it. Tanks, aircraft, and warships fought World War II and figured in the prewar arms trade. But guided missiles grew out of that war and soon enabled armies, air forces, and navies to extend the destructive reach of their respective weapon platforms by tens to hundreds of kilometers with increasing accuracy and decreasing risk to human crews as time passed and technology evolved.

Defining Generations of Missiles

The extensive evolution of military missile technology over the course of four decades (which is detailed in appendix B) suggested that it could be useful to apply the notion of technology generation, as used for aircraft and tanks, to missiles. For this study, military missile types were sorted into at most four generations. Because missiles are more task-specific than combat aircraft, generational comparisons are most meaningful within specific types. One can readily demonstrate that an older-model infrared-guided air-to-air missile that can home in only on hot engine exhaust is less advanced than a newer one capable of engaging targets from any angle, coming or going. It is more difficult, however, to assert that a particular air-to-air missile is more advanced than a particular antitank missile. The two types use very different guidance systems and warheads, require very different levels of maneuverability, and have very different purposes. Representative examples of different types and generations of missiles are listed in table 1.10.

Comparing the Trade in Attack and Defense Missiles

The security studies and arms control communities have long debated whether weapon systems can be categorized as offensive or defensive in character. Discussed at greater length in chapter 2, the general community consensus has been that they cannot be so categorized; that virtually any weapon can be used to support an offensive strategy even if its immediate purpose is defensive; and that virtually any weapon can be used to defend, even if its principal purpose is attack. However, in terms of their specific tactical functions, surface-to-air missiles (SAMs), air-to-air missiles, and antitank guided missiles (ATGMs) are more defense-oriented than not, while air-to-ground, ground-to-ground, and antiship missiles are more attack-oriented than not. That is how they will be treated here, for analytical purposes.

With a caveat that the data on deliveries of missiles are by and large less accurate than data on missile launchers or other major weapon delivery systems, defense-oriented missile deliveries to developing countries have outnumbered deliveries of attack-oriented missiles by about nine to one. Because the volume of missiles estimated to have been transferred is rather large (over a quarter-million units), that still represents roughly 30,000 offensive weapons. Nearly half of those were cruise missiles. As we discuss the overall patterns of transfers in this section, these proportions should be borne in mind, as well as the fact that an unknown number of these missiles have been expended in training, testing, and combat.

Table 1.10 Missiles Entering Service, by Generation and Year, Examples

Type	First Generation	Second Generation	Third Generation	Fourth Generation
Air Defense (SAM, AAM)	AIM-9A Sidewinder (56) AIM-7A Sparrow (56) R-530 (63) AA-2 Atoll (59) FIM-43A Redeye (65) SA-7 Grail (66)	AIM-9H Sidewinder (70) AIM-7C/E Sparrow (58/69) R-550 Magic (75) AA-7 Apex (70) Seacat, Tigercat (63) Blowpipe (74) MIM-23A HAWK (59) SA-2 Guideline (58) SA-14 Gremlin (78)	AIM-9L Sidewinder (76) AIM-7F/M Sparrow (81/83) Skyflash (78) AA-8 Aphid (75) R-440 Crotale (71) Rapier (73) FIM-92A Stinger (81) Javelin (UK) (85) Improved HAWK (82) SA-6 Gainful (67) SA-16 Gimlet (87)	AIM-120 AMRAAM (91) AA-10 Alamo (85) VT-1 (92) Improved Rapier (87) Starstreak (UK) (94) MIM-104 Patriot (82) SA-10 Grumble (8X)
Anti-Armor	SS-10 (FR) (56) BO-810 Cobra (GR) (61) Vigilant (UK) (63) AT-1 Snapper (60)	Milan (74) Swingfire (69) BGM-71A TOW (70) FGM-77A Dragon (73) AT-3 Sagger (65) HJ-8 Red Arrow (82)	Milan-2 (84) HOT (77) BGM-71C I-TOW (81) AT-5 Spandrel (77)	MR TRIGAT (00) BGM-71D TOW-2 (85) AGM-114A Hellfire (85) Javelin (US) (94)
Cruise	SS-12 Anti-ship (75) AS-1 Kennel (61) SS-N-2 Styx (59)	Sea Killer (70) Hai Ying-2 (76) SS-N-9 Siren (69) SS-N-2 Styx-C (67)	MM-38/40 Exocet (73/78) Otomat I/II (76) AGM-84A Harpoon (77) C-801 (China) (85) SS-N-22 Sunburn (80)	
Other Air-to-Ground	AGM-12B Bullpup (59) AS-11/12 (60)	AGM-65A Maverick (72) AS-20/30 (62)	AGM-65D Maverick (83) AS-30L (88) AS-14 Kedge (80)	AGM-65G Maverick (91)

Patterns of Missile Diffusion

Eleven different models of first-generation guided missiles entered service in the arsenals of industrial states in the second half of the 1950s, along with a half-dozen second-generation models, which dominated the field through the 1970s (figure 1.18).[51]

Figure 1.18 Missile System Models Entering Service, by Generation and Time Period

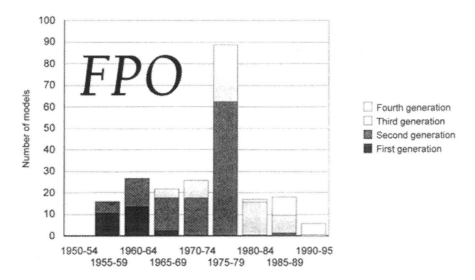

The latter 1970s were particular boom years for production of new designs, both second and third generation, while the most advanced, fourth-generation missiles made their appearance in the latter 1980s. Overall, more than 200 missile types and models that entered military service between 1955 and 1995 eventually showed up in developing countries' arsenals.

Figure 1.19 shows how the four generations of missiles proliferated into those arsenals, counting just dates of *first* delivery of new missile models to any given recipient. That is, if Afghanistan received deliveries of Soviet AA-2 Atoll air-to-air missiles

Figure 1.19 Rates of Guided Missile Proliferation to Developing Countries, by Time Period

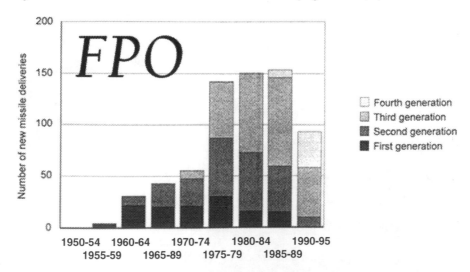

in 1979, 1980, 1982, and 1985, only the 1979 delivery is counted for purposes of figure 1.19; subsequent deliveries just added more of the same item. Only five developing countries received guided missiles in the 1950s (Israel, the Philippines, South
Africa, Pakistan, and Taiwan). By the early 1960s, about 30 first deliveries were made
to 15 countries, from either the USSR or Western suppliers, but not both. By the early
1980s, about 150 first deliveries reached 42 countries, of which six (India, Pakistan,
Iran, Iraq, Libya, and Argentina) were clients of both East and West. Third-generation
missiles increasingly dominated first deliveries from 1980 onward. Fourth-generation
missile types appeared in quantity only in the 1990s.

During the Cold War, Western suppliers were the first to transfer missile systems to
developing countries, sold them to many more countries than did the USSR, and sold
them with much shorter delay. Figure 1.20 depicts first deliveries of missiles over time,
for Western suppliers and for the Soviet Union/Russia, subdivided into short (not
more than five years after entry into service) and longer delays. Western (mostly American) missiles initially went to urgent Cold War clients, who received essentially brand-
new equipment. While the United States was heavily involved in the Vietnam War,
developing state clients received older model missiles, perhaps because the newest
models were needed and being used in quantity by U.S. forces themselves. After the
war ended, short first-delivery delays once again dominated Western missile trade.

The Soviet Union made one brief foray into the quick-delivery market in the early
1960s, to leftist-nationalist clients like Cuba, Indonesia, Egypt, Iraq, and India. Thereafter, the Russians became more cautious missile proliferators, waiting longer to make
first-delivery. States receiving new Soviet equipment were primarily in the Middle
East (Syria, Egypt, Algeria). Smaller-volume Russian missile sales in the 1990s show
relatively greater emphasis on early deliveries once again, as Moscow seeks to market
its latest high-tech wares to any willing customer.

Figure 1.20 Relative Delay in First Delivery of Missiles, by Time Period and Supplier

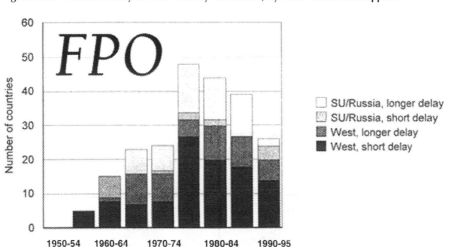

Note: Short delay = five or fewer years after entry into service with first producer state.

Finally, with so many missile models chasing customers in developing countries, I was curious to confirm the degree to which missile proliferation focused on the leading arms recipient regions. In fact, the "density" of missile proliferation—the average number of missile models imported by states in a region—varied by a factor of three across the six major developing regions. Between 1955 and 1995, states in the Persian Gulf and Middle East imported an average of 15 models each (see figure 1.21). At the other end of the scale, South American countries imported just five models apiece over the same 40-year period.

Figure 1.21 also allows one to gauge the proportion of missile types transferred to each region by each major arms supplier. The United States and France dominate transfers to the Persian Gulf, while Soviet/Russian models dominate the Middle East (outside Israel). In Northeast Asia, Taiwan and South Korea import mostly American technology, while North Korea received most of its missiles from the USSR. The distribution of supply in South Asia reflects both a succession of Soviet/Russian arms deals with India, and Pakistan's successive supply relationships with the United States, France, and China.

As missiles become smarter and more lethal, their unit cost has of course risen. The longer-range, independently maneuverable missiles used to attack aircraft, tanks, warships, and hardened ground targets tend to be the most costly. On the other hand, if a $200,000 smart antitank missile can destroy a $4 million tank with high probability in one shot, that could make heavy armor a less attractive option for many countries. Similarly, a $1 million air-to-air missile able to bring down a $40 to 60 million fighter with similarly high probability is a relative bargain that may cause some countries to think twice about the airframe investment. This vulnerability of large military platforms like tanks, ships, and aircraft to smart munitions is what's driving aviation engineers toward "stealthy" designs that are hard to track and target. Similar worries have driven tank designers in a different direction. Since tanks and armored troop carriers

Figure 1.21   Density of Missile Proliferation, by Region and Supplier, 1950–1995

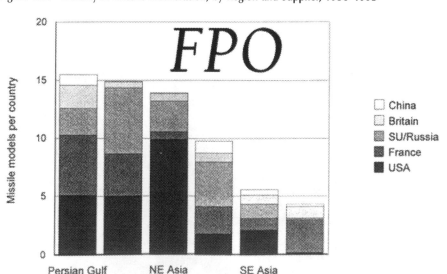

are often acquired as targets at fairly close range (a few kilometers), by infrared or direct eyesight, ground vehicle engineers try not so much to hide their vehicles as to make them harder to destroy, most recently using so-called active/reactive armor (explosive charges on the hull and turret of a tank that detonate outward instantly when they sense an incoming projectile, to disrupt its trajectory). Over the long haul, smart missiles coupled with even smarter command, control, communications, computers, and intelligence spell trouble for large fighting machines.

Naval engineers face particularly complex problems when it comes to keeping a big metal object floating in a vast stretch of water from being a homing beacon for missiles that "see" their targets in the frequencies of radio waves, microwaves, infrared, and visible light. They also have to worry about ships as radar signal reflectors, bouncing back signals from a cruise missile's target seeker.

Naval Cruise Missiles

The growing prevalence of missile-equipped ships in the naval arms trade has been a cause of concern for blue-water (oceangoing) navies everywhere, because cruise missile-armed ships, patrol craft, and airplanes can pose significant risks to major warships. In 1982, during the Falklands War, the Royal Navy lost the destroyer *Sheffield* to a French-made Argentine Exocet cruise missile. During 1987–1988 patrols in the Persian Gulf to protect commercial shipping, one U.S. frigate (the *Stark*) was damaged by Iraqi Exocets. (Missiles are of course not the only risks posed by close-in naval warfare. During the same Persian Gulf patrols, another American frigate, the *Roberts,* was torn nearly in half by a mine, while a mine punctured the hull of the helicopter carrier *Tripoli.*)

Figure 1.22 charts the proliferation of naval cruise missiles by region. First deliveries leapt upward in the latter 1970s, but the regional distribution is surprisingly even. The Middle East and Persian Gulf accounted for most new deliveries in the early

Figure 1.22   First Deliveries of Naval Cruise Missiles by Region

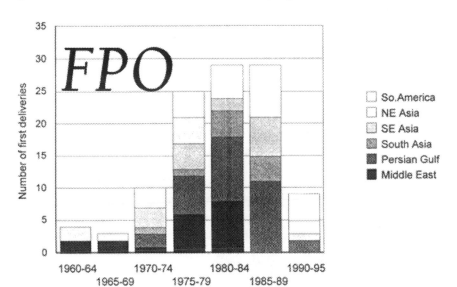

1980s, and Persian Gulf demand remained high in the late 1980s. Somewhat unexpected is the steady importation of new models of cruise missiles by countries in South America, but more than in most developing regions, South America's countries operate blue-water warships, and many of them are missile-equipped.

Altogether, 46 developing countries had taken delivery of naval cruise missiles by the mid-1990s, but the last *new* recipient, Bangladesh, was added to the list in 1989, and it was the first addition in five years. That is, while new *models* continue to appear in developing states' armories, the proliferation of naval cruise missiles to new *operators,* which progressed rapidly and steadily from 1971 through about 1983, was largely completed by then. All coastal states in the Middle East and North Africa, the Persian Gulf, South Asia (except Sri Lanka), Southeast Asia (except Burma), Northeast Asia, and South America (except Uruguay) own cruise missiles, and roughly 60 percent of them operate advanced, third-generation models. Missile inventory upgrades continued in the 1990s, as fig. 1.22 indicates, but even the pace of proliferation of new models decelerated markedly in the 1990s.

Is there a threat to Western interests inherent in the proliferation of these weapons? Most of the countries that operate them are a match only for one another and even the continuing acquisition of cruise missiles by current owners will not change that fact very much. In the event of a war that involved Western powers, cruise missile-equipped aircraft, in particular, could pose a temporary danger to an expeditionary force, as Argentine aircraft did to British forces in the Falklands/Malvinas War.[52] But that danger existed mostly because the British lacked sufficient ability to screen out Argentine fighter-bombers or to destroy Argentine air bases on the mainland. No engagement involving American forces would lack such capabilities.

More pertinent might be the risks of indirect or inadvertent engagement in somebody else's conflict, for example, a conflict that threatened shipping through international straits or enclosed seas. The *Stark* was attacked by mistake, as was an Iranian airliner shot down over the southern Persian Gulf by an American missile cruiser one year later. In such instances, defensive capability is more important than the ability to deter (which can't prevent accidents or mistakes) or the ability to counterattack (which may be politically infeasible). Western navies have been working on cruise missile defenses for some years, fielding a variety of decoys, high-rate-of-fire Gatling guns, and fast, short-range antimissile missiles. None of these defenses has as yet been tested in high-intensity combat. They might be most sorely tested if the missile launch platform were a submarine undetected before its missiles broached the surface, three to five minutes from their targets.

The International Arms Market in the Late 1990s

The Cold War dynamic that inspired a substantial fraction of arms transfers to developing countries has disappeared, but politics continues to play a role in the arms market, especially from Washington's perspective. The United States continues to use arms sales as a means of building and sustaining political bonds with selected recipient governments, particularly Israel, Egypt, and a number of Persian Gulf states. Conversely, it has pressed for international arms and technology embargoes on states like Sudan, Libya, Iraq, and North Korea because of their support for terrorism, their sheltering of terrorists, or their aggressive histories and substantial capabilities in conventional and/or mass destruction weapons.

Most other arms suppliers in the 1990s pursued arms sales for economic reasons, in a market that, by the mid-decade, seemed to have bottomed out at about $40 billion a year (of which $20 to 25 billion worth was sold to developing countries). That market was dominated by the United States with 40 to 45 percent of total deliveries, but it dominates a market with a remarkably narrow base. Since 1980, the ten biggest importers of arms in the developing world have accounted for 40 to 50 percent of major arms transfers worldwide and for 75 to 85 percent of transfers to the developing world. Membership in the top ten has varied somewhat over time. Only three countries (Saudi Arabia, Egypt, and India) made the top ten in 1980, 1985, 1990, and 1995, as well as 1998 (the latest year for which data is available at this writing). But the five yearly snapshots together capture just 22 countries (see table 1.11).

The purchases of two major buyers, Egypt and Israel, are, moreover, paid for with American aid funds in an elaborate system of transfer payments that nets the U.S. economy very little. Depressed petroleum prices are cutting into oil producers' arms-buying power. Indeed, a majority of American arms sales are once again being made to developed industrial states. Japan, which buys all of its imported arms from the United States, is Washington's main industrial customer with $9 billion in purchases between 1994 and 1996. The United Kingdom is second with about $6 billion in the same period.[53] Moreover, a glance at the list of leading arms importers among developing states for 1995 shows two (Malaysia and Thailand) that were hit hard by the Asian economic crisis of 1997–1998.

The Clinton administration has done what it could to encourage U.S. arms exports, so as to ease the defense industry's shrinking pains. Its long-debated policy on conventional arms sales, issued in February 1995, noted the impact of exports on defense jobs and authorized U.S. government agencies and representatives abroad to facilitate foreign access to U.S. defense firms once a major sale has been approved by the U.S. government.[54] Moreover, the National Defense Authorization Act for fiscal year 1996 created something called the Defense Export Financing Program (DEFP), a sort of export-import bank for arms sales. DEFP is authorized to issue several billion dollars per year in loan guarantees underwriting overseas weapon sales. Congress excluded most Middle Eastern countries from eligibility for such loan guarantees but made exceptions for Israel and Egypt.[55]

European arms producers have faced the same smaller arms markets as their American counterparts but have been unable to consolidate their business to the same degree, hampered by often-conflicting national defense policies. Despite the Maastricht Treaty and the European Union's rhetoric of unity, nationalist tendencies remain strong, especially in defense. Moreover, the union's founding Treaty of Rome exempts defense industries from common market rules, making it particularly difficult to form real multinational companies (as opposed to project-related consortia to which national firms make individual contributions). The pace of privatization and industrial consolidation in Europe varies from country to country, with Britain, Italy, and Germany much further along than France, whose defense industry remains largely government-owned.[56] Countries' differing arms export policies also make multinational arms sales problematic. Germany's policies are far more restrictive than those of France, for example, and France has insisted that any new multilateral companies' exports not be constrained by such foreign restrictions.

In early December 1997, political leaders of Britain, France, and Germany nonetheless asked their defense companies to come up with a plan for a multinational European

Table 1.11 Top Ten Developing Arms Importers at Five-year Intervals, 1980–1995, plus 1998

	Frequency of Appearance in Top 10	In Top 10 for 1998?
Saudi Arabia	4	Y
India	4	Y
Egypt	4	Y
Iraq	3	
Syria	3	
Libya	2	
South Korea	2	Y
UAE	2	Y
Pakistan	2	Y
Iran	2	
Kuwait	1	
Malaysia	1	
Taiwan (ROC)	1	Y
Thailand	1	
Afghanistan	1	
Algeria	1	
Angola	1	
Chile	1	
Indonesia	1	
Israel	1	Y
North Korea	1	
Viet Nam	1	

Importer rankings, at five-year intervals, in descending order of arms deliveries

1980	1985	1990	1995	1998
Libya	Iraq	India	Egypt	Taiwan (ROC)
Syria	India	Afghanistan	South Korea	Saudi Arabia
Iraq	Syria	Saudi Arabia	Saudi Arabia	Israel
India	Saudi Arabia	South Korea	Malaysia	UAE
Israel	Libya	Iran	Taiwan (ROC)	Singapore
Saudi Arabia	Egypt	UAE	India	South Korea
Algeria	Iran	Pakistan	Kuwait	Pakistan
Egypt	Angola	Iraq	Thailand	India
Viet Nam	North Korea	Syria	Chile	Egypt
Indonesia	Pakistan	Egypt	UAE	Qatar

Source: SIPRI Arms Transfers Project, "The 50 leading recipients of major conventional weapons." Internet: http://www.sipri.se. Updated February 2000.

aerospace conglomerate, and to do so in three months. Such a plan was to be the first step in a long process of rationalizing Europe's defense industry. Its leaders realize that without such an effort, Europe's industry will be left behind by that of the United States, but as of mid-1998, France remained adamant and progress remained illusory.[57]

While all numbers on China's defense industry are "dubious," as Britain's *The Economist* concluded in a 1997 review, that industry probably employs at least three million people.[58] China has been focusing primarily on upgrading its own defense capacity to world-class standards and has been exploiting a certain synergy with the

Russian Federation. China has money that Russia needs, while Russia has the technology and expertise that China lacks. So far, it has bought three squadrons of Russian Su–27 fighters and plans to build more under license. China's navy has bought several Russian Kilo-class submarines and plans to build more, and it has bought at least two modern, Russian-designed Sovremenny-class destroyers. On April 23, 1997, the two countries' presidents signed a "Joint Russian-Chinese Declaration About a Multipolar World and a New International Order." The United States downplayed the significance of the accord but China's neighbors publicly professed a certain unease.[59]

Russian export agencies beating the bushes for new cash clients had some success by mid-decade, having gotten the hang of marketing their wares. In 1995, the value of Russian firms' contracts even temporarily exceeded the value of U.S. government-to-government sales. Large clients like India (which signed a total of $2.5 billion in new agreements in 1996 with all of its arms suppliers) contributed to Russian firms' mid-decade gains, but Russian sales are unlikely to expand much further. The market is highly competitive, Russian production costs are rising for reasons noted earlier, and several previously promising Asian customers are offline for a while. The most optimistic of Russian forecasts anticipated annual sales of about $7 billion in 1997, but preliminary figures from Russia's state-owned arms export agency, Rosvoruzhenie, suggested sales of just $2.6 billion, down 25 percent from 1996.[60] Russian political gridlock and slow economic reform also present serious problems for Russian enterprises trying to convince foreign customers that they will be able to count on Russian firms for technical support and parts over the 10- to 30-year lifetime of a major weapon system.

Arms control made an ironic contribution to the arms trade with the developing world in the 1990s, spurring rather than constraining weapon transfers as a byproduct of political and military reconciliation in Europe. In the early- to mid-1990s, a large number of heavy weapons, primarily ground forces equipment, had to be destroyed or shipped out of the area covered by the 1990 Treaty on Conventional Forces in Europe (CFE). The treaty, discussed further in chapter 5, established concentric zones of force limitations from western Europe's Atlantic coast to the Ural Mountains that separate the western and eastern parts of Russia. Some treaty-limited equipment could legally be transferred to NATO countries such as Greece, Turkey, or Spain that lay largely outside the zones of limitation, in a process known as "cascading." Instead of paying to destroy excess weapons that did not cascade, NATO CFE signatories offered them to buyers outside Europe at little more than scrap value.[61] As a consequence, in the mid-1990s the market for new, front-line tanks and armored personnel carriers was constrained to the few states willing and able to pay top price for new equipment, but the market for component upgrades boomed.

Countries taking delivery of "scrap" equipment could count on visits from firms specializing in system upgrades, as could countries that obtained arms years ago at a deep discount or as outright grants and wanted to modernize. Many developing countries availed themselves of these firms' services, ending up with more capable weapons for a fraction of the cost of buying new models. Upgrade packages were offered by some novel cross-national partnerships that included companies whose home states used to be Cold War adversaries.[62]

Typical upgrades include improved radars, communications, guidance systems, weapons, and night-fighting capability for aircraft; improved targeting, armor, guns, and ammunition for ground vehicles; and new radars and missile systems for naval vessels.[63] The upgrade market at the end of the decade was brisk, and its ability to give new life and new capabilities to older platforms, including the sorts of communications and data systems that make a modern fighting force effective, suggest that it will remain brisk for some time.

As a result, some of the most salient elements of the "revolution in military affairs" that the U.S. military is counting on to give it battlefield dominance in future conflicts are going to be appearing in forces that the United States may one day have to engage. Of course, technology alone does not make a military effective. It also needs good doctrine, training, and leadership; unit cohesion under wartime conditions; and good logistics, maintenance, and medical support to help it endure those conditions. In the meantime, however, the basic technology of twenty-first century combat continues to spread.

Conclusions

Nearly fifty years of politically and economically motivated arms sales have saturated the testiest parts of the developing world with lethal heavy armaments, serving suppliers' and recipients' purposes but spreading increasingly sophisticated military technologies to more and more users. The proliferation of military technology and the continuing evolution of smart munitions are together making large, traditional weapon platforms increasingly vulnerable to attack and destruction. As a result, in the most advanced industrial countries, military planners and engineers have been turning much of their attention to the problem of making such platforms—the warships, fighter jets, bombers, and tanks of industrialized states' armed forces—less conspicuous targets.

Nonetheless, the United States and a few other countries have gotten so good at the classic forms of heavy mechanized warfare that direct military competition with them may be pointless. Evolving technology may make direct competition unnecessary, however, in all but the most open terrain. In other settings, striving to make life on the battlefield (or life back home) difficult for the major military powers is a far from pointless enterprise. Effective defense might now be had, owing to the advance of technology, without adopting the heavy, armored force structures that have dominated European, Middle Eastern, and Korean militaries for the past several decades. Intensive communications, longer-range weapons for older aircraft, aerial surveillance drones, smart antiarmor munitions, and increasingly capable and cheap computer support offer other opportunities both for national defense and for reprisal. So while the arms market is unlikely ever to regain the size, form, and content that it had in the 1970s and 1980s, it will continue to have an impact on international relations and national interests, partly through the proliferation of major platforms, partly through the provision of the latest components and support for those platforms, and partly through enablement of new forms of warfare.

Still, the arms market in the developing world is driven not so much by recipients' needs to counter the policies or actions of the major powers as it is to deal with regional contingencies, neighboring states, and internal security problems. The legitimacy, stability, and effectiveness of governance in developing states, how their minorities are treated, and how effectively their national economies are integrated with the outside world will have as much to do with states maintaining effective national security as will the status of their armed forces' equipment. Chapter 2 looks in more detail at the sources of insecurity within and among developing states, and thus at the sources of demand for arms, and also examines the role that arms transfers may play in contributing to that insecurity and to the outbreak of conflict.

CHAPTER 2

Arms Transfers and Regional Security

Only for the largest powers—and maybe only for the United States—is national security globally determined. For nine-tenths of humanity it is largely locally determined and focused. The rhythms of the larger international system did strongly influence the security of some developing states during the Cold War, as the opposing sides scrambled to block one another and back developing country clients. The United States paid a heavy price trying to keep South Vietnam out of the communist fold. The Soviet Union paid a similar price in Afghanistan to keep it in. Neither was successful. The United States used guns as well as money to sustain larcenous rulers like the late Mobutu Sese Seko of Zaire (now the Democratic Republic of Congo) because they kept the lid on their countries and kept Soviet influence to a minimum. Mobutu and other autocratic rulers sitting atop unstable states welcomed such largesse and parlayed threats from one side and its clients into arms from the other side. The whipsaw was often effective for the ruling regime but did little to improve basic governance or the lot of the average citizen.

Today, military security threats emanate not from the international system at large but from the next-door neighbors or from within. Ethnic or religious animosities keep a number of developing countries at odds, and many are weak states with limited capacity to deliver goods and services to often-burgeoning populations. Many also have militaries that institutionally predate the state itself, remnants of imperial or colonial rule that were the strongest institutions of government when the current state was formed and that in many instances remain the strongest institution today. This unsettled sociopolitical landscape was the arms trade's major market in the late twentieth century and its likely major market in the twenty-first century as well.

The purposes of this chapter are to lay out the internal security problems and regional security dynamics that help generate developing states' demand for modern arms; to apply two important concepts from international relations theory—deterrence and the security dilemma—to the security situation of developing states; and to assess the impact of arms imports on developing states' economic and political stability, the stability of regional arms balances, and the propensity of political crises to tip into conflict. I also use some simple but powerful statistical measures to examine in greater depth the relationship between arms transfers and both internal and external conflict in the developing world since 1970. I conclude from that analysis that propensities toward conflict and proclivities to import conventional arms are positively but not closely related. Conflict behavior and arms imports seem wired into states' behavior in parallel, as it were, rather than in series: clip the wire to one and the other may keep on going.

Problems and Pathologies of New States

With the exception of some persistent interstate feuds in the part of the world that stretches from the eastern Mediterranean to the vale of Kashmir in southern Asia, internal conflict has been the predominant problem for developing countries. A fundamental lack of governmental legitimacy, and primary group loyalties that are directed neither toward the state nor the government, plague many developing states and leave their ruling elites preoccupied with their own security and the state's cohesion. With Cold War overlays lifted and the military aid and political support that they brought to ruling elites largely gone as well, internal tensions have come front and center in several security complexes, sometimes erupting into civil war, as in Liberia, Somalia, or the Great Lakes region of central Africa. Where the stakes have been neither ethnic nor religious but political in nature, and fueled in part by Cold War interests, the end of the Cold War has enabled civil strife to cool, as in Nicaragua or El Salvador. But throughout the developing world, there remain substantial obstacles to progress in building the cohesive states, responsive governments, and progressive economies that help reduce internal conflict. Developing countries came late to the state-building process but are expected to accomplish it faster and more humanely than their ex-colonial overlords did back home. In the headlong rush to do so, they grasp at tools, like modern armaments, that promise enhanced security, but the question is: security for whom?

The Pressures of State Making

Mohammed Ayoob at Michigan State University has looked long and hard at the structure of what he calls the "security predicament of the Third World," which arises from the "twin pressures of late state making and . . . late entry into the system of states." The first pressure compels new states to compress into a few decades what may have taken centuries to unfold in places like Europe. State making involves consolidating political authority over a defined territory and its people (historically, by war, if necessary); maintenance of order (policing) within that territory; and extraction of resources from both territory and populace (taxation) to support both war making and policing and to maintain the apparatus necessary for routine public administration. The more primitive the stage of state building, the more coercive the strategies employed to accumulate and concentrate power "in the hands of the agents of the state."[1]

Europe, the cradle of the modern state system over four centuries ago, evolved slowly from the absolutist monarchies that replaced feudal society. Over time, the monarch's subjects became a people with a common history, legal system, language, and often religion, evolving a common national identity or even what today might be called ethnicity. In the case of Germany, Italy, and even France and England, the "nation" evolved from such historical habits of loyalty to and identification with the state, and not the other way around. Finally, and gradually, came the evolution of representative institutions "dictated by the need to co-opt into the power structure new and powerful social forces that emerged as a result of the industrial revolution." By and large, however, the old ruling structures did not give in gracefully to these republican pressures. The American Revolution to split the 13 American colonies away from the British Crown, and the French Revolution, which overthrew the Bourbon monarchy, were the opening rounds of two centuries of on-again, off-again democratization and reaction in Europe.

Nonetheless, the three phases of state-building—dynastic monarchy, evolution of national identity, and evolution of representative institutions—were completed before the era of modern mass politics.[2] Developing states, on the other hand, emerged into an international system in which the norms of mass politics and individual rights already were ascendant, and are expected "to perform the tasks of state making in a humane, civilized, and consensual fashion" that is unlike the process that unfolded in Europe. Moreover, they are expected to substitute, for what was in Europe a "largely unpremeditated and uncoordinated evolutionary process," a policy process with predetermined goals to be reached on a "ridiculously short" timetable. The great experiment in central planning that was Soviet Russia served for a time as a seemingly successful role model for many former colonies that needed to bootstrap themselves into statehood quickly, but that model caused at least as much bloodshed, in the USSR and elsewhere, as Europe's unplanned progress. Moreover, its failure, in both human and economic terms, is an object lesson to leaders in developing states who are prone to hold the reins of government and the economy too tightly. But open, competitive markets and politics still need management, lest they decay into anarchy. Successful management requires respect for the rule of law, and laws widely seen as fairly structured and enforced. It requires, as well, internal consensus about government legitimacy, and that is a stumbling block for many developing states.

A Shortage of Good "Security Software"

Lacking internal cohesion, many developing states and regimes suffer from chronic insecurity, which Ayoob considers the defining characteristic of Third World states, even those such as China, India, or Brazil with significant amounts of material capability.[3] A large military and the guns, bombs, tanks, and other accouterments of measurable military capability that come along with it give a state only part of what it needs to be secure. Together with such hardware it needs good "security software," or political legitimacy, political integration, and policy capacity, which together determine how national values are defined and perceived, how national resources are allocated, and how national policies are chosen and implemented. A high level of legitimacy—broad agreement about the government's right to rule and about how it rules—fosters citizen loyalty. Most developing governments don't have it. Faltering legitimacy increases the likelihood of internal challenge, and most regimes respond to such threats as though they were threats to state security rather than to the security of the regime itself.[4]

Political integration means the evolution of commonly shared values and interests among a country's populace. Many developing countries lack this element of security software. Often-arbitrary colonial borders became the equally arbitrary borders of the new states of Africa and Asia. Those borders frequently bisected ethnic groups that might well have preferred a state of their own and under the right circumstances might fight to achieve it, much as the Jewish people did in founding the state of Israel in 1948. Once postcolonial states began to populate the United Nations in significant numbers, however, their leaders made sure that the principle of self-determination could not be used to threaten their new, rather precarious, political positions. Thus, the landmark UN General Assembly resolution that recognized colonial self-determination as a basic human right (UNGA Res. 1541 of 1960) did not further extend rights of self-determination to minorities within the borders of independent states.[5] That has not kept dissatisfied groups from claiming rights to self-determination, and the patchwork of communal interests and identities that marked many developing states at birth has proven quite resistant to amalgamation. Under certain circumstances, those differences can explode into civil war. Consider the futile, decades-long struggles against state authority and repression by the Kurdish peoples living within the borders of Turkey, Iraq, Syria, and Iran, the equally long struggles of Tamil separatists fighting the government of Sri Lanka, or the Muslim-Christian war in Lebanon (1975–1989).

Policy capacity finally relates to how a government detects and processes data from its environment (what it "sees" and what it makes of that); how it creates policy and allocates resources; and how it determines national behavior.[6] A government may aspire to see its world more clearly and objectively but lack the resources to do so. It may profess to create policy and allocate resources efficiently and fairly but be driven in reality by the whims of the leader or the consensus of the junta, by considerations of kinship, or by opportunities for corrupt self-enrichment. A government may aspire to change national behavior by public debate but fall back instead on propaganda. It may aspire to deliver public services competently, efficiently, and flexibly, but achieve none of the three. In an environment where legitimacy is weak and integration lacking, those elements of policy capacity that are best armed may dominate policymaking. Their own institutional interests will receive high priority, but in a framework of low political legitimacy, and hence minimal intrinsic loyalty to the government in power, the coercive potential that militaries and police forces represent, can turn against the government just as readily as it moves against the enemies of that government.

Human Rights and Treatment of Minorities

New states with arbitrary borders, whose governments command the hardware but not the software of national security, and command the bodies but not the allegiance of their peoples, have been expected to observe international human rights norms in their policies and actions. These norms arose out of the "successful functioning of industrialized, representative, responsive states of Western Europe and North America," which collectively set the standard for "the humane and civilized treatment of citizens." But when these bellwethers set the standards that others were expected to follow, they had finished with state-building, were relatively affluent, and "unconditionally legitimate" to their citizens, so they could afford to "adopt liberal standards of state behavior," knowing that societal demands would not run counter to state interests or jeopardize state institutions.[7] To expect developing states to adhere fully to such norms may be too much to expect, but they may have little choice but to try.

States with ethnically mixed populations—and there are many—can take three basic approaches in dealing with their minorities: balancing ethnic demands, as Malaysia has done; trying to assimilate minority groups and ignore demands for group rights, as Turkey has done; or trying to wipe out or drive out minorities, as the Serbs tried to do in Bosnia and Kosovo. Given the number of ethnically mixed states in the world, the third approach is a recipe for "a flood of virtually uncontrollable violence."[8] The second approach merely pushes the problem underground, whence it may later erupt, as did Kurdish violence in Turkey after 1984. Balancing demands may therefore be the unavoidable alternative for states that wish to avoid chronic unrest, and a share (and a stake) in governance may be the only way to produce visible balance. Democracy, therefore, is not just a laudable goal but a political precondition for legitimate state structures and popular regimes. Governing elites are coming to realize that they can't "build credible states and legitimate regimes without guaranteeing minimum civil and political rights and at least some political participation." In the absence of democratic institutions, dissent can easily turn violent. The tragedy for many developing countries is that requirements for enhancing overall state security may clash with requirements for maintaining the security of the ruling regime and are thus resisted by it. Such regimes, it seems, would rather go down fighting and take the state with them, than surrender power peaceably. One hopeful exception may be Nigeria, which, after a succession of increasingly rapacious military rulers, has an elected president who has pledged open and democratic rule.[9]

Security for Whom?

"security for whom," contrasting the interests of the individual, the "nation," the ruling regime, and the state.[10] The nation is the dominant group whose self-identification is seen as the "basis for expression of legitimate political identity and power" within the state. For example, the United States government legitimately represents Americans, not males or white people, black people, or any other people of particular ancestry or class. (Social-democratic states in Europe whose redistributions of domestic wealth have reduced class differences substantially are only now learning how to adapt to this notion of multihued nationality. Other countries fight or reject it. Japanese citizenship, for example, is accorded only to people of one particular racial and ethnic heritage.[11]) The ruling regime is the elite that holds the highest offices of state and commands its institutional machinery. The state is both an international actor in sovereign control of a chunk of territory and a set of institutions that holds the right to coercive use of force to maintain internal order and settle disputes. The security interests of these three entities and of individuals within society may not be congruent. Concentrated ownership of most economic capacity in the country, for example, might benefit the ruling elites, as it did former Indonesian president Suharto and his friends and family, but not ordinary citizens or the state, at least not in the long run, as 1997–1998 economic and governmental crises in Indonesia demonstrated.

In previous eras, such weak, self-conflicted political entities would have been absorbed by other, stronger states looking to expand. Under current international norms, such top-down state-building by territorial accretion is ruled out. "Weak states," argues Job, "have a guaranteed existence"[12] unless, like the Soviet Union or East Germany, they choose to dissolve themselves or, like Yugoslavia, cannot hold their fractious parts together. Somalia has lacked a central government since 1991 but has yet to be absorbed by any other state. Afghanistan has been riven by civil war for decades and may yet split along ethnic lines either despite or because of efforts of the now-dominant Taliban to impose a uniform and draconian version of Koranic law.[13] Neither place is at risk of absorption by its neighbors, who have troubles of their own and don't need to adopt more. Only the export of disorder—including narcotics—from such troubled lands might invite outside intervention, to raid poppy fields or terrorist bases but not to rule, thus giving local anarchy a kind of guaranteed existence, too.

As the principal institutional tools supporting weak regimes' hold on power, police and military forces gain considerable political clout. In some cases, military leaders themselves seize political control, a common occurrence throughout the developing world. In other cases, the military serves as the real power behind nominal civilian rulers. In still others, as will be seen in chapter 3, it arrogates to itself (or may even be legally assigned) the constitutional role of "guardian of the state." As such, it may not often rule, but it sets the parameters of government policy and acts when those boundaries are breached.

The Dynamics of Regional Security Complexes

We all know, in general terms, what people mean when they refer to a "region": a piece of geography of limited extent that is part of a larger whole. The Northwest is a region of the United States; Scandinavia is a region of Europe. But when talking about the particular sets of relationships among states that interest us here—arms and antagonism—it helps to be more specific about the definitions that we use to group states together. What makes it more important, for example, to group Egypt with the Middle East rather than with Africa, the continent on which it sits? British international relations (IR) scholar Barry Buzan has argued that the deciding factors should be not just physical proximity but also "established patterns of amity and enmity and an established distribution of power."[14] He calls such a collection of states a "regional security complex." (A complete region by region discussion of the security complexes that are the focus of this study can be found in appendix A.) Buzan posits a rough global hierarchy of such complexes. A lower-level complex is one comprising states whose power does not extend much beyond regional borders; this is the sort of complex to which most developing countries belong. A higher-level complex counts among its members one or more great powers whose interests reach beyond their immediate region. Europe or North America would be examples of higher-level complexes (the latter comprising Canada, the United States, Mexico, and elements of the Caribbean). Members of higher-level complexes intervene in the affairs of the lower levels, thereby altering the relative importance of local and global dynamics for security relations at those lower levels.[15] But the spread of military technology and "western-style sociopolitical organization" has altered relations between higher and lower complexes, eliminating many of the grossest differences in military and political capacity that in past centuries enabled European states to impose their will on much of the rest of the world.[16]

Power can shift within a security complex because of the disintegration or merger of member states; because of differential rates of economic growth or the introduction of new technologies; or because of outside powers' intervention into member states' affairs.[17] Only serious outside intervention, however, what Buzan calls the "overlaying" of a higher-level on a lower-level security complex, can fundamentally alter the latter's security relations. The security problems such intervention is intended to address are sometimes eliminated but often only repressed. After World War II, Germany and Japan were subject to complete overlay by the victorious Allies, their war parties were destroyed, their military options were constitutionally curtailed, and they were tied into American-led political-military alliances that entailed American guarantees for their security In the 1990 breakup of Yugoslavia, on the other hand, the international overlay on the Balkans security complex was tentative and tenuous, the war parties largely remained in power, conflict has recurred periodically, and by mid-2000 the end was not yet in sight.

Higher-level overlays are of course not always benign, as the history of empire, whether European, Ottoman, or Soviet, attests. In the aftermath of empire, new security complexes must sort themselves out, and the sorting is often not pretty. The Balkans and the Middle East were the northern and eastern marches of the Ottoman Empire in the seventeenth century, carved off piecemeal into other European empires over a 200-year period, until those empires, too, dissolved and their various colonies and protectorates ran free.[18] Both of these complexes have seen multiple conflicts in this century, some of which have drawn major outside powers close to or over the brink of war, as well.

Higher-level overlays are not necessarily political-military in nature. The entire process of economic globalization that attends the opening of developing countries' markets to foreign trade and investment also entails vulnerability to global economic turbulence and to potential regional meltdown—as occurred in East Asia in late 1997—when national or regional economic practices fail to meet international standards. International Monetary Fund structural adjustment packages impose such standards in return for money needed to keep a troubled country's economic institutions from collapse. On the other hand, the heavy, political-military overlay that characterized the Cold War has largely vanished, and whatever local insecurities it helped to suppress in the name of the larger cause have since been free to manifest themselves.

States within a security complex perceive and react to the intensity or immediacy of threats to their security along six related dimensions. These include *specificity* (how sharply defined is the potential threat?), *physical proximity* (is it right on the border or quite a ways away?), *timing* (if it doesn't exist now, how soon might it materialize?), *probability* (how likely is that to happen?), *damage potential* (how bad could it be if it did happen?), and *history* (has it happened before?). A rather diffuse potential threat, in a state a thousand miles away, that might or might not materialize for ten years is unlikely to stimulate advance planning and defense building unless the early signs of trouble match patterns that resulted in serious damage to the same country in the past.[19] On the other hand, a clear threat nearby that could cause trouble soon and that looks like past sources of trouble could stimulate an early and energetic counteraction.

A country's external security is most directly affected by the military capabilities and political intentions of its immediate neighbors and by the recent history of their interactions. Terrorism and espionage excepted, effective military threats require the potential for direct physical interaction, that is, they require a direct "force interface."[20] A historically unique feature of the Cold War was the existence of a force interface between two powers—the United States and Soviet Union—that were physically far removed from one another. Most interfaces are closer and more direct (for example, India and Pakistan, or North and South Korea, whose forces face each other across narrow, "temporary" dividing lines.)

Each country, Buzan argues, looks for different indications of threat based upon its particular history, geography, and vulnerabilities. Russians and Poles worry about land invasion. Americans worry about surprise attack. Japanese worry about economic strangulation. History and proximity are relatively fixed elements in every country's evaluative framework, but of course history is subject to interpretation. A country's sense of the power or proximity of a threat also depends in part on its own sense of military power and political integrity. Few developing states have much in the way of either, and those that have one may lack the other, leaving most perennially insecure. The urge to acquire arms is a common result.

Each regional security complex in the developing world has unique characteristics. In Latin America, a dominant military-political tradition and nineteenth-century social structures lasted well into this century and combined with chaotic domestic politics to keep national militaries close to political power or actually in control of government. This tendency was only reinforced by twentieth-century communism's challenge to such traditional structures.[21] The collapse of communism, the military's failure as an economic administrator in many countries, and occasionally its failure on the battlefield discredited Latin military rule and led to a transition to civilian rule by the late 1980s. In the intervening decades, Latin militaries acquired such things as submarines, cruisers, aircraft carriers, and supersonic jets that had only passing relevance to national security but considerable relevance to institutional military pride.

In sub-Saharan Africa, as A. F. Mullins has observed, most states became independent before acquiring the normal trappings of sovereignty, either the ability to defend their borders or a sense of popular "identification and affiliation with government."[22] Although independence occurred, for the most part, in an environment where immediate border challenges were nil, the ruling claims of new sub-Saharan governments were especially weak and open to internal challenge, increasing the value of a strong and loyal, politicized military able to meet such challenges.

Arab nationhood has been defined perhaps more by what Arab states oppose (such as the state of Israel) than by what they support. Personalized governance, benign and otherwise, has been the rule in most Arab countries for most of their independent existence. One thinks not of parliamentarians or governing structures so much as one thinks of Gamal Nasser, Anwar Sadat, and Hosni Mubarak of Egypt; Saddam Hussein of Iraq; Hafez Assad of Syria; Moammar Qaddafi of Libya; and the House of Saud in Saudi Arabia. Such narrowly based regimes naturally turn to military forces for support and support them in turn. They also have, to some extent, a built-in need for an external enemy to divert public attention from, and to excuse failures of, governance. Egypt lost Israel as an enemy in 1979 but focused on Libya as a replacement. Saudi Arabia may be able to afford the loss of Israel as an enemy if the Middle East peace process finally bears fruit, but it will still have Saddam's Iraq and radical Shi'ite Iran to contend with. Radical Islamist movements challenge some Arab regimes. Security forces have been called upon to deal with a xenophobic terror campaign in Egypt, full-scale insurrection in Algeria, and separatist forces in southern Sudan fighting the government's drive to impose Koranic law on believers and nonbelievers alike.

In South Asia, India is more democratic than most developing countries and has a better tradition than most of separating the military from politics. However, it exemplifies the state driven by national self-image to acquire the sort of arsenal that generates global respect and befits a would-be regional hegemony and continues to have serious problems with minority populations desiring autonomy, if not independence: India must deal with the long-standing problem of Kashmir and with other sectional rebellions of varying intensity.

While most of the troubles faced by developing states stem from the weakness of their internal political consensus and capacity for effective, even-handed governance, most of the theory developed by Western scholars to explain and predict the behavior of states discounts internal problems and focuses on external relations. Does that leave us with any good models to use in explaining or predicting the behavior of developing states, especially with regard to arms acquisitions, that aspect of their behavior that concerns us most here?

Western Security Models and Arms Transfers

The ideas and models of Western, primarily American, international relations theory, whether "realist" or "liberal," derive primarily from European historical experience, notes University of British Columbia scholar K. J. Holsti. They concern themselves with territorial and power competitions among coherent political entities historically prone to the use of force internationally, both against one another and against overseas lands and peoples brought under their collective suzerainty through the eighteenth and nineteenth centuries. From that historical experience, and from observation of the behavior of Eastern and Western blocs during the Cold War, American international relations (IR) scholars in particular spun a web of theories about state behavior that emphasized relative military power and the number of significant power nodes or "poles" in the system of states (unipolar, bipolar, multipolar). Very little of this theorizing, argues Holsti, has much applicability to developing countries.[23] With the exception of the handful of long-running interstate fights, which upon close examination look more like intercommunal feuds (Jews and Arabs, Hindus and Muslims, Sunni and Shi'ite Muslims), the developing world's problems are largely internal in nature and rooted in the sorts of weak social and political cohesion just discussed.

Holsti's argument is compelling in large part, but there are two important elements of security-related IR theory that do seem important to regional security complexes in the developing world. One is the theory of deterrence and the other is the concept of the security dilemma. Deterrence theory evolved in the context of the U.S.-Soviet nuclear standoff but soon came to be formulated in generic terms, involving the following components: a threat delivered to prevent something from happening, obvious capability to carry out the threat if the forbidden action is taken, demonstrated will to carry out the threat, and communication of the threat to the state/group against which it is targeted. On the target's part, deterrence requires vulnerability to the threat, attention to efforts to communicate it, and credence that the threat will be carried out if the threatened-against action is taken and be damaging enough to outweigh any gains from that action.[24] Deterrence can fail to work if any of these components are missing.

The security dilemma can apply whenever entities are uncertain of their security and work to increase it by building up what they may regard as defensive military capability. The objects of their concern (other states or groups) may respond by further building up capabilities of their own. Moreover, they may take the original build-up to be a sign of hostile intent where none in fact existed. Thus the dilemma: do not build up and remain insecure, or build up and risk reactions on others' parts that restore insecurity and may leave one even worse off than before.[25]

It matters whether leaders get the model right in formulating their national defense strategies. If I am a defense-minded leader with limited information about the intentions of neighboring countries, I may guess that they are motivated by the security dilemma and make efforts to improve communications and build friendlier relations to dispel the uncertainty on which the security dilemma feeds. If I guess right, everyone can live in peace at lower levels of military spending. If I guess wrong, however, and the neighbor really needs to be firmly deterred from aggressive behavior, I risk going down in history as Neville Chamberlain II at a twenty-first-century Munich.[26] Turning the problem around, if I guess that a neighboring leader needs to be deterred but he or she is really in thrall to the security dilemma, then the firm warnings and military buildups that I undertake to reinforce deterrence may instead confirm the other's worst fears, trigger an arms spiral, and generate hostile intent where none existed before.[27]

Are these models of relevance to developing states and their security problems? Holsti would argue not, inasmuch as Western IR theory is a sovereignty-based construct that assumes strong (in the sense of legitimate) states and has little to say about the "presovereign" situations of most developing countries and their predominantly domestic troubles.[28] Yet whether or not places like Syria, Egypt, Israel, India, Pakistan, and other participants in regional feuds have all the traditional attributes of sovereign states, they are treated as such by the rest of the international system. Moreover, their serious shortcomings as states do not prevent them from contracting with other states and private firms for armaments, or from being threatened by equally presovereign neighbors, or from experiencing the security dilemma, especially in cases where mutual ill-feeling stifles effective interstate communications.

For most developing states, external threats will tend to take the form of spillover effects from conflicts in neighboring states. These could include incursions by armed groups seeking respite from fighting (or support from cross-border ethnic kin), smuggling to finance the fighting, and refugee flows. Deterrence might come into play if the neighboring government is in fact the proximate source of the ripple effects (deliberately "cleansing" itself of minority groups, for example).

The security dilemma can also apply within states. Holsti notes that, in many developing countries, the government is the major internal threat to citizens' well-being. Since 1945, such governments have killed far more of their own in internal wars than they have lost in interstate conflict.[29] Communities within such states thus may face a dilemma with respect to their own rulers: take repression lying down, or invite greater repression by taking up arms and fighting. During the Cold War, such a decision to fight tended to attract the attention of one or both superpowers. The United States was inclined to provide governments the arms and training that they wanted to fight such insurrections. The USSR was inclined to support self-described national liberation movements that could cause trouble for friends of Washington and perhaps create eventual allies for Moscow inside a regional security complex and in the halls of international organizations.

Ruling elites in such situations face what Brian Job calls an "*in*security dilemma." Their concerted efforts to shore up their internal positions tend to make those positions even less secure, because the coercive methods used to shield a narrowly based, illegitimate regime or to substitute for competent administrative capacity will successively alienate people whose loyalty the elites both want and need to sustain their rule. But not taking such punitive actions also threatens their rule by permitting the rise of disloyal opposition.[30]

Where government fecklessness rather than government action has been the problem, argues MIT's Barry Posen, especially in the "wreckage of empires" like Yugoslavia and the former Soviet Union, the security dilemma may become especially acute between groups.[31] Whether such groups misperceive one another's intentions or see them clearly, they may feel especially vulnerable in an environment that increasingly mirrors the anarchy of the international system. One or more groups may be genuinely malevolent, as in Bosnia-Herzegovina, a fragment of Yugoslavia that dissolved into genocidal bloodshed in the early 1990s. All that is needed to ignite violence in such a situation is a "tipping event"—such as the referendum on independence that triggered fighting in Bosnia—or a balance of power and opportunity favorable to the malevolent party that is likely to shift against it over time. A tipping event that also alters the balance of power against the malevolent party may thus be especially dangerous.[32]

Such an event occurred with respect to Kosovo, a southern province of Serb-ruled Yugoslavia with a population that was 90 percent ethnic Albanian. In early 1997, many Albanians' life savings were lost to a failed government-sponsored financial pyramid scheme. Angry mobs overran government buildings, including arsenals and public order in Albania fell apart. At least a half-million weapons disappeared from those arsenals, and many found their way—very cheaply—across the border and into the hands of the "Kosovo Liberation Army," a coalition of separatist groups favoring an independent Kosovo. With ready availability of mostly small arms, the loosely organized KLA began a campaign of harassment and assassination against Yugoslav paramilitary police forces who were largely of Serbian ethnic extraction, and who retaliated with increasing brutality, not just against the guerrillas but against the ethnic Albanian population of Kosovo.[33] The conflict attracted the political and military attention of the United States and its European allies in the North Atlantic Treaty Organization (NATO), who were already busy keeping the peace a few miles away in Bosnia. In the spring of 1999, a year after serious trouble in Kosovo began and following two failed rounds of peace talks, NATO began a concerted bombing campaign against Yugoslavia to force a peace in Kosovo, and Serb forces began "cleansing" Kosovo of its entire ethnic Albanian population. Soon, NATO and the United Nations were keeping the peace for the duration in Kosovo as well.[34]

Had Albania not suffered the meltdown that freed up cheap arms, the KLA's actions in Kosovo very likely would have remained relatively low-key, and Kosovo would have remained, for at least some while longer, a merely unpleasant rather than genocidally dangerous place in which to live. Thousands of lives might have been spared, and billions of dollars in property damage in Kosovo and across Yugoslavia might have been avoided. This is not to argue that Yugoslav president Slobodan Milosevic and his armed forces do not bear major responsibility for the conflagration in Kosovo, but to emphasize the destabilizing effect that sudden influxes of armaments can have in tense situations. Serbs consider Kosovo the cradle of Serbian nationalism and sacred ground, owing to a decisive defeat suffered there at the hands of imperial Turkish forces in 1389.[35] Whether or not one sympathizes with the Kosovar separatist cause, the KLA's newly found strength proved just enough to seriously alarm the Serb-dominated Yugoslav government, police, and army, but not enough to deter them, let alone defeat them. Had the KLA gained access to heavy weapons instead of small arms, the situation might have been different, but heavy weapons require much more training and unit coordination to operate, are much more visible, and need roads on which to move. The KLA would have had to bring its arms across border roads closely guarded by Serbian forces. An attempt to contest control of Kosovo on the Serbs' terms with a poor replica of Serbian forces, without benefit of air cover, would have failed.

Arms Transfers and Recipient/Regional Stability

In the jargon of international security studies, events like the sudden availability of armaments in Kosovo that shake up the political-military status quo are "destabilizing." Throughout the Cold War, "stability" was the main objective of American foreign policy in the Third World, just as "containment" was its main objective in Europe and around the Soviet periphery. Stability connoted calm, control, and the absence of trouble, all desired ends when policy's aim was to avoid political change, because change carried the risk of bringing into power groups unfriendly to the West and thus, by extension, friendly to the USSR or China. Today, Western interests in developing regions are largely commercial, and Western investors seek politically and economically stable locales in which to invest. Western firms and governments are also the principal suppliers of major conventional weapons to those regions. Opponents of the arms trade argue that it undermines developing economies, while supporters argue that arms transfers can reinforce political stability in key countries. Opponents also argue that the arms trade can contribute to regional arms races and that certain types and configurations of armaments can make political-military crises more likely to tip into conflict. Let's look at each argument in turn.

Economic Stability

The dominance of economics and trade in 1990s U.S. foreign policy gives economic stability abroad a value that is equivalent to that of political stability during the Cold War. Indeed, the latter may not even be possible without the former. Economic stability might be defined, somewhat crudely, in terms of what is *not* happening in a country: annual inflation is not rapidly destroying wages and life savings; banks are not failing due to inept or corrupt investment practices; armies of job seekers are not fleeing from rural to urban areas; and people already in the cities are not taking to the streets to demand work or public support. The government, consequently, is not at risk of violent or other extralegal replacement due to any of these factors.

What impact can arms transfers have on economic stability? Since arms sales these days are mostly retail, arms imports can hurt a country's balance of trade, run up its tally of external debt, and divert public resources from other areas of expenditure that may contribute more directly to the country's ability to compete in the global economy. But the average developing country spent only about 8 percent of its annual military budget on imported armaments in 1993. Since 1993 was a low-water mark for the arms trade, increase that fraction by half for a more typical year, to 12 percent. That still leaves 88 percent of military spending in developing countries directed toward things like salaries, operations and training, and building and equipment maintenance. Moreover, when economic crises occur, states can more easily delay offshore arms purchases than they can cut domestic military spending (they may need the army to weather the crisis, and would want it to be well-disposed toward the ruling regime). When the economic crisis hit East and Southeast

Asia in 1997, pending international arms deals were among the first obligations that governments sought to shuck.[36]

Whether the costs of imported arms (both initial acquisition costs and ongoing costs of operation and maintenance) are worth it to a recipient country, its government, its political leaders, or its military depends, of course, on what threats the acquired arms will counter, what political coalitions they will cement, and (in too many countries) whose pockets the transactions will line. Corruption is a perennial problem in developing and transitional economies, where the dividing line between private and public goods and authority is as yet not strongly drawn, and the rule of law may take second place to personalized governance and the accumulation of family wealth. Since 1995, Berlin-based Transparency International (TI) has been compiling and publishing annual surveys of perceptions of corruption by "business people, risk analysts, and the general public," with 1998 coverage of 85 countries.[37] That is, TI's index does not measure corruption directly but instead measures beliefs about corruption. However, perceptions affect confidence in government and the investment choices of international business.

Of 47 developing countries included in the survey, 37 have below-average index scores (see table 2.1); that is, on the whole, developing countries are perceived to have much greater problems with corruption than developed states. On the ten-point scale that TI uses, ten is "least corrupt." The median index score (half of the countries score higher and half lower) for all 85 countries in the survey is 4.2, whereas the median score for developing countries is 3.3. However, some countries still classed as developing (Israel, Chile, Botswana, and Costa Rica) scored well above the overall median.

If governments with poor TI index scores also tend to buy more arms, we might be able to start building a case for corruption as a driver of the arms trade with developing countries. The TI index rates fewer than half of the states whose arms buying habits we tracked in the charts in chapter 1, and misses a few major buyers like Saudi Arabia or Kuwait, but it does include 12 of the current top 20 buyers. Moreover, while the index, as its creators note, may not include the most corrupt countries in the world, it does include several ranked about as low as the scale can read. In short, while not a scientific sample, it contains a broad cross-section of the countries whose policies and behavior are of interest here.

Is there a relationship between arms imports and corruption? A simple scatterplot comparing corruption scores with the three most recent years of arms import data from the U.S. Arms Control and Disarmament Agency (figure 2.1) suggests that there is no appreciable relationship. Three-year average values for arms imports from 1994 to 1996 were converted into logarithms for ease of plotting. (A value of 1 on the X-axis corresponds to $10 million in arms imports per year; a value of 2 corresponds to $100 million per year; a value of 3 corresponds to $1 billion per year; and so on.) Countries paying a lot for imported armaments and countries paying relatively little can be either quite corrupt or not very corrupt at all. So arms transfers seem at least not to induce enough corruption to regularly affect perceptions of national honesty. Nor, if one uses the sketchier data from the first TI survey in 1995 and compares it with arms deliveries in 1996 and 1997 to states covered by that survey, is there much association at all between preexisting levels of corruption and amount of later deliveries.[38]

Do developing countries nonetheless spend money on arms imports that they ought to be spending on basic economic development? During the Cold War, one

Table 2.1 Transparency International's Corruption Perception Index, 1998

Country Rank	Country	1998 CPI	Country Rank	Country	1998 CPI
1	Denmark	10	43	South Korea	4.2
2	Finland	9.6	44	Zimbabwe	4.2
3	Sweden	9.5	45	Malawi	4.1
4	New Zealand	9.4	46	Brazil	4
5	Iceland	9.3	47	Belarus	3.9
6	Canada	9.2	48	Slovak Republic	3.9
7	Singapore	9.1	49	Jamaica	3.8
8	Netherlands	9	50	Morocco	3.7
9	Norway	9	51	El Salvador	3.6
10	Switzerland	8.9	52	China	3.5
11	Australia	8.7	53	Zambia	3.5
12	Luxembourg	8.7	54	Turkey	3.4
13	United Kingdom	8.7	55	Ghana	3.3
14	Ireland	8.2	56	Mexico	3.3
15	Germany	7.9	57	Philippines	3.3
16	Hong Kong	7.8	58	Senegal	3.3
17	Austria	7.5	59	Ivory Coast	3.1
18	United States	7.5	60	Guatemala	3.1
19	Israel	7.1	61	Argentina	3
20	Chile	6.8	62	Nicaragua	3
21	France	6.7	63	Romania	3
22	Portugal	6.5	64	Thailand	3
23	Botswana	6.1	65	Yugoslavia	3
24	Spain	6.1	66	Bulgaria	2.9
25	Japan	5.8	67	Egypt	2.9
26	Estonia	5.7	68	India	2.9
27	Costa Rica	5.6	69	Bolivia	2.8
28	Belgium	5.4	70	Ukraine	2.8
29	Malaysia	5.3	71	Latvia	2.7
30	Namibia	5.3	72	Pakistan	2.7
31	Taiwan	5.3	73	Uganda	2.6
32	South Africa	5.2	74	Kenya	2.5
33	Hungary	5	75	Vietnam	2.5
34	Mauritius	5	76	Russia	2.4
35	Tunisia	5	77	Ecuador	2.3
36	Greece	4.9	78	Venezuela	2.3
37	Czech Republic	4.8	79	Colombia	2.2
38	Jordan	4.7	80	Indonesia	2
39	Italy	4.6	81	Nigeria	1.9
40	Poland	4.6	82	Tanzania	1.9
41	Peru	4.5	83	Honduras	1.7
42	Uruguay	4.3	84	Paraguay	1.5

Note: Shading indicates median score for developing countries.
Source: Transparency International, Internet: http://www.transparency.de, 1998.

would have had to pose a counterpart question: does security assistance (in the form of money, training, and infrastructure assistance) accelerate such development? In the late 1980s, Nicole Ball, then working at the National Security Archive (a private, nonprofit group in Washington, D.C.), published an extensive analysis of military spending in and military aid to the Third World. She concluded that until govern-

Figure 2.1   Corruption and Arms Transfers

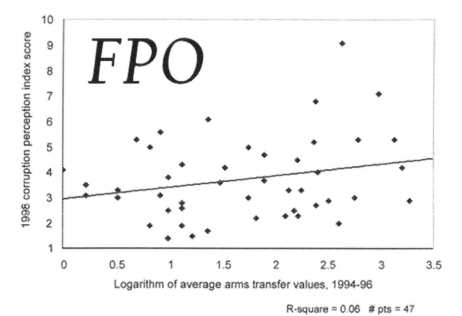

Source: Transparency International, *1998 Corruption Perception Index*. Internet: *http://www.transparency.de*. Accessed October 1998.

ments in developing regions found a way to reduce economic inequalities within their societies, those governments would continue "to arm themselves against their own people as well as potential external enemies."[39] In a project created more or less as a result of Ball's findings, an interdisciplinary research group based at Michigan State University set out to determine the relationship between military spending and economic performance among developing countries. It mostly concluded, however, that "more research is required"—that is, the results were ambiguous and country-dependent.[40]

In the early 1990s, the United Nations Development Program took the uncharacteristically bold step of devising and publishing a "Human Development Index" (HDI), a measure that compared how well UN member states met the needs of their peoples for such things as health and education. One might expect that those developing countries that spend the most on their militaries, or the most on imported weapons, would rank worst on the HDI, but that is not the case. Across all of the countries in this study, arms import expenditures in the same time period from which data was drawn for the HDI show a modest positive statistical relationship between human development levels and arms imports (a correlation of +.26, with only a 1 percent probability of being spurious). Correlation between military spending per capita and HDI is similarly positive and slightly stronger (a correlation of +.43). The best fit between that spending and the HDI is actually nonlinear in nature: expenditures tend to be disproportionately higher in countries with higher human development indices (see figure 2.2), suggesting that countries that can afford high military spending per capita can also afford to spend money on human

Figure 2.2 Military Expenditures per Capita and Human Development

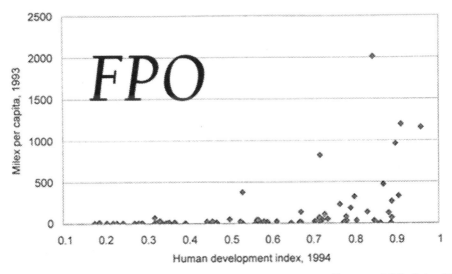

R-square = 0.188 # pts = 91

Source: United Nations, *Human Development Report, 1994*. U.S. ACDA, *World Military Expenditures and Arms Transfers, 1993–94*.

development. While these relationships are not very strong (only about 7 to 19 percent of the variation in one number can be accounted for statistically by variation in the other), countries' levels of human development and their spending on the military and on arms imports are not strongly at odds, either. Indeed, in the first half of the 1990s, only two of the top 20 arms importing countries—Angola and Pakistan—had human development index rankings below the median for developing countries. (Pakistan ranked 134th of 174 countries and Angola 165th.)[41]

Could or would developing countries devote greater resources to human development if their arms imports and overall military spending were lower? They might, but there is no guarantee that funds would be moved from military to civilian accounts on a one-to-one basis. In today's privatizing, globalized economy, cuts in military spending may respond to market pressures and demands from the International Monetary Fund that governments cut spending overall rather than shift it to the social sector.

Political Stability

Maintaining political stability in developing regions was a principal U.S. foreign policy objective during the Cold War, and its first line of defense against the creeping red stain of communist influence. Indeed, not just political stability but also political *stasis* was the order of the day, even though stasis, in the longer run, undermined stability by preventing small, adaptive changes that might ultimately have spared places like Nicaragua both revolution and U.S.-inspired counterrevolution. Nonetheless, change was viewed as a chink in the armor, an invitation to communist meddling in countries' political affairs, and an avenue of Soviet influence. To avert such risks, the United States strongly supported "authoritarian" anticommunist regimes, sometimes for decades, without regard to their records on civil or political rights. Arms transfers shored up political stability almost by definition if they contributed to the power and prestige of rulers and of the militaries who helped them keep the lid on their countries.[42]

In the post–Cold War 1990s, promoting political stability remains an important goal of American policy but the canonical source of instability is no longer the communist menace. Political stability is now seen to depend on governments' abilities to meet the needs and demands of their publics while at the same time respecting the human rights of ethnic and religious minorities as well as dominant groups. That is, stability for the purpose of keeping undesirable political influences *out* of other countries has given way to stability for the purpose of enticing desirable outside influences *into* the country, through lowered trade barriers, expanded avenues for capital investment, reduced risk of sudden investment loss, movement toward democratic governance with free and fair elections, and openness to the values exported by hundreds of nonprofit, nongovernmental organizations (NGOs) that deal in emergency relief, conflict resolution, economic development, or the training of democratic political cadres.

Arms transfers have much less of a politically stabilizing role to play in this new, marketizing environment. They may continue to play an influential role, however, in interstate relations within the more contentious security complexes, where arms acquisitions remain a competitive enterprise.

Arms Competitions, Stable and Otherwise

The notion of a stable arms race, like most concepts in contemporary American analyses of international security affairs, arose from the seemingly endless accumulation of conventional and nuclear armaments by the United States, the Soviet Union, and to a lesser extent, their allies during the Cold War. In the context of such a competition, stability describes a rate of acquisition (measured by money expended or destructive capacity acquired) that does not escalate over time. Such competitions have been characterized as either "quantitative" (piling up numbers of weapons) or "qualitative" (racing to acquire the best technology).

Quantitative Competitions

Harvard's Samuel Huntington published an early and influential work on arms races that pointed to escalating *quantitative* competitions as the ones to worry about. Ever-increasing expenditures on arms to avoid falling behind the adversary in military capability could drive one or the other side to war before it was driven to bankruptcy or capitulation.[43]

Thirty years after Huntington wrote, 40 years after the start of the Cold War, Soviet president Mikhail Gorbachev and his colleagues, facing severe economic problems at home, eschewed war and tried to avoid bankruptcy by implicitly capitulating in late 1988. Gorbachev withdrew Soviet forces from Europe, removed short-range nuclear weapons from Soviet navy ships, and gave other clear signals that the Cold War competition was over without officially running up a white flag. Gorbachev's actions were in fact consistent with the outcomes of a large majority of major power arms competitions over the previous century and a half, which ended in de facto capitulation rather than war.[44]

Might developing states choose differently? Recent research by David Kinsella at American University has suggested that the most destabilizing arms transfers to security complexes of developing states have been those made by anti-status quo suppliers (especially the Soviet Union) to anti-status quo recipients, whereas "weapons flows from the United States . . . seemed to do no further harm." That is, the political attitudes of suppliers and recipients had an important impact on recipients' tendencies to engage in conflict, regardless of the quantities of arms transferred.[45]

In the Middle East, from the mid-1950s onward, Western and Soviet arms supported mutually hostile clients and contributed to a rising regional balance of power. Fueled in part by cheap outside arms, relations between Israel and its Arab neighbors were punctuated by increasingly destructive wars in 1956, 1967, and 1973. A cool peace broke out between Israel and Egypt after Egyptian president Anwar Sadat's trip to Jerusalem in 1977, followed by the U.S.-brokered Camp David accords and a bilateral peace treaty that cost Sadat his life, two years later, at the hands of Islamic militants. Subsequently, American arms and financial assistance to both countries were intended to help cement the peace, make both feel more secure against third-party threats, and keep Soviet influence in Egypt at a minimum.

Kinsella's work would tend to support the U.S. argument that continuing arms aid to *both* Israel and Egypt since their 1979 peace treaty has been neither wasteful nor dangerous. However, as Nicole Ball suggested in her earlier analysis, without a careful look at the opportunity costs, such as alternative uses for the more than $50 billion given the two countries under this program, that conclusion would be premature.[46]

Qualitative Competitions

Was Huntington right that "qualitative arms races" or competitions in technology (the building of prototypes, for example, rather than active forces) are less costly and therefore less dangerous than competition in quantity of arms? Looking back at several decades of East-West arms acquisitions, George Downs, writing for a National Research Council report on the risks of nuclear war, concluded that competition in technology is expensive rather than cheap (consider the $2 billion cost of a "stealth" bomber) and that qualitative competitions tend to become quantitative at some point.[47] High-quality weapons must be produced or purchased in quantity if they are to have any military value, unless the *entire* point of the competition is symbolic. The United States and Soviet Union did upgrade their forces continuously with new technologies and new weapon systems (and the weapons they sent to Third World clients improved in quality over time as well). Even after the end of the Cold War, the habit of fielding new forces has been hard to break in the United States, where new, stealthy fighter-bombers, attack helicopters, submarines, and warships are slated to enter production in the new century.

That the East-West competition in armaments did not lead to general war as Soviet technology lagged that of the West is attributed by many analysts to the looming presence of nuclear arsenals on both sides.[48] The West spent several trillion dollars, however, on conventional forces despite its nuclear arsenals, both to convince the other side of the seriousness of its resolve to defend itself—by nuclear means, if necessary—and (somewhat contradictorily) to avoid rapid escalation to nuclear arms in the event of East-West war.

This Cold War dynamic is not particularly comforting to opponents of nuclear weapons proliferation or to advocates of conventional arms control among developing countries, as it vests major deterrent value in weapons of mass destruction (WMD), and major symbolic value in highly capable conventional forces. This is an expensive path for any country to tread, let alone developing countries with limited resources. Cold War logic suggests, however, that states locked in confrontation may be driven toward such force structures and, unfortunately, supporting evidence may be found in the Arab-Israeli dispute, in Iraqi and Iranian WMD programs, and in the nuclear weapons programs of India and Pakistan. WMD programs may not substitute for conventional arms but instead may stimulate their acquisition as continuing evidence of political resolve, while major conventional arms imbalances may stimulate the acquisition of WMD.

Crisis Stability

Crisis stability refers to the relative propensity of tense international situations to tip into conflict. Crises that can be tipped easily by small actions (whether deliberate or mistaken) are unstable. Patterns of arms acquisitions and military deployments believed to be conducive to such tipping are considered "crisis destabilizing"—that is, they are believed to make conflict a more likely result of crisis.

Cold War strategic planners used to worry about whether U.S. bombers and missiles could survive a surprise nuclear attack, because logic and quantitative modeling strongly suggested that vulnerable forces undermined crisis stability. (Inability to survive surprise attack, it was believed, could tempt the other side in a crisis to launch a preemptive attack in hopes of catching vulnerable forces at their bases.) Over the years, hundreds of billions of dollars and rubles were devoted to programs to mitigate potentially destabilizing vulnerabilities in nuclear force postures and command and control systems.

Striking first is not just an option cherished by aggressors, however. A state that considers itself outnumbered and outgunned may launch preemptive attacks in a crisis to get the drop on its adversaries rather than give them the option of landing the first blow. Such logic drove Israel to attack the mobilizing forces of its Arab neighbors in 1967, to deliver a knockout blow against the Egyptian air force and send an armored blitz into the Sinai, efforts that succeeded in driving back Egyptian forces hundreds of miles, to the bank of the Suez Canal, in a matter of days.[49]

By the same token, the ability to strike back against attack is an option that few militaries would willingly give up. Over the course of the long standoff in Europe, some analysts urged Western governments to convert their forces to a "defense only" posture capable of denying victory to Soviet forces but incapable of offensive operations. They argued that war avoidance in Europe depended in part on NATO *not* posing a threat to Eastern European territory comparable to that posed to Western territory by forward-deployed Soviet forces, an argument that assumed the equipment requirements of defense and offense to be radically different. It is not clear that they are.[50]

Indeed, the utility of most weapon systems for defensive as well as offensive purposes makes the task of devising stability-enhancing weapon control regimes that much more difficult. The least stable situation may be one in which short-warning weapon systems are possessed in quantity by an anti-status quo state with an offense-oriented military doctrine and strategy.

Short-warning Weapon Systems

Weapon systems perceived as capable of striking with little advance warning or of giving great advantage to the side that strikes first may undermine crisis stability or feed a security dilemma. However, they may also allow a status quo power to hold off a known aggressor like Saddam Hussein by threat of preemptive action. In such circumstances, they may lend credibility to deterrent threats. But some short-warning systems have fewer redeeming values than others.

Modern strike aircraft or long-range cruise missiles using low-altitude, nap-of-the-earth flight paths can cross national borders with little advance warning. Israel used such a low-altitude stratagem to attack Iraqi nuclear facilities in 1981. India and Pakistan both have supersonic attack aircraft and Pakistan's urban/industrial areas are close to its border with India. But some of the same aircraft are equally capable of mounting a defense against aerial attackers. They are dual purpose.

Long range missiles have fewer obvious defensive applications. A cruise missile is a small, self-piloted aircraft. It may be as big as an aircraft or compact enough to fit in a submarine torpedo tube (with wings folded). The most sophisticated, long-range (1,000-kilometer-plus) cruise missiles can duplicate a strike aircraft's terrain-following path and hit a preassigned target with precision. Long-range cruise missiles thus far remain a possession solely of the United States and Russia. The United States used them extensively in the 1990s, against Iraq during and after the Gulf War, against Serb forces in Bosnia and Yugoslavia, and against targets in Afghanistan and Sudan in retaliation for attacks by terrorist groups hosted in those countries. Lightweight, fuel-sipping engines and sophisticated, programmable guidance systems have been the main hurdles to other countries building long-range cruise missiles. With the availability of satellite-based navigation signals from the U.S. Global Positioning System (GPS), however, anything on or near the earth can be accurately positioned in three dimensions, even while moving at high speed. Public signal channels provide position accuracy of about 100 meters and encoded military channels enable accuracies of less than 20 meters.[51] Compact, basic GPS guidance systems can be purchased on the commercial market. While the engines used in American missiles may be difficult to replicate, there are other suppliers of sophisticated missile engines, like Russia, suggesting that in due course, longer-range cruise missiles will be appearing in more countries' arsenals.

Effective warning of attack by low-flying aircraft and cruise missiles can be difficult. Ground-based surveillance radar is limited in its line of sight by the curvature of the earth and may give as little as one minute of warning of a low-level attack. Airborne early warning systems can provide about 15 minutes' notice of aircraft approaching a country's border at supersonic speed, but such systems are sophisticated, expensive, and as a result used by only a few countries. Without such warning, defense forces have little chance to mobilize and intercept attackers.

A ballistic missile follows a parabolic arc from launcher to target, like a precision-crafted rock. Ballistic missiles have been spreading slowly into more countries' arsenals. They are also difficult to intercept but for different reasons than those for cruise missiles and aircraft. A missile like the Soviet-designed Scud, with a range of 300 kilometers, has a flight time of less than five minutes, reaches a peak altitude of roughly 120 kilometers, and is moving at 2,000 meters per second when it reenters the atmosphere on final approach to its target. Atmospheric friction slows it down quickly, unless its warhead is highly aerodynamic and designed to separate from the missile body. In that case, the warhead may still be moving at roughly 1,000 meters per second when it comes within range of air defense systems, while atmospheric buffeting may break up the trailing missile body, dazzling air defense radars with a shower of scraps.[52] Both the warhead's speed and the radar-interfering scraps make interception of the warhead difficult. Warhead reentry speeds increase directly with missile range (intercontinental-range ballistic missile warheads can approach their targets at seven kilometers per second). Even the United States has experienced difficulty developing defenses against ballistic missiles of longer range than the Scud.

Thus far, most short-range ballistic missiles have carried relatively inaccurate inertial guidance systems, making them useful mostly against broad targets like urban areas (a basic Scud will land within about a half-mile of where you point it). Iran and Iraq both fired Scud and Scud-derived missiles against one another's cities during their eight-year war, and Iraq attempted a similar blitz against Israeli and Saudi cities during the Gulf War. While a 700 to 800 kilogram high explosive missile warhead can be deadly, a long-range ballistic missile is a very costly way to deliver conventional explosives. A modern fighter-bomber can carry five to ten times as much ordnance, and it can be used more than once.[53]

For ballistic missiles carrying chemical, nuclear, or biological warheads, however, inaccuracy is less of a handicap; WMD have much larger destructive radii. This enhanced destructive potential with WMD warheads reinforces the ballistic missile's reputation as a terror weapon and has underpinned one of the few global conventional arms control efforts to date, the voluntary Missile Technology Control Regime, discussed at some length in chapter 4.

Offense-oriented Forces and Strategies

Even the great aggressors of history have claimed defensive motives for their actions (*lebensraum,* security of oil supplies, a buffer zone to protect the motherland), but offense dominated their military strategies. From the pounding cavalry of Genghis Khan's Golden Horde, to the *blitzkrieg* that rolled over Europe and western Russia in World War II, and the Operational Maneuver Groups that likely would have spearheaded Soviet advances into Western Europe in World War III, offense has always been the aggressor's friend. But it has also been the defender's revenge: Maginot lines and other defenses incapable of forward movement cannot eject aggressors from conquered territory. Hitler's armies were pushed out of Russia by the same sorts of forces that pushed into Russia: heavy armor and attack aircraft. Saddam Hussein's army was pushed out of Kuwait in much the same way. The very flexibility of most modern weapon systems, which can be used to attack or to defend more or less interchangeably, make intentions more difficult to divine from knowledge about capabilities and place a premium on intelligence and clear communications.

Rather than the performance characteristics of an individual weapon system, what matters to the stability of a tense regional standoff is the "aggregate impact of all weapon systems in a given arsenal," and "the prevailing doctrine that determines its use."[54] Indeed, national strategy and military doctrine may be the keys to making a military establishment offensively or defensively oriented. Strategy articulates national goals and doctrine expresses these goals in operational terms.[55]

For the community interested in controlling conventional armaments, this conclusion offers both bad news and good news. It is bad news in that it suggest that efforts to label certain types of weapons "particularly destabilizing," must be carefully crafted and closely focused, not broad-brush. It is also bad news because changes in doctrine can make any country's military more threatening to its neighbors without changes in force structure. Such changes in doctrine may not be visible to neighboring states, who may either be unpleasantly surprised one day or who may plan for the worst only to find they have wasted a lot of money that better lines of communication might have saved. Unfortunately, good communications and mutual trust both are in short supply in a number of regional security complexes.

Doctrine's dominance can be good news inasmuch as doctrine can be changed to make forces less threatening without requiring that countries junk their existing arsenals and invest in costly new weapons. Instead, new doctrine requires reeducation and retraining of personnel, new methods of operating, revised field exercises, and perhaps altered structures of command and control—changes in security "software." Where states struggle with security dilemmas, improved political communications—either direct or facilitated by third parties—might facilitate doctrinal change. Where deterring a hostile neighbor is a bigger problem than the security dilemma, doctrinal change and arms control may take a back seat to the search for allies within and beyond the region. Arab states in the Persian Gulf region, for example, have inked basing and other agreements with the United States to counter perceived threats from Iran and Iraq.[56] Finally, where hostility is based partly in mutual grievances—as with Pakistan, India, and the disputed territory of Kashmir—better communications and willingness to stop provoking the other side may lead over time to doctrinal adaptations that reduce the likelihood of conflict in times of tension. Such confidence and security building measures will be discussed at some length in chapter 5.

Do Arms Transfers Contribute to Conflict?

Sophisticated weapons have made their contribution to the body count in wars in the developing world by increasing what A. F. Mullins dubbed "soldier productivity," or the unit destructive power of armed forces.[57] But have arms transfers substantially increased the levels of conflict between or within developing countries? If they have, then limiting such transfers should be a concerted goal of international public policy. If they have not, or the linkage is weak, then limiting transfers may have little impact on levels of violence and those interested in conflict prevention and resolution should look elsewhere for leverage. In this section, we compare developing states' current situations with the past in Europe around which much of the arms-and-conflict literature has been built, and then do some modeling of arms transfer and conflict interaction in the developing world.

Differences Between Developed and Developing Regions

As noted earlier, studies of arms buildups or competitions primarily among major European states have found the relationship between such buildups and conflict to be tenuous. But three interrelated elements of arms and security suggest that there might be a different answer for regional security complexes of the developing world. Their members depend more heavily on outside sources of armaments; their regions may be subject to overlay by the members of higher-level security complexes; but despite the overlays, very few members of the developing world enjoy the sorts of security guarantees that major allies of the United States enjoyed during the East-West standoff.

Although the capacity for building basic armaments is spreading, most developing countries still import all but the most basic of weapons, and they are years if not decades from producing the most sophisticated technologies already on the market.[58] Developing states' ability to provide for their security is therefore dependent on decisions made in foreign capitals and board rooms. But the existence of an international arms market—open, gray, and black—also means that with enough cash or a politically interested benefactor, states and groups with security concerns or ambitions can arm themselves much faster and more heavily than they could if they had to design and produce the arms on their own. New capabilities that alter local or regional force balances may appear without much warning, giving the state or group receiving them serious new damage potential rather quickly. Under some circumstances (for example, where a peace has been kept by a tenuous local balance of power), acquisition of substantial new capabilities might even be the tipping event that triggers conflict. In general, relationships between arms and conflict might be expected to be more volatile in the developing world than relationships found in historical case studies of largely self-supplied states.

Second, the existence of a larger system of international politics operating above and beyond regional politics has made late-twentieth century interactions among states such as Israel, Syria, Jordan, and Egypt quite different from the nineteenth- and early twentieth-century relations of the major European powers. The latter states functioned, to all intents and purposes, within a self-contained international system subject to no other power's decisive influence. Developing states' interests and relations, on the other hand, have been overshadowed by the interventions of larger powers whenever those powers decided that their own national interests required it (for example, the United States in Latin America, France in West and Central Africa, and both superpowers everywhere during the Cold War).[59]

Both of these differences made the Arab-Israeli dispute, for example, far bloodier and longer-running than it might otherwise have been, as both sides had access to large amounts of arms and the occasional direct military support of the superpowers. On the other hand, Israel has survived nearly 50 years of collective Arab hostility because of outside support and would likely have succumbed to that hostility without it. Much the same could be said of the Muslim community in Bosnia-Herzegovina in southeastern Europe. Outside assistance, however limited initially, helped the "Bosniacs" resist heavy military pressure from their Serb and Croat neighbors and to survive as a distinct community.[60] An outside overlay thus may alter not only the time line of a conflict but also its outcome.

Third, a particular kind of overlay existed during the Cold War, consisting of the United States extending a metaphorical "nuclear umbrella" over its allies, offering implicitly to retaliate in kind against any state that used nuclear weapons against those countries. Among developing states, only South Korea enjoys comparable patronage. No such security guarantees are forthcoming for states like Israel or Pakistan that feel threatened by the conventional military power and ambiguous or hostile intentions of their neighbors. The technologies of several kinds of WMD are now 50 to 80 years old. That, plus the growth of significant commercial chemical, nuclear, and expected biotechnology industries around the globe make the diffusion of these technologies difficult to halt despite an array of treaties and informal supplier arrangements with that objective.[61] States that face highly unfavorable conventional arms imbalances may find WMD an attractive solution to their security problem.

The Iran-Iraq War (1980–1988) provides an example of all three factors at work to pernicious ends. The dictates of Cold War competition and the ambitions of megalo-maniacal leaders in both countries combined to stimulate substantial flows of weapons to each. When the United States lost its military gamble with the shah of Iran in 1979, most of the Western world came to view neighboring Iraq as the cork that would bot-tle up Iran's revolutionary mullahs, much as the United States and Britain viewed Stalin as a counterpoise to Hitler. The grinding war that resulted when Saddam Hus-sein's troops attacked Iran in September 1980 was sustained both by unprecedentedly large imports of money and arms and by deployment of Western naval power in the Persian Gulf to protect the petroleum shipments of states supporting Iraq.[62]

Despite this support, Iraq turned to chemical arms to counter Iran's advantage in military manpower and to deal with internal unrest. Iraq soon became the first state in decades to use lethal chemicals on the battlefield and the first ever to turn lethal chemicals on its own citizens, in this case, Kurdish villagers. Only two years after the cease-fire in this conflict, Iraq wheeled its forces southward and invaded another neighbor, Kuwait, stimulating an American response, a devastating military defeat, and the discovery in the aftermath of that defeat of yet more Iraqi WMD programs. Had Saddam Hussein delayed his new war for a year or two, U.S. and allied forces in the Persian Gulf might have faced not only Iraqi tanks but also the threat of Iraqi nuclear weapons, either aboard Baghdad's shaky missiles or trucked into Kuwait City to hold the city hostage indefinitely.

In sum, the interactions between arms supplies, force balances, political-military volatility, and levels of conflict in the regional security complexes of the developing world are likely to be more complex than similar interactions among European pow-ers in previous eras. Outside influence, particularly in the areas of conventional arms, political support, political opposition, and military intervention, is much more preva-lent. While modern developing states may suffer no more inner turmoil than Euro-pean states at a similar stage of state-making, they do so in an era of accelerated change, accelerated expectations, and global news coverage. That very coverage, how-ever, may accentuate the negative to such an extent that open warfare seems en-demic. To get a more balanced view of the extent of turmoil, and the contributions that arms transfers may make to it, we have to step back for a broader view and look at the data in comparative fashion.

Doing the Numbers

Finding and comparing patterns in the mass of data generated by the arms trade and contrasting them with patterns of internal and external conflict in the developing world requires that we turn to numerical measures and statistical analysis. The follow-ing sections describe measures and patterns of conflict, chart changes in conflict against changes in arms acquisitions, and, finally, track arms transfers and conflict over time. In the process, we will be looking for answers to the following questions:

- Do arms transfers increase levels of conflict?
- Do increased levels of conflict generate arms transfers?

- Are these effects region-specific or more general in nature?
- Do these effects change over time?

What follows is not time series analysis, in which one set of multiyear data is compared with one or more comparable series of data. Such analysis is difficult to justify with multiple cases (countries) whose patterns of behavior are known to vary considerably. "Pooling" their data streams to make one time series for global arms deliveries and another for global conflict could mask rather than reveal patterns of interaction. So instead of time series analysis, I will be taking a sequence of snapshots of arms transfers and conflict and comparing them in search of change and continuity.

Periods of Analysis

I begin the quantitative analysis in 1970, the start of phase two in the post–World War II arms trade. That transitional phase marks the shift from mostly grants in aid from the United States to cash-and-credit sales, rapidly rising amounts of Soviet arms for developing countries, and rapid commercialization and expansion of the trade. The year 1970 is also a good starting point because all but a handful of developing states (9 out of the 106 included in this study) were independent by then, providing a mostly common set of cases over the entire period of analysis, which runs through 1995.[63]

I subdivide these 25 years into five-year periods consistent with the phasing of the arms trade laid out in chapter 1 (the 1970–1974 transitional period; 1975–1979, 1980–1984, and 1985–1989, which together make up the third phase of Cold War competition and commercialization; and 1990–1995, post–Cold War). I use five-year periods rather than something shorter (annual data, for example) because the arms transfer business is uneven and cyclical. Countries may receive shipments in odd-numbered years but not even-numbered years, as a quirk of contracting. Their policies and attitudes may be no different in the odd and even years, but a model based on yearly data would predict such differences. A three- to five-year moving average would have filled and smoothed over such data gaps, but then delivery numbers for any given year would be linked to data for preceding and succeeding years, and data for each time period would as a result be contaminated by data from the ones preceding and following it. Such a model would not differentiate cleanly between them. So instead of either annual data or moving averages, I opted for the half-decade time periods. The same intervals were used by Ted Robert Gurr and his colleagues in their long-running evaluation of communal conflict.[64]

These five-year periods fit the phasing of the arms trade, as noted, and also match the phasing of a number of significant conflict events. In that sense, they may predispose the model to pick up relations between arms and conflict, so that strong relationships might be discounted a bit for that reason. Weak relationships, on the other hand, may be even weaker than indicated by the quantitative results.

The time periods match conflict phases in the following ways: 1970–1974 encompasses major wars in the Middle East, South Asia, and Southeast Asia; 1975–1979 marks a relative lull in major interstate warfare in the developing world; 1980–1984 includes the first half of the Iran-Iraq War and the first Reagan administration, whose policies emphasized carrying counterrevolution to the developing world; 1985–1989 marks the second half of the Iran-Iraq War; 1990–1995 is marked by a sharp decline in interstate conflict and the resolution of some Cold War-influenced internal conflicts.[65]

Coding Armaments

The arms data used in the quantitative analysis reflect total numbers of weapons received by each developing country during each five-year period. A more refined analysis might weight the relative fighting power of each weapon (using something akin to the Weapon Effectiveness Indices calculated by the U.S. Army), but that would also introduce vast data problems, inasmuch as key weapon performance characteristics may be closely held by some governments. The army indices used by some conflict modelers are based on declassified 1979 U.S. government documents. Dozens of new weapon systems have entered the arms trade since then, and the knowledge and resource base is lacking outside of government to rate those weapons authoritatively. Rather than introduce such a complicating factor, I opted to use the basic delivery data, but not quite the raw data. Because the statistical models that I use below (and discuss at greater length in appendix C) could be thrown off by the disparities in scale between the conflict data (which has just six levels) and arms deliveries (which range up to several thousand units per year to a given country), I opted instead to use the common logarithm of the arms transfer data, as I did for the values in figure 2.1. Thus, the actual numbers representing arms deliveries that I used in the models ranged from zero to about 3.87 (which is the common log of 7,375, the number of heavy weapons imported by Iraq between 1985 and 1989).

Arms categories counted for the analysis of *external* conflict were heavy combat aircraft, missile-armed helicopters, medium and heavy tanks and those land vehicles derived from them (recovery vehicles, bridging vehicles, and the like), armored personnel carriers (APCs), infantry fighting vehicles, and specialized vehicles derived from them; and air defense guns and missile launchers.

On the assumption that internal conflict generates lesser or perhaps just different requirements for arms, categories of arms transfers included for the analysis of internal conflict were light attack aircraft, transport helicopters, and armored cars (wheeled vehicles that carry fewer troops than APCs but are better suited to urban settings).

Coding Conflict

In looking for appropriate measures of external and internal conflict in the developing world, I reviewed several other, related research efforts and the conflict scales that they used to measure conflict. The scales used in this analysis are hybrids of those other efforts. (A comparative assessment of four other conflict scaling efforts may be found in Appendix B.) The external conflict scale that I devised differentiates six levels of strife between states:

- no conflict (assigned value: 0);
- tension (assigned value: 1) resulting from proximity to conflict, assigned to states that were neighbors of states involved in some level of interstate fighting;
- small, cross-border clashes (assigned value: 2), such as the raids staged from Lebanon into Israel, and Israel's tit-for-tat reprisals; also includes small air duels;
- limited cross-border operations of brigade size or larger (assigned value: 3) such as Israel's Operation Litani in 1978 against the PLO in Lebanon;
- relatively small-scale conventional conflict (assigned value: 4), such as the 1982 Falklands/Malvinas War between Argentina and Great Britain, or Israel's Operation Peace for Galilee, when Israel reentered Lebanon and engaged in di-

vision-level combat with Syrian land and air forces in and over Lebanon alone; and

- medium-scale conventional war (assigned value: 6), as between Iran and Iraq (1980–1988), confined to a single security complex but involving the full national resources of the states involved and potentially threatening the survival of at least one of them.

The most devastating wars in the developing world have done considerably more damage than conflicts closer to the threshold criterion of 1,000 or more battle deaths that projects such as the Correlates of War have used to define a war. On the other hand, a relatively brief, relatively bloodless conflict like the Falklands War is far removed from long, bloody conflicts like the Indochina or Iran-Iraq wars, or even from short, bloody conflicts like the October 1973 Middle East War. Reflecting this gross difference, and my intent to treat this measuring stick as a quasi-interval scale, I separated the deadliest sorts of external conflicts from the rest by assigning them a scale value of six. (For the validity of interval-level assumptions, here, see the discussion of Joshua Goldstein's scaling tests in appendix C.)

For internal conflicts, I adapted Ted Robert Gurr's "rebellion" scale.[66] The resulting scale codes levels of internal strife as follows:

- no reportable activity (scale value: 0);
- attempted coup, politically motivated riots, occasional terrorist act (scale value: 1);
- successful coup, terror campaign, politically motivated extrajudicial killings (scale value: 2);
- state of emergency, "dirty war," guerrilla war in part of the country (scale value: 3);
- nationwide insurgency, internally based (scale value: 4);
- sustained, nationwide insurgency with cross-border bases (scale value: 5);
- large-scale civil war entailing the near-collapse of government and economy and, in most cases, generating large numbers of refugees (scale value: 6).

In creating this scale, I separated Gurr's category of "protracted civil war" into two categories of lesser and greater destructiveness. An example of level five conflict would be the contras' war against Nicaragua. Examples of level six conflicts would be the civil wars in Lebanon (especially 1975–1989) or Angola (1974–present).

To create an interstate conflict data set suitable for this analysis, I merged relevant portions of the Militarized Interstate Disputes (MID) data set from the long-running academic Correlates of War project with a data set of my own devising. (Both sets, and the reasons and methods for combining them, are detailed in appendix C.) Each developing country received a conflict score in each five-year time period that corresponded to its *worst* experience of conflict during that period, on the assumption that the worst experience exerts a greater influence on policymakers than does average experience.[67] Thus the model expects policymakers to act according to that worst-case experience. This assumption accords with the well-known tendency within defense establishments toward worst-case planning.

Graphing Conflict over Time

The number of states involved in violent interstate disputes between 1970 and 1994 is shown in figure 2.3. The upper line plots the number of states involved in disputes

Figure 2.3 Developing States and External Conflict

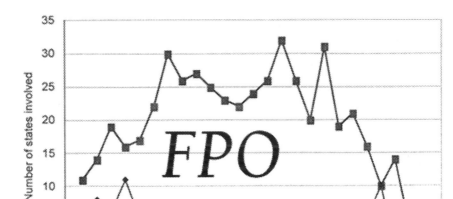

involving uses of force short of war (scale values 2 and 3). From the early to mid-1970s, this number triples; it hovers near 25 for the next eight years and gyrates in the latter 1980s before dropping significantly to around 5 states in 1993–1994.

The lower line in figure 2.3 represents the number of countries involved in interstate wars (scale values 4 or 6). The two peaks, in 1973 and 1991, reflect the 1973 Middle East War and the Gulf War, in which coalitions of states fought on one side of the conflict (the Arab states against Israel, and the U.S.-led multinational coalition against Iraq). Peace breaks out in 1976 but doesn't last long. From 1977 through 1990, an average of five developing countries were involved in a regional war (the odd number comes from Afghanistan's war against the USSR; Afghanistan is one of the few countries rated as being engaged in civil and interstate wars simultaneously, to reflect the Soviet intervention). After the 1991 Gulf War, interstate war in the developing world dropped to zero (and stayed at zero until the Ethiopia-Eritrea border war of 1998–2000). In total, no more than one-third of developing states experienced recordable conflict. Just under one-third were scored as experiencing "tension" due to proximity to their other conflicts, and half to two-thirds experienced no external conflict at all.

Internal conflicts were coded and scaled in a similar manner. Figure 2.4 shows the number of states experiencing three levels of internal violence between 1970 and 1995. The upper line represents more serious insurgencies that invest a substantial portion of a country's territory but fall short of full civil war (scale values 3 or 4). An example would be Colombia's perpetual struggles with guerrilla groups and drug cartels, or Peru's struggles with the guerrilla forces of Tupac Amaru and Sendero Luminoso. The number of states afflicted with such insurgencies increased slowly to the mid-1970s, where it plateaued; rose slightly to plateau at around 20 states in the mid-1980s; and dropped at the end of the Cold War only to rise briefly in 1993–1994 before

Figure 2.4 Developing States and Internal Conflict

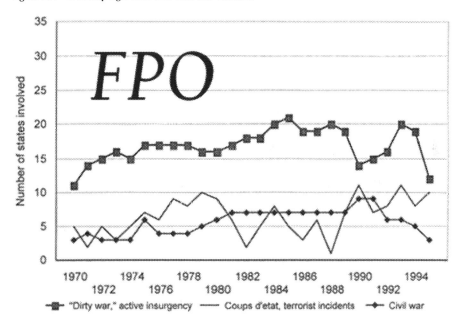

dropping again to 1970 levels. The unadorned line represents the number of states suffering from relatively low-level violence on the order of coups d'etat or sporadic terrorism (scale values 1 or 2), showing a somewhat cyclical pattern that was actually higher on average in the 1990s than earlier. (Because internally directed activities at lower levels of violence may not be picked up as readily by international news media and published conflict compendia, this category may well be underreported.) Finally, the dark string of diamonds in figure 2.4 represents full-scale civil wars, which afflicted an average of seven developing states in the 1980s and seemed to be tailing off by the mid-1990s.

Plotting Changes in Conflict Against Changes in Arms Transfers

When searching for patterns of association between variables, one can usually learn quite a bit just by plotting the data. Two different approaches illustrate that relationships between arms and conflict can be tricky to pin down and that patterns of change in conflict behavior and arms deliveries are not especially consistent with one another.

Plot the rise and fall of states experiencing recordable conflict (scale value 2 and higher) over time and the value of international arms transfers on the same chart and the pattern match is striking (figure 2.5). The two curves suggest a close relationship between arms transfers and conflict. If the two variables are plotted directly against one another (figure 2.6), the data points line up in reasonably linear fashion, and would be even more linear but for the two data points toward the left side of the graph. They represent data for 1993 and 1994, when arms delivery value remained relatively high even though numbers of states experiencing conflict was low. Much of that value represents arms delivered to states in the Persian Gulf in reaction to the Gulf War and the threat from Iraq (and, to some extent, Iran). Knowing the number of

Figure 2.5   Arms Transfer Values and External Conflict

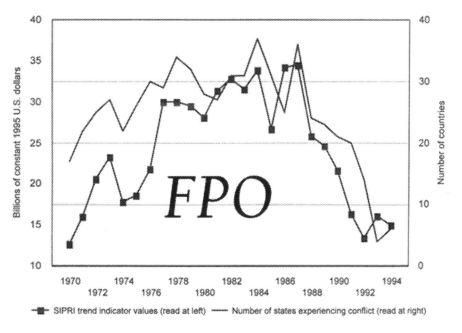

Figure 2.6   Number of States in External Conflict vs. Value of Arms Trade

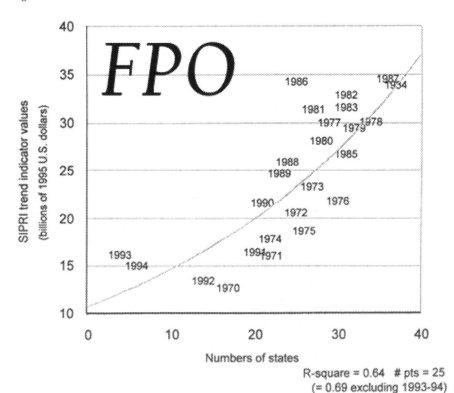

countries experiencing conflict helps us to account for about 60–70 percent of the value of the arms trade from 1970 to 1994.[68] This, too, seems like a reasonably good fit.

However, these comparisons use aggregate values of arms delivered and total numbers of countries in conflict. They do not match arms to recipients. What happens if instead of an aggregate measure over the whole period, we look at five-year time slices and match countries experiencing conflict, as above, with the value of arms deliveries? If arms deliveries and peak experience of conflict really are closely related, then this limited selection of cases should highlight the relationship. But these more explicit comparisons generate a much less compelling set of numbers.[69] For 1970–1974, conflict accounts for just 23 percent of the value of the arms trade to the developing world; for 1975–1979, essentially zero percent; for 1980–1984, 30 percent; for 1985–1989, 53 percent; and for 1990–1994, 24 percent. The fit between arms deliveries and conflict is not looking nearly as good.

Let's take another cut at the problem, measuring not absolute values for conflict or arms deliveries but *changes* in those values from one time period to the next, and examine all developing countries, not just those experiencing conflict in any one time period. If changes in conflict levels and changes in amounts of arms delivered were strongly related, then a scatter diagram of arms and conflict would, much like figure 2.6, show a pattern of dots running from lower left to upper right. At the lower left would be those countries experiencing a great decrease in conflict level, and a corresponding decrease in arms imports, from one five-year period to the next. At the upper right would be countries experiencing great increases in both conflict and arms imports. At the lower right would be countries that experienced an increase in conflict but imported fewer arms; and at the upper left would be countries that experienced a drop in conflict but increased their arms imports nonetheless. Toward the center would be countries whose status had not changed much on either dimension.

The results of plotting changes in conflict levels against changes in arms deliveries are depicted in the next four charts. Figure 2.7 compares changes between the first and second five-year periods, figure 2.8 between the second and third periods, and so on. The further from the center of the plot, the greater the change in status from one time period to the next.

None of the scatter diagrams show much correlation overall between changes in conflict and arms deliveries. Countries experiencing large changes in conflict status may show little or no change in arms deliveries, while countries with little or no change in conflict status may show large changes in weapons delivered. In short, for the developing world at large, there is no clear and consistent pattern of change. Each of the outliers in the scatter diagrams has its own particular story. Many of those stories involve conflict or its resolution, but the implications for arms imports are (literally) all over the map.

Looking at the outliers in figure 2.7, Saudi Arabia, Libya, and Iran all undertook substantial oil-revenue-funded arms buildups in the latter 1970s, accounting for the large increases in their import levels by comparison to 1970–1974. South Vietnam finally lost its bid for independence in 1975, and the flow of American weapons stopped abruptly. Egypt and Israel found relative peace in the late 1970s and imported far fewer arms by comparison to the massive deliveries each received before and during the 1973 October War. India and Pakistan, which fought a brief, sharp war in 1971 but kept the peace later in the decade, each reduced arms imports.

Figure 2.7   Comparing Changes in External Conflict Level and Arms Deliveries, 1970–1974 to 1975–1979

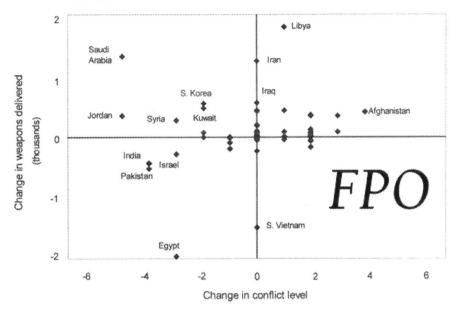

Figure 2.8   Comparing Changes in External Conflict Level and Arms Deliveries, 1975–1979 to 1980–1984

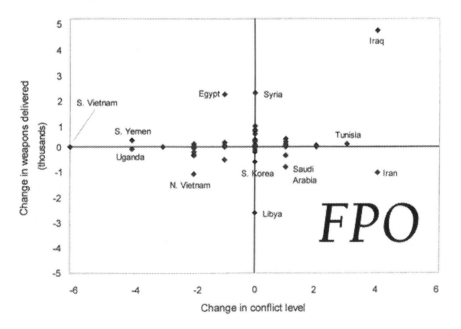

States in the upper left quadrant of figure 2.7, though experiencing less overt conflict than in the first half of the 1970s, could all still be characterized as having tense relations with their immediate neighbors. Indeed, someone who would interpret interstate relations in the Middle East and on the Korean peninsula as more strongly influenced by deterrence than by the security dilemma might consider the combination of increased armaments and decreased conflict levels to be exactly what a properly functioning deterrent relationship would generate: peace through strength.

That objective fairly well characterizes U.S. policy regarding Egypt and Israel after their 1979 peace treaty. Between the late 1970s and early 1980s (figure 2.8), Egypt became a major American arms client, and the combination of arms and peace puts Egypt in the upper left part of the chart. Syria accelerated imports of Soviet arms. While it did not find itself in full-scale war with Israel, neither did its conflict situation improve further in the 1980s. Syria was heavily involved in Lebanon's civil war in both the late 1970s and early 1980s, when it fought a limited war with Israel in and over Lebanon.

The Syrian case illustrates the importance of remembering that all the data points in figures 2.7 to 2.10 represent relative rather than absolute data: a country like Syria may be engaged in limited warfare in both the earlier and later time periods used to create one of these charts, and so it rates a zero—no change—in conflict level. Increases in Syrian armaments were more likely motivated, instead, by peace between Israel and Egypt (since it could no longer rely upon Egyptian help in a future clash with Israel) as well as by continuing clashes with Israel in Lebanon.

Iraq's position in the far upper right corner of figure 2.8 suggests that Saddam Hussein may have underestimated what it would take to seize a bit of Iranian territory in 1980, since Iraqi arms imports were much higher in 1980–1984, during the war, than in 1975–1979, the years leading up to the war. Revolutionary Iran, meanwhile, lost access to the substantial arms pipeline from the United States and had to find new suppliers after 1978, so it actually imported fewer arms after the war started than beforehand.

The Iran-Iraq War continued into the late 1980s and both combatants increased their arms imports substantially in that period (see figure 2.9). Iraq increased its already-high rate of acquisitions as it tried to counter Iran's superiority in manpower. Panama was caught by surprise in late 1989 by invading U.S. troops looking for strongman Manuel Noriega: no time to buy arms. Egypt, Israel, and Syria all imported far fewer weapons in the late 1980s, even though conflict levels in the region remained stable; Syria, by the late 1980s, was encountering Moscow's growing reluctance, under Mikhail Gorbachev, to provide arms below cost to its previous customers in the developing world. Short of grants from other, oil-producing Arab countries, Syria could ill afford to keep importing arms at its historical rates at retail prices.

Experience of the Gulf War helped to make Kuwait and Saudi Arabia the two developing states with the greatest increases in arms imports in the first half of the 1990s (figure 2.10). Egyptian and Syrian conflict ratings ticked upward due to their direct participation in the Gulf War. Iran and Iraq both imported far fewer weapons after their long war ended in 1989, but Baghdad's high conflict rating remained unchanged because of its invasion of Kuwait in August 1990. Its steep drop in arms imports in the 1990s reflects the effect of a United Nations arms embargo applied at the time of the invasion and still in place at the end of the decade.

Figure 2.9 Comparing Changes in External Conflict Level and Arms Deliveries, 1980–1984 to 1985–1989

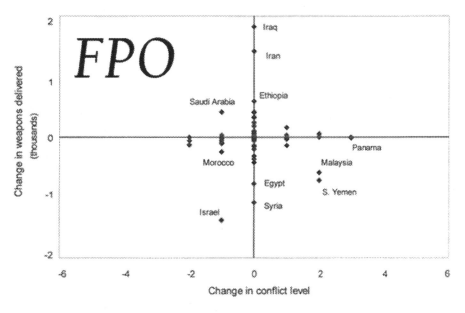

Figure 2.10 Comparing Changes in External Conflict Level and Arms Deliveries, 1985–1989 to

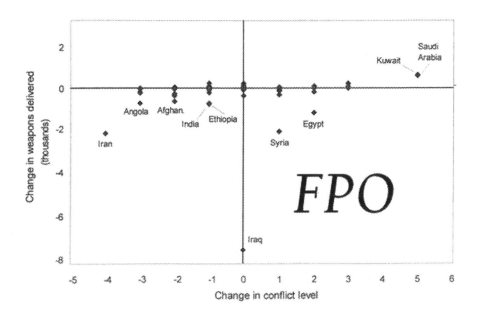

In sum, external conflict and arms transfers in the developing world can be connected, but the connections, even for the outliers in the data, do not suggest a simple and straightforward relationship between changes in conflict levels and changes in levels of arms deliveries. (If the major outliers are removed from the analysis, moreover, the basic shapes of the scatter diagrams do not change substantially. The relationship between conflict and arms becomes no simpler.)

Path Analyses of Arms Transfers and Conflict

The scatter diagrams facilitate useful insights about how conflict levels and arms imports co-vary, but they give us an incomplete picture and would be an inefficient means for generating a more complete one. Additional diagrams could tell us more about, say, current conflict and prior transfers, or current transfers and prior conflict, but we could not easily link the various relationships together.

To do that, I used a "path model," a way of testing influence relationships among variables. Using prior assumptions about how variables relate to one another, one diagrams the hypothesized relationships and then uses a statistical technique (multiple regression) to test them against the data on hand (in this case, data about conflict, arms transfers, and membership in regional security complexes). Path models can be used either to test the relative strengths of the causal relationships among variables, if one is sure that all relevant potential causal variables have been included in the model, or to "forecast" outcomes without necessarily attributing causality. Since this model most surely does *not* incorporate all potentially relevant causal factors (including unidentified variable[s] that might be the cause of both conflict and the quest for armaments), it is advisable to take the latter, less-assertive approach, discussing association rather than causation.[70] Since, as I noted earlier, my objective here is not to determine the overall causes of conflict in the developing world but just the relative influence of arms transfers, I will try to avoid use of cause-and-effect terminology in discussing the results of the path analysis.

The division of the data into half-decade time periods lent itself to creation of a multistage model. A one-stage version is depicted in figure 2.11. To begin with, each possible path between variables is drawn in, with lines symbolizing linkages between the variables and arrowheads indicating the hypothesized direction of influence.

Arrow direction is clear when one variable represents an earlier time period—the future cannot influence the past (at least not until someone invents time travel). In what directions the arrows really are within time periods, however, is a toss up. It is equally plausible, for example, to posit that war creates demand for immediate delivery of weapons (represented by the down arrow between conflict and deliveries in figure 2.11), as it is to posit that sizable, rapid deliveries contribute to the outbreak or sustenance of war (represented by the up arrow).[71] Regression helps to winnow out the weakest relationships among the variables (those that prove to be statistically insignificant) and thus to reduce the number of paths and arrows in the final model.

Since some types of heavy weapons are well known to have been drawn disproportionately to particular regional security complexes (tanks to the Middle East, for example), I introduced dummy variables (with values of just zero or one) to represent regional security complexes.[72] Accounting for states' membership in particular security complexes permits one to discern whether conflict and weapon deliveries exert influence upon one another that is more powerful than the influence exerted by geographic location.

Figure 2.11 Path Model Template, One Stage

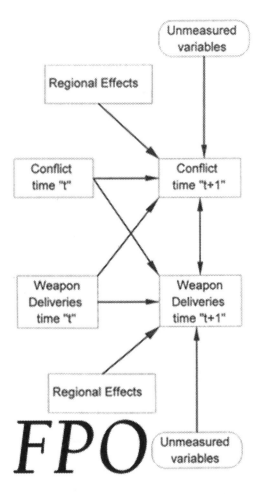

Finally, for completeness, the finished model indicates the aggregate influence of "unmeasured variables"—factors and influences *not* explicitly included in the model—as a reminder that neither conflict nor arms transfers are fully explained here.

Separate regressions were run using each conflict and weapons delivery variable from 1975–1979 onward as the dependent variable.[73] For those runs the arrows' target was the dependent variable. For analysis of relationships in the next stage of the model, the same variable was used as a predictor (independent variable). The analysis started with the earlier time periods and worked forward. So, for example, "Weapon Deliveries 1975–1979" first served as a dependent variable, with "Conflict 1970–1974," "Weapon Deliveries 1970–1974," "Weapon Deliveries 1975–1979," and "Regional Effects" as its predictors. Subsequently, it served as a predictor of conflict and weapon deliveries in 1980–1984. In the final, multistage model, a number of arrows are missing, reflecting links that proved statistically insignificant and were as a result dropped from the model. Details on this process of crafting and culling may be found in appendix C.

The summary statistic indicating how well the model predicts the dependent variable is the proportion of variance in that variable accounted for by model. It is the square of the correlation between the actual value of the dependent variable and the model's estimate of that value ("R-squared," for short).

The numbers associated with each arrow in the model indicate the relative strengths of the links between variables. Called path coefficients, they represent the amount of change produced in the dependent variable (the arrow's target) by a *unit* change in a particular independent variable (the arrow's origin), and can vary from −1.0 to +1.0. For example, an arrow with a path coefficient of .33 from arms deliveries in one time period to conflict in the next would indicate that a unit increase in arms is associated, on average, with a one-third-unit increase in conflict. Equivalently, a three-unit increase in arms is associated with a full unit increase in conflict (that is, one notch on the external conflict scale).

The arms data used in the model are the common, base−10 logarithms of actual arms deliveries. A country that received a mix of 100 tanks, troop carriers, and aircraft over a five-year period, for example, would have a data entry of 2.0 (representing 10^2 weapons) for that period. A country that received 1,000 weapons over the same period would have a data entry of 3.0. So a three-unit increase in weapons delivered would be 10^3 or *1,000 weapons*. In other words, the model would be suggesting that changes in countries' levels of conflict are associated only with *major* changes in arms deliveries. In fact, that is about the order of magnitude of relationship that we find globally between deliveries of heavy arms and levels of external conflict in the developing world.

Heavy Arms and External Conflict. The completed multistage model links the data from the respective five-year time periods like freight cars in a train. The model for heavy weapon transfers and external conflict is depicted in figure 2.12. In general, weapon deliveries are better accounted for than are conflict levels, as reflected in the higher reported R-squares for weapon deliveries in the 1970s and 1980s. That is, with minor exceptions, weapon deliveries do not do that good a job of explaining conflict.

The lack of any negative linkages in figure 2.12 is striking, however. *Increased* arms deliveries are nowhere associated with *decreased* conflict, nor are *decreased* arms deliveries associated with *increased* conflict. If arms imports do not strongly increase conflict, neither do they seem to provide a barrier against it, which might cast some doubt on the deterrence model, which associates peace with greater military strength (at least among countries supportive of the political status quo). Conversely, these results suggest at least a weak role in developing country security complexes for the security dilemma (since increasing arms are associated with at least some increased tension). Finally, the results suggest that measures to reduce the volume of arms deliveries to developing regions might have an impact on levels of tension and thus may be worth pursuing, but they should not be expected to produce miracles of transformation in regional relations in and of themselves.

Regional variables in the model are more strongly associated with levels of arms transfers than with levels of conflict—that is, the arms buying behavior of countries in a regional security complex tends to be more consistent than those same countries' conflict behavior. (Only regions with statistically significant path coefficients are shown in figure 2.12). Thus, while membership in the Persian Gulf security complex is associated with weapon deliveries in both the 1970s and 1990s, it is not associated

98

Figure 2.12   External Conflict and Heavy Weapons

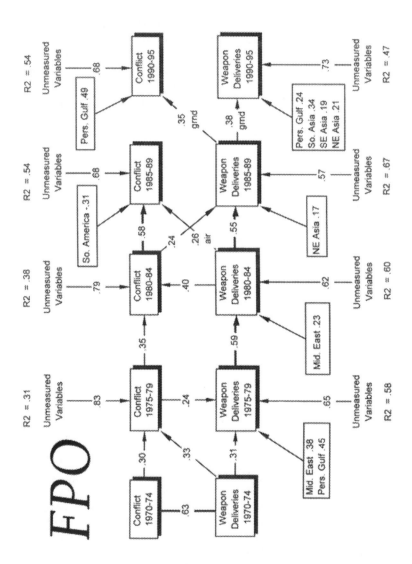

with conflict until the 1990s. Before then, only Iran and Iraq were at war, while their neighbors were merely hedging their bets with arms purchases. In 1990–1991, Operation Desert Storm directly involved every country in the complex with the exception of Iran. South America is the only other region associated with conflict, and that association is negative, indicating that in the late 1980s South America was substantially more peaceful than other regional security complexes in the model.

Of the regional variables, the Persian Gulf and Middle East so outdistanced other regions in arms deliveries in the 1970s that they are the only security complexes to have significant linkages with weapon deliveries in that time span. After the shah of Iran's buying spree was ended by his ouster, the Persian Gulf loses significance in the model until the Gulf War and its aftermath revives it. The Middle East, long associated in most people's minds with substantial deliveries of arms, was still significantly associated with arms deliveries in the early 1980s, but the path was weaker and it lost significance entirely thereafter. It faded out not because countries in the region stopped buying arms completely but because of growing global arms deliveries in the 1980s to *other* developing regions. Indeed, the market was sufficiently well distributed in the 1980s so that taking all regional variables out of the model reduced its basic explanatory power (R-squared) by not more than 3 percent, a tiny amount. In the late 1970s and early 1990s, regional arms patterns were more differentiated, and regional variables added 10–20 percent to explanatory power. Differentiation in the 1990s was driven by deliveries to India and Pakistan, the growing "tigers" of East Asia, and the Arab states in the Persian Gulf. As a final note on regional influence, removing all regional variables from the models for weapon deliveries did nothing to increase the association between arms deliveries and external conflict in any time period.

Note that paths pointing from conflict to arms deliveries are especially sparse. The two links that survived the statistical winnowing process are weak, with path coefficients of just .24. In other words, a roughly four-unit increase in conflict level (say, from peace to a small-scale regional war) is associated with a bit less than a one-unit increase in arms deliveries, or about nine weapons ($10^{.96} = 9.12$). That's not much of an association, globally. Clearly, countries caught up in full-scale war, like Iran and Iraq in the 1980s, or Egypt, Israel, and Syria in 1973, or the two Vietnams until 1975, do try to obtain substantial quantities of arms to sustain or increase their fighting abilities. But countries not at war have also accumulated large arsenals (for example, Libya, Saudi Arabia, Taiwan, South Korea, India, and Pakistan). Their direct experience of war may be more than five years past (which was true for each of these states in the 1980s) and thus was not captured by this particular model, or their acquisitions may be anticipatory. But the truly feuding countries in the developing world are sufficiently small in number that their effect—even on a model such as this one, which tries to account for regional variations in arms buying behavior—is washed out in the volume of other states' data. Moreover, as we saw earlier with the scatter diagrams, even these major arms recipients show no consistent pattern of changes in their conflict levels and arms acquisitions.

Indeed, the most consistent linkages in the model are those between conflict and conflict, from one time period to the next, and from arms deliveries to arms deliveries. The paths between conflict variables and the paths between weapon delivery variables tend to grow stronger over time until the 1990s. In so doing, they also grow stronger than the links between conflict and arms. These results suggest that states become stuck in conflict and habituated to arms acquisitions, but in parallel rather than in highly interdependent or interactive processes. However, the modest positive links between *prior* arms deliveries and *current* conflict in three out of four time periods depicted in figure 2.12 suggest that arms imports may contribute to keeping states stuck in conflict. Because this model looks at these relationships in the aggregate, rather than in terms of "conflict dyads" (pairings of grumpy states), such conclusions are tentative.

But could it be that, as James L. Payne argued in a late-1980s book on the role of militaries in the affairs of states, that rational, threat-responding reasons for building militaries and buying arms are supplemented and maybe even dominated by "non-rational" reasons for doing so? Might countries build outsized military institutions to serve crusading ideologies or warrior cultures, and might such militaries see to their own institutional well-being by continuing to accumulate power, partly in the form of armaments? Since consistent, detailed spending data or information on insider political-military influence relationships can be hard to come by, Payne argued that a workable measure of military influence should use visible military attributes as its unit of accounting. Since the great majority of countries' militaries are manpower-intensive, the ratio of military personnel to the population at large would be a truer indicator of military burden (and, indirectly, military influence) than would counting or otherwise trying to measure a state's inventory of armaments.[74]

If such a measure of military burden (what Payne dubbed the "force ratio," or armed forces per thousand population) were applied to the path model in figure 2.12, would it have much impact on the results? The argument would be that militarism, rather than membership in any particular regional security complex, was the more appropriate predictor of countries' arms buying.

I reran the model for weapon deliveries between 1975–1979 and 1990–1995, running the force ratio in lieu of the regional variables, and also running them together.[75] For deliveries in 1975–1979, R-squared dropped 22 percent without the regional variables and the force ratio was statistically insignificant. For that period, being a part of the Middle East or the Persian Gulf was thus much more closely associated with arms buying habits than were countries' force ratios. There was no similarly steep loss of explanatory power when regional variables were left out in 1980–1984, but the force ratio barely passed muster in statistical significance, with a weak path coefficient of .13.

For weapon deliveries in 1985–1989, the force ratio worked a little better, with a path coefficient of .17 without regions in the model, and .16 with them included. For deliveries in 1990–1995, when regional influences on arms deliveries increased once again, the force ratio made a statistically significant contribution to the model, with or without regions included, but without them, explanatory power dropped 6 percent from the original model. When both force ratio and regions were included, explanatory power gained 8 percent over the original. But there's a catch: in the 1990s, the force ratio has a negative path coefficient. When prior arms buying behavior and region of origin are both included in the model, it seems to say that countries with higher force ratios bought *fewer* arms. That result was influenced in part by two countries with high force ratios that imported no heavy weapons in the 1990s: Iraq (under UN embargo) and Uganda (with high manpower to fight multiple, low-tech insurgencies). Plotting the data on 1988 force ratios against 1990–1995 arms deliveries (figure 2.13), it is clear that there is no native trend in the data; the raw correlation between the two variables barely registers ($r = .13$). Countries with relatively low force ratios bought all number of arms in the 1990s, from none at all to nearly a thou-

Figure 2.13 Arms Deliveries vs. Armed Forces per 1,000 Population

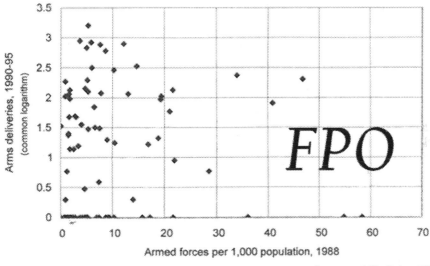

R-square = 0.02 # pts = 106

sand items. Countries with high forces ratios were similarly scattered in their habits. My inclination is to view the negative coefficient as an artifact of the model that really has no traction out there in the world.

In short, militarism as Payne defined it doesn't account very well for developing countries' arms buying patterns. It may very well be that some other measure of military influence would have greater explanatory power. At the moment, however, the best that can be said is that arms, like nicotine, appear to be addictive. Approaches to breaking the addiction are likely to be even more complicated for the collective case of a state and its security institutions than for an individual. Indeed, the best that might be expected would be to protect third parties from the regional equivalent of "second-hand smoke," which is the subject of chapter 5 and confidence building measures.

Medium Arms and Internal Conflict. While interstate conflict among developing states has been a worrisome and deadly phenomenon over the past quarter-century, levels of interstate conflict have been on the decline for some years. Internal conflict (recall figure 2.4) showed no such trend, however, except at the most violent end of the spectrum: the numbers of civil wars had declined slightly and perhaps temporarily by the mid-1990s. Serious internal conflict creates refugee flows whose presence can burden, if not destabilize, neighboring countries, and groups fighting for control of a state frequently base their operations just across the border. In the case of Rwandans in Zaire from 1994 to 1996, the two groups (refugees and fighters) were intermixed, a mixture that eventually led to renewed fighting and the overthrow of Zaire's long-time ruler, Mobutu Seso Seko.[76]

This section looks at the relationship of deliveries of medium weapons (light attack aircraft and attack-trainers, transport helicopters, light artillery, wheeled armored personnel carriers, and armored cars), the sort well-suited to counterinsurgency warfare

Figure 2.14 Internal Conflict and Medium Weapons

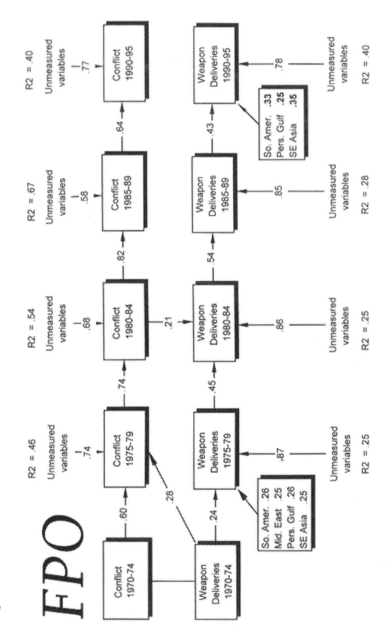

and urban warfare, to levels of internal conflict. This class of arms is distinct from the personal or portable, crew-served weapons (automatic rifles, heavy machine guns, mortars, and antipersonnel land mines) that are usually referred to as "light" weapons in recent literature on regional conflict and arms control. I do not address deliveries of heavy weapons in this discussion, on the assumption that main battle tanks and heavy combat aircraft are both ill suited and too expensive to use in internal strife.

Internal conflicts tend to be longer-lived than interstate wars.[77] Coups d'etat may be relatively brief individually, but a country may suffer a string of them over time, as have a large number of developing states. Low-level insurgencies and even civil wars can drag on for decades. As a result, we would expect to see fairly strong ties linking measures of internal conflict from one time period to the next. In fact, that is what we do see when the relationships between armaments and internal conflict are portrayed in a multistage path model like the one for external conflict (see figure 2.14). The habit of conflict is sufficiently strong, and sufficiently distributed around the developing world, that regional effects on internal conflict are statistically insignificant in all time periods. Moreover, the path coefficients from conflict to conflict grow stronger through the 1980s, beginning to weaken in the 1990s as the number and intensity of internal upheavals began to taper off.

There are similarly strengthening paths from one period of arms deliveries to the next, with a similar falling off in the 1990s. Unlike heavy weapons, whose historical delivery patterns predict future deliveries passably well, the explanatory power of medium weapon deliveries is quite weak (R-squared of roughly .25) and their patterns less consistent.

Fig. 2.14 indicates regional influence on medium weapon deliveries similar to that for heavy arms. Regions have a differential impact on deliveries in the 1970s and 1990s, but none in the 1980s. South America, Southeast Asia, the Persian Gulf, and the Middle East imported most medium weapons in the late 1970s and, minus the Middle East, in the 1990s. Finally, there are just two, relatively weak links between weapons and conflict. I would have expected to see more arrows pointing from conflict to arms deliveries, reflecting governments' responses to insurgent threats.

Clearly, many other factors influence internal conflict and the acquisition of medium arms than are measured here. Countries with lower-level internal strife may not need helicopters and armored cars to deal with it, may be able to deal with lower-level insurgencies from stocks on hand, or may make greater use of jeeps, trucks, mines, mortars, and other relatively simple and durable crew-served weapons not tallied in this data set.

States facing full-blown civil wars may seek heavy rather than medium weapons. Angola, for example, acquired tanks, heavy artillery, and heavier jet fighter aircraft in its long fight with Jonas Savimbi's insurgent Union for the Total Independence of Angola. Aggregate data testing determined, however, that heavy weapon deliveries are not significantly associated with internal conflict in any time period. The world's Angola conflicts are the deadly anomalies.

This path model demonstrates the tenacious nature of much internal strife, and suggests that its sources are not found in deliveries of medium-capability weapons. Nor does internal strife particularly stimulate such deliveries. As the first part of this chapter suggested, the internal troubles of developing states are both deep and complex. This brief section suggests, further, that arms control per se is not a solution to those troubles.

Conclusions and Observations

The human race is a complex and combative species. Most states' decisions with respect to armaments, when parsed in some detail, probably would be found to be some admixture of ambition, fear, pride, uncertainty, institutional inertia, and well-placed financial incentives. What this chapter suggests is that conflict per se is not an adequate explanation for the arms trade, nor is the arms trade an adequate explanation for conflict within and among developing states.

Does this mean that the United States and its friends (and competitors) should pay no attention whatsoever to who sells what to whom in the conventional military arena? Not necessarily. Competent states that adhere to basic international norms of interstate relations and (more controversially) international norms in treating their own peoples, should have access to the arms that they need for self-defense in a still-dangerous world. How much do they need? In today's market, that amounts to asking how much they can afford, and that will be determined country by country.

It is, of course, the less competent states that do not adhere to international norms of behavior that are the source of most concern to the international community. Their internal troubles attract the distracted attention, and occasionally the intervention, of the developed world. When they appear to be acquiring weapons of mass destruction, the developed world also sits up and takes note. Demand for such weapons is often greatest, as noted earlier, where cross-border hostility is coupled to a hard-to-right imbalance in conventional military power and the survival of the smaller power(s) is at stake. So in the end, we return to the question of how to produce a stable relationship where an unstable one currently exists. For answers, we must drop below the level of regional aggregates and look at particularly instructive case studies. These are the subject of the next chapter.

CHAPTER 3

Getting Down to Cases

C ountries buy and build major conventional weapon systems for a host of political, military, and economic reasons that vary by region and country, from defense against external threats, to control of internal minorities, to national image-building and military institution-building, sometimes by armed forces with too much political power and too little accountability. The specific cases examined in this chapter lie along a broad arc from western to eastern Asia that accounts for more than 80 percent of the arms delivered to developing states in the 1990s.[1] That arc begins with Turkey, which anchors Asia's western end and serves as both a geopolitical and cultural bridge between Asia and Europe. Although a member of the North Atlantic Treaty Organization (NATO) and the Organization for Economic Cooperation and Development (OECD), and thus technically not a developing country, Turkey's location is pivotal, as it has been through history, and the problems and forces it contends with—national identity, an Islamic revival, and pressures for ethnic autonomy—are shared by, and bear upon the stability of, regional security complexes to its west, east, and south. The forces of secularism and Islamism also vie for influence in Turkey's large eastern neighbor, Iran, but in a kind of mirror-image fashion. Where the government in Ankara fights to preserve its secular state, Iran's clerics fight to preserve their theocracy. To the east of Iran lie Pakistan and India, a pair of mismatched twins still angry over the botched operation to separate them at birth. That anger is now nuclear-armed. As India contends with Pakistan, it also keeps a wary eye on China, with which it shares contested mountain borders and which it views as its major competitor for preeminence in Asia. (About one in three human beings lives in either India or China.) The politics and defense policies of China exert major influence, in turn, on the policies of the states on its periphery in Southeast and Northeast Asia. Of these the largest and, at the close of the 1990s, most fragile was Indonesia with 200 million people—including 8 million ethnic Chinese—spread across a 13,000-island archipelago.

Together, these cases illustrate the full range of conditions that influence developing states' decisions to purchase or produce major conventional arms. They also illustrate the limits of unilateral power-balancing with conventional armaments where the size and power of one's adversary is very great. The countries of Asia vary in population by as much as 400 to 1, and in territory by as much as 10,000 to 1. Facing such extremes, states may seek safety in alliances, in mutual arms restraint, in mutual trust-building, in an external champion, or in an unconventional balance—that is, one based upon weapons of mass destruction.

Turkey

Turkey sits athwart the ancient crossroads between Europe and Asia. For six centuries, the empire of the Ottoman Turks controlled much of the modern Middle East, Greece, and the Balkans, where it left behind Muslim communities in Bosnia and Albania and a Serbian nation whose cultural legends are built around resistance to and defeat at the hands of "the Turk." The Balkans slipped away from the Ottoman Empire before World War I. Opting to join the losing side in that war cost the Ottomans their lands in the Middle East (modern Lebanon, Syria, Iraq, Jordan, Israel, the rest of Palestine, coastal Saudi Arabia, and Egypt). Allied forces occupied Constantinople (Istanbul) in 1919 and Greek troops occupied the Smyrna (Izmir) region of western Anatolia—the Turks' core territory. The following year, Anatolia was partitioned by the 1920 Treaty of Sèvres. The Greeks, who had been Ottoman subjects for nearly 400 years until some broke free in 1830, were given islands in the Aegean and the area around Smyrna where more than a million Greeks lived. Britain received control of Cyprus, plus the Dardanelles and the Bosporus, the narrow passages from the Mediterranean to the Black Sea that British and Commonwealth forces had failed to conquer during the Great War.[2]

That campaign had failed thanks in part to the generalship of one Turkish officer, Mustafa Kemal, who after the war was seized with the notion of building a smaller, stronger, unified, and secular Turkish state. Sèvres and the occupations helped Kemal to rally support to his cause. An ill-fated decision by Greece in 1921 to attempt the conquest of more Turkish territory further spurred Turkish nationalism. Securing arms from the struggling new Bolshevik regime in Russia, Kemal beat back the Greeks, won new borders for Turkey, and declared a republic in 1923.[3] At the time, every other Muslim area from Morocco to the Dutch East Indies, partly excepting Iran, was a European colony or protectorate. The European model had produced powerful, expansive, secular states that had overrun or outlasted the old Islamic empires and sultanates. The Western state model was a winner. Moreover, Islam was and is a universalizing religion and Mustafa Kemal (thereafter known as Atatürk, "Foremost Turk") wanted a strong state with *Turkish* nationalism, not a more broadly based religion, as its fundamental bond. Without direct foreign role models, Turkey embarked on a massive project of "political and social engineering [that] was unique in its time and continues to be unmatched today in its comprehensiveness."[4] Secular law supplanted Islamic law (*sharia*). Women were forbidden to wear the veil and men forbidden to wear the fez. The alphabet was Latinized. A one-party state until after Atatürk's death in 1938, Turkey evolved thereafter into a fractious multiparty democracy with periodic lapses into military rule.

Domestic Politics and Insecurity

Today, Turkey remains a multiparty democracy with close ties to the West as a member of NATO (since 1952) and as a new member of the European Customs Union—its only Muslim-majority member. Turkey is a similarly "singular" member of the Muslim world, which is a source of both internal and external tension as Turkey tries to live in both worlds. As Bilkent University scholar Duygu Sezer has observed,

> because Turkey is neither a fully Westernized nor an orthodox Islamic society but a modernizing, Europe-oriented country with an Islamic substratum, its domestic regime has

been inherently exposed to shifting counter pressures from the external environment. The ideology of Westernization and pro-Westernism has been anathema in the regions surrounding Turkey for a very long time—except belatedly in Greece. . . . [5]

Maintaining Atatürk's vision has not been easy. Within the structures of the secular state, the Turkish military has a singular role as guardian of the vision. The Turkish constitution sets the military above the country's political leadership, and as the Platonic guardian of the state it has not hesitated to seize power or nudge governments aside when either secularism or national unity seemed threatened.[6] The military is both part of society's fabric and apart from it: 18-month conscripted service for soldiers means that most Turkish males have some military service background, but officers are housed apart from the rest of society. Regulations forbid praying in uniform in public. "Defamation and belittling of the state's security and military forces" is punishable under Article 159/1 of the Penal Code.[7] Ethnic nationalism and Islamic fervor draw the military's ire in roughly equal proportion, except when one can be set against the other.

Anatolia's population has never been 100 percent ethnic Turk. At modern Turkey's founding, after the wars of intervention, there were mass relocations of Greeks from Turkey and Turks from Greece (1.3 million and 400,000 respectively). This population shift followed by just eight years the Ottomans' attempted final solution of the "Armenian problem."[8] And there remain today, primarily in southeastern Turkey and scattered through northern Syria, Iraq, and Iran, roughly 20 million ethnically and linguistically distinct peoples who call themselves Kurds, whose presence in the region long predates the influx of the Turks themselves. A Kurdish state was promised by the Treaty of Sèvres, but the promise went unfulfilled. A rump Kurdish state established by Soviet forces in northwestern Iran in 1945–1946 became one of the new Cold War's first contests of political will between Moscow and Washington. Moscow backed down and the "Mahabad Republic" disappeared.[9]

Roughly half of the present-day Kurdish population lives in Turkey, but until 1990 its ethnicity was not officially acknowledged and use of the Kurdish language was banned (broadcasts in Kurdish remain banned).[10] Since 1984, the Turkish police, gendarmeries, and military have been waging an internal war against a violent group of Kurdish separatists (the Kurdistan Workers Party or PKK). Any expression of sympathy or support for separation is an indictable felony.[11] A half-million Turkish Kurds have been displaced from 1,500 villages destroyed in army operations in the southeast. Some of the displaced live in shanty towns on the edges of the conflict zone while others have moved to the fringes of Turkey's burgeoning cities. In its anti-Kurd operations, the army has enjoyed remarkable autonomy, at one time preventing the country's deputy prime minister from entering one province to investigate charges that the army was razing Kurdish villages suspected of cooperating with the PKK. The U.S. State Department's 1996 report on human rights cited "serious human rights problems" related to Turkey's campaigns against the PKK, noting both the forcible displacements of population reported in the press and the widespread use of torture.[12] The PKK in turn has burned villages that cooperate with the government and has been responsible for multiple cold-blooded killings, including the deaths of those who have tried to quit the movement, or their families. As usual in such wars, the common people whom both sides want to control suffer the most.[13]

Islam, the Turkish state's other main concern, is influential in Turkey's rural areas and among the growing numbers of urban poor, many displaced by the anti-PKK campaign. In 1996, the first coalition government led by an Islamist party was formed under Prime Minister Necmettin Erbakan, leader of the Islamic Welfare Party. Less than a year into his term, Erbakan was urged to resign so as to "protect democracy" from being drawn into "endless darkness"—an allusion by Turkey's chief of military intelligence to Islamist rule.[14] Erbakan had made prominent visits to Libya and Iran and promoted, among other things, more Islamic education within Turkey. His government was supplanted by a secularist coalition in July 1997. Six months later, his party was disbanded by order of Turkey's Constitutional Court.[15] It was soon reborn as the Virtue Party, although seven former leaders of Welfare were barred from political activity by court order for five years.

Welfare had sought to increase the role of Islam in Turkish life and governance. In a society in which politicians are generally assumed to be corrupt, it had a reputation for "honest and impartial administration in the cities and towns where it is in power" and did seem interested in the welfare of those it represented. Its elected officials occupied some 400 mayors' offices, including those in the two largest cities (Istanbul and Ankara) and throughout the Kurdish-majority southeast. The appeal of honest and competent governance is not likely to fade as urban crowding increases. Roughly 70 percent of Turkey's population will reside in cities by 2000. About 60 percent is under 20 years of age, reflecting its rapid growth. Unemployment at the end of the 1990s was running as high as 40 percent. An underemployed, poorly housed, low-income urban class represents a substantial pool of potential voters for Welfare/Virtue.[16] The "resurgence of Islam as a source of identity for Turkish-Kurds" suggests, in turn, convergence of the military's two main fears.[17]

Disturbing reports hit the press as the new decade began, indicating that the state, or elements of it, had enlisted the aid of domestic terrorists who call themselves Hizbullah (or Party of God, ostensibly unrelated to the group of the same name that attacks Israeli targets in southern Lebanon). Dozens of brutalized bodies thought to be those of Kurdish politicians, intellectuals, and rights advocates were exhumed from mass graves in or near Hizbullah safehouses at widely separated sites. Some viewed the evidence uncovered by late winter 2000 as indicative of the fate of thousands of other persons whose murders or disappearances in southeastern Turkey since the start of the anti-PKK war remain unresolved. Remarks made in the context of these discoveries by former prime minister Tansu Ciller, that "terror was the top issue, and . . . we had to do whatever was necessary" to contain it, and by President Suleyman Demirel that the state may "deviate from routine when higher interests require it" did little to assuage critics of the government's possible use of terror to fight terror.[18]

Regional Threats, Opportunities, and Uncertainties

Until Gorbachev and *glasnost,* the Soviet Union and communism were the major external threats to Turkey's experiment. Turkey joined NATO in response to those threats and was welcomed in turn as a barrier to the southern expansion of Soviet influence. The Soviets generally leapfrogged Turkey in their efforts to build patron–client relations with front-line Arab states (those actively engaged in the struggle against Israel), but for the United States, Turkey was valuable real estate from which to monitor the Soviet missile and space program in the days before all-seeing, all-hearing reconnaissance satellites. Turkey benefited in turn from substantial U.S. economic and military assistance.

Now, in place of the Soviet Union, there is a jumble of new states that are still getting their bearings. Several are dealing with domestic unrest, some of it very serious and right on Turkey's northeastern border: Georgia is split by a separatist movement, while Armenia and Azerbaijan are deadlocked in their war over an Armenian-dominated enclave (Nagorno-Karabakh) within Azeri territory. Further afield, not only war but also peace in Afghanistan potentially threatens the stability of Turkmenistan and Tajikistan, if the likely victors of Afghanistan's long civil war, the fundamentalist Taliban, seek converts in other states. The stability of Russia itself remains a cause for concern in Ankara, and the bloody fighting in Russia's southern Chechen Republic stirred that concern.

In the early 1990s, Ankara sought to exploit its new access to the Turkic-speaking republics of Central Asia, whence the Turks of Turkey came in the eleventh century, seeing an opportunity to influence their political development and hoping to "shape a group of mutually reinforcing, like-minded states." Because the newly independent states' physical infrastructure and economies tended to be tied to Moscow, it would have taken substantial and sustained effort to reorient them. Turkey's leaders soon realized that they did not have the resources needed to sustain the initiative.[19]

The wars in the former Yugoslavia drew Turkey's close attention. Ankara contributed troops to international peacekeeping efforts there, first as part of the United Nations Protection Force in Bosnia (1993–1995) and then with the NATO Implementation/Stabilization Force (1995–). Turkish jet fighters helped NATO patrol Bosnia's UN-declared "no-fly zone," and in 1999 joined NATO's air strikes against Serbia, and the subsequent NATO peacekeeping force in Kosovo.

It is one of the complex political ironies of this region of the world that while forcefully putting down its own Kurdish separatist movement, Turkey has hosted U.S./NATO forces patrolling the skies of northern Iraq to protect Iraqi Kurdish areas from Iraqi army incursions. Support for the 1991 UN embargo on Iraq that followed the Gulf War has cost Turkey a great deal in lost trade, and the government has looked the other way regarding certain kinds of cross-border smuggling. Turkish truckers regularly head for northern Iraq to swap flour for diesel fuel in Mosul, selling the diesel for a factor-of-ten profit back in southeastern Turkey. In late 1994, up to 40 million gallons of diesel a day were crossing the border. The proceeds supported the extended families of the 60,000 largely Kurdish truckers in the otherwise depressed region hard hit by the sanctions against Iraq and the war against the PKK. Iraq in turn has gained food supplies and hard currency, and the Iraqi Kurdish faction that controls the border areas, the Kurdish Democratic Party (KDP), has earned a substantial income from transit duties imposed on the contraband. The KDP also has cooperated with the Turkish army in its periodic forays into northern Iraq to hunt down PKK guerrillas and destroy their base camps.[20]

The war against the PKK has also affected Turkey's relations with Syria. Long known to harbor PKK forces and their leader, Abdullah Ocalan, Syria was forced to disgorge both in October 1998. Turkey's National Security Council had met in late September, announced that it was considering "economic, political, and military sanctions" against Syria, and moved 10,000 troops and their equipment close to the two countries' common border. On October 4, Turkey's chief of staff said the two countries were in an "undeclared state of war." Later that month, Ankara announced that military exercises would be conducted in the area in early November, involving 50,000 troops. Syria got the hint and on October 21 signed an accord pledging that it "will not allow the PKK to receive military, logistic or financial support or to carry out propaganda on its soil." Ocalan fled to Moscow and his troops decamped for Iraq. On November 8, 25,000 Turkish troops moved into northern Iraq to engage the guerrillas.[21] Ocalan meanwhile turned up in Italy, which refused to extradite him because of Turkey's death penalty. After two months, Ocalan left Italy and, after being refused entry by several European countries, made his way to Greece. The Greeks, nervous about sheltering Turkey's public enemy number one, flew him to their embassy in Nairobi, a city that happened to be swarming with American FBI agents investigating the August 1998 bombing of the U.S. embassy there. U.S. intelligence tipped off the Turkish government, which sent a unit of commandos to Nairobi. After two weeks, Ocalan left the embassy, persuaded that a flight out to Western Europe awaited. Instead, Turkish commandos seized him and brought him home for trial.[22] Convicted and sentenced to death, Ocalan offered to end the PKK's armed struggle and his group took up the refrain. The Turkish government refused to negotiate, and by its actions in early 2000—including the brief arrest of several popular Kurdish mayors—still appeared more inclined to suppress Kurdish nationalism than to seek a modus vivendi with it.[23]

Turkish Armaments

As a member of NATO, Turkey is subject to the constraints of the Conventional Forces in Europe (CFE) treaty, unlike its Middle Eastern neighbors, but those neighbors are smaller and poorer and can no longer buy high-tech arms at bargain-basement rates from the Soviet Union. The Turkish military, on the other hand, has benefited from the "cascade" of arms removed from the old central front in Europe under CFE. For each weapon received, Turkey must dispose of an older model, but its arsenals have been substantially modernized at low cost as a result of cascading.

Turkey is a firm supporter of the international nuclear nonproliferation regime. Ukraine's ratification of the Nuclear Non-proliferation Treaty has eased Turkish proliferation concerns looking northward. To the south, however, Iraq remains unrepentant about its nuclear, chemical, and biological weapon programs. Iran is reputed to be seeking (while denying) a nuclear weapon capability, and Syria is known to have a chemical arsenal. All three have 200-km-range Scud ballistic missiles. Iraq may be busy recreating longer range versions now that United Nations arms inspectors have left the country, and Iran has tested a medium-range missile. Syria, Iraq, and Iran all share borders with Turkey, which makes Turkey's membership in NATO, and the security guarantees that go along with it, that much more important. The very pervasiveness of ballistic missiles in the region also gives Turkey an interest in ballistic missile defenses.

Arms Industry

Elements of Turkey's armaments industry date back to the Ottomans but the most modern elements, like co-production of U.S.-designed F-16 fighter jets, date only to 1983 and a major new effort, spearheaded by Turgut Ozal (later prime minister and president) to promote Turkish self-sufficiency and export potential in military equipment. Sentiment favoring self-sufficiency initially grew out of the international arms embargoes that followed Turkey's 1974 invasion of Cyprus on behalf of the ethnic Turkish minority there and from the 1980 military takeover of the government. However, the Reagan administration's willingness to downplay the latter in the interest of building higher security walls around the Soviets' "evil empire" led to major new Turkish reliance on advanced U.S. military technology and to substantial growth in the size and sophistication of Turkey's arms industry.

The industry is closely tied to the Turkish military through the Ground Forces, Naval, and Air Force Foundations, merged in 1987 as the Turkish Armed Forces Foundation, and to the holding company OAK, which is funded by tithes on officers' salaries. The latter gives the officer corps a stake in military production that is entirely separate from that of the civilian-run government.[24] Within the Ministry of National Defense, the semi-autonomous Undersecretariat for Defense Industries (UDI) oversees arms production and controls the Defense Industries Support Fund (DISF), established in 1986. DISF receives resources from the military foundation, plus 5 percent of income tax revenues, 10 percent of the taxes on fuel consumption, and tax revenues from cigarette, alcohol, and tobacco consumption. As the Turkish economy grows, so do resources for arms production, although high inflation rates have undercut real revenues.[25] In 1995, the Turkish government inaugurated a "buy Turkish" program to discourage imports, partly because of inflation-sapped exchange rates and partly because Western arms exporters, including the United States and several Nordic countries, blocked exports to Turkey or expressed reservations about imported weapons being used, inadvertently or not, against Kurdish civilians in the government's campaign against the PKK.[26] The head of UDI estimated that owing to the limited size of Turkey's defense market, viable defense producers had to be able to export 60 percent of their production, making export markets key to military import substitution.[27] However, since most high-end Turkish military production is under license from other producers, end-user constraints may limit how and where excess production might be sold.

Turkey produces a wide array of defense equipment. For its ground forces, it builds U.S. Blackhawk and French Super Puma transport helicopters, Stinger portable surface-to-air missiles, antiaircraft guns, and various trucks and other vehicles. For its air force, it builds the F-16, Italian-designed jet trainers, and Spanish-designed light transport aircraft. Turkey's shipyards build German-designed frigates, submarines, and fast patrol boats.[28]

In April 1996, the Turkish military released a 30-year military modernization plan estimated to cost $150 billion that called for acquisition of hundreds of new combat aircraft and helicopters and 14 new frigates.[29] Prime Minister Erbakan's Welfare-led government was slow to ratify the modernization plan, once in power. After four months of delay, Foreign Minister (and political competitor) Tansu Ciller signed a decree launching the first part of the plan, primarily for local production of attack and transport helicopters, while Erbakan was on tour in Africa.[30]

Arms Imports

Local production notwithstanding, Turkey has been taking in substantial numbers of arms from abroad in the 1990s. The CFE cascade process has brought in hundreds of tanks, armored personnel carriers, and heavy self-propelled artillery. Radars, guns and missile systems for its license-produced frigates are imported, as are the air-to-air and air-to-ground missiles that equip its F-16s.[31] Under a $630 million defense cooperation deal with Israel signed in December 1996, Israeli firms have upgraded Turkey's older F–4E jet fighters to Israel's Phantom 2000 standard. Israel has also bid on a Turkish requirement for four airborne early-warning aircraft along with a number of American firms and would like to be the chosen provider of 1,000 new main battle tanks (most of which would be built in Turkey).[32]

Human rights issues have impeded some but not all recent U.S. sales to Turkey. In spring 1996, the White House delayed the sale of ten AH–1W Supercobra attack helicopters in response to objections by human rights groups, which led to cancellation of the sale by Turkey in October and to the opening of negotiations with a number of other suppliers to co-produce its next generation attack helicopter. Delivery of three former U.S. Navy Perry-class frigates to Turkey was also delayed.[33] On the other hand, the Congress let pass the first export sale of the Army Tactical Missile System (ATACMS) to Turkey in late 1995. The Defense Department's memorandum on the sale certified that it would "not adversely affect . . . the military balance in the region," and used identical language when announcing a sale of 40 ATACMS to Turkish rival Greece in June 1996.[34] Originally, Turkey wished to coproduce many more of the missiles, but the United States refused to release the rocket motor technology, citing the Missile Technology Control Regime. The 165-kilometer-range missile cannot reach very far into neighboring states' territory from Turkey, except for Cyprus, which it can reach with relative ease. Maker Lockheed-Martin received an order for 72 missiles with cluster munition warheads in November 1996.[35]

Turkish Dilemmas

Turkey has faced a brutal opponent in the PKK, but the government's equally brutal campaign against it and its only grudging recognition of Kurdish ethnic identity detract from the cause of a secular, democratic country and raise questions in the West about whether Turkey should ever be part of the European Union. Such questions in turn kindle Turkish resentment. Yet democracy without respect for human rights is an empty framework, and efforts to quash dissent and discontent may only drive more Turkish voters to cast their ballots for Virtue, or for the party that succeeds it when *it* is banned. By standing ready to intervene again should voters grow too enamored of Islamist parties and the care with which they attend to the public weal, the Turkish military may be doing democracy—and ultimately Turkey itself—no favors.

The West's dilemma is that neither a Turkey whose military rules with an iron fist nor a Turkey that turns to seventh-century *sharia* to solve twenty-first-century problems is a particularly desirable partner. The former is likely to be chronically unstable and thus unable to attract the investment it needs to grow and to reopen its political system, while the latter is more likely to turn away from Europe and the Westernizing experiment of the last 75 years. Much as Marxism-Leninism failed to snuff out Russian Orthodox Christianity in 72 years of trying, Turkey's determined experiment in secular rule has failed to suppress or supplant either Islam or Kurdish nationalism, and the harder it tries, the more radical the Islamic challenge it is likely to produce, and the less Turkish its Kurdish minority will feel. Turkey is the world's oldest running experiment in building Westernized political institutions in an Islamic culture. The longer it tries to sustain the original vision that modernity and democracy require the sidelining or suppression of both Islam and ethnic diversity, the bigger the crunch if the experiment fails.

Iran

Much as Anatolia and the Bosporus have been the meeting points of East and West for centuries, Persia (Iran) has been a transit route, its history a succession of domestic empires and foreign invasion. Today, Iran is as a pivotal bogeyman in U.S. foreign policy as Turkey is a pivotal ally, but just like Turkey, Iran is a country of much greater complexity than casual observation—or categorical demonizing—would suggest.

The Roots of the Islamic Republic

While America's collective memory goes back to George Washington and 1776, Persia first became a power to be reckoned with around 550 B.C. under Cyrus the Great. For two centuries, he and his descendants ruled an empire stretching from the Nile to the Indus River in the south, and from the Bosporus to the Onyx River, east of the Caspian Sea, in the north. Alexander the Great added most of it to his Hellenic Empire in 330 B.C., but the Romans never managed to penetrate farther eastward than the mountains that form the current Iraq-Iran border. Persia flourished for centuries before falling to Arab conquerors a few years after the death of Mohammed in the mid-seventh century. Arab domination ended less than a hundred years later when armies sweeping out of northeastern Persia established the Abbasid dynasty that presided, from Baghdad, over Islam's "golden age."[36]

Persia was overrun again in the mid-eleventh century by the Seljuk Turks on their way west, but the Turks were content to let local representatives rule in their stead. Not so the Mongols (1217–1265), who pillaged, burned, and massacred millions. Persia was not stitched back together for 250 years and did not recover its pre-Mongol population levels until the mid-twentieth century. The stitching was done in typical fashion by military conquest, this time at the hands of a mystical strain of Islam known as Sufism and a Sufi order called the Safavids. By the early sixteenth century, about the time Cortez and his troops were bringing Christianity and measles to Mexico, the Safavids had secured control of Persia and enforced Shia Islam as the state religion.[37]

Shi'ism has remained largely non-institutionalized through much of its history: "Imprecise judgments concerning knowledge of Islamic law arrange religious authority figures in a fluid hierarchy" wherein rankings are "a loose function of popular esteem and eminence." Shias believe that the Koran's teachings are interpretable by man, so Shia clerics traditionally combine roles of theologian and judge, interpreting "all religious, social, and political issues facing the Islamic community." But while Shi'ism theologically is the "great equalizer of society and the mighty shield protecting the masses against the elite," in practice it is "authoritarian and elitist," and its hierarchy, while imprecisely defined, nonetheless commands the obedience of the faithful.[38]

As a minority sect within Islam, moreover, Shi'ism evolved the protective practice of *taqiyeh,* or "concealment in defense of the faith." Historically, it has meant concealing the truth "in any situation in which that seems propitious . . . [Thus today's] Islamic Republic suffers the abiding suspicion among many foreign governments experienced in the Iranian philosophy and practice of taqiyeh that pronouncements of policy on everything from economics to terrorism may be concealment of the truth."[39]

The other pillar of Persian order, the monarchy, reflected the strongly patriarchal nature of Persian culture and family life, including its tradition of mistrusting non-kin.

The Persian/Iranian concept of the monarchy "never really evolved," concluded historian Sandra Mackey, from the first days of Cyrus the Great to the last days of the Pahlavis in 1978. "Except for two popular movements that exploded in 1906 and 1951, the monarchy remained absolute, and [in] its own view, infallible" until it was destroyed, only to be replaced by a different sort of one-man rule under the Ayatollah Khomeini.[40]

Iran and the West

From the nineteenth century onward, the principal trespassers into Persian affairs were Western (or Westernized) powers: Russia, Britain, and the United States. As weak as China in the same period, similarly ruled by a tribe from the outer provinces (the Qajars), Persia relinquished mineral rights to Western states and companies. A 1906 reform movement resulted in the convening of the first Majlis (advisory council), but its real power lasted only five years, when it was overthrown with the assistance of Russian troops.[41]

In 1921, Reza Khan, a military officer and head of the aptly named Cossack Division, set about restoring order. He was appointed prime minister in 1923 at about the time that Atatürk was founding a new republic in neighboring Turkey. Urged by reformers to emulate Atatürk, Reza Khan instead cut a deal with the clergy to support his new dynasty in return for his keeping democracy at bay. He soon outraged the clergy, however, by decreeing a secularist regime—Kemalism without the voting parts. He changed the country's name to Iran, built up its infrastructure, and banned the veil. His police were every bit as vigilant in removing veils as more recent Islamic militants have been in enforcing their use.[42]

In 1941, Iran was once again overrun by foreigners—the Russians in the north and the British, soon backed by the Americans, at the head of the Persian Gulf, which the British already dominated through their oil concessions. Together, the three foreign powers built a vital supply line to Soviet forces battling the Nazis. A powerless Reza Shah abdicated in September 1941, handing the reigns of government, such as it was, to his son, Muhammed Reza.

After the war, the younger shah became increasingly reliant on American support to fend off the Soviet-backed Tudeh Party on his left, clerics on the right, and nationalists in the center. The secular National Front, led by the aging and quirky Mohammed Mossadeq, dominated the Majlis. Upon becoming prime minister in 1951, Mossadeq ordered the British, including oil industry engineers and technicians, to leave the country. Without the foreigners, the oil industry fell into ruin. Mossadeq's clerical allies deserted him, and by 1953, the Tudeh Party was mounting a strong challenge to his rule. This time, the coup de grace to a tottering nationalist regime was delivered by a handful of American agents with money to pay street demonstrators.[43]

After Mossadeq's downfall, resistance to the shah's state shifted from secular to religious channels. Never hesitant to put down dissent, the shah created more of it by his ruthless use of force. His strong ties to the United States allowed the opposition to associate him with the country's long history of foreign domination. A 1964 U.S.-Iran Status of Forces Agreement granting American personnel immunity from Iranian law evoked fiery condemnation from the clerical opposition's most articulate and charismatic leader, Ayatollah Ruhollah Khomeini. Arrested the day of his speech, he was sent into exile, spending most of the next 15 years amidst the holiest shrines of Shia Islam, in southern Iraq.[44]

An imperial police state with some of the outward trappings of democracy, Iran continued to accumulate social-political stresses throughout the 1960s and 1970s. From October 1971, when the shah ceremoniously commemorated the 2,500th anniversary of the Persian monarchy, until his overthrow seven years later, his rule relied almost entirely upon an extensive security apparatus whose forces leaned heavily, in turn, on torture as an instrument of control, engendering violent organized opposition. "To [the shah], critics were dissidents and dissidents were traitors."[45] Eventually, even the shah came to recognize that his policies were counterproductive, but by then the counterpressures were such that the effort to relieve them with a new round of reforms only let loose the deluge.

As in the Soviet Union, Iranian *glasnost* was not enough. In a 14-month period preceding the shah's final self-exile from Iran in early 1979, more than 10,000 Iranians died and 40,000 were injured. In late October 1979, he entered the United States for cancer treatment. On November 1, U.S. national security adviser Zbigniew Brzezinski met with moderate Iranian secular leaders in Algiers, inflaming extremist sentiment in Iran and contributing, in historian James Bill's estimation, "both to the taking of American hostages and to the destruction of moderate influence in the revolution."[46] The American embassy in Tehran was seized by Iranian militants three days later, and U.S.-Iranian relations entered into a death spiral from which they have yet to fully recover.

The Islamic Republic at War

In September 1980, Iraqi president Saddam Hussein attacked the new Islamic Republic of Iran (IRI), nominally to reclaim territory promised Iraq in a 1975 agreement made with the late shah. But the Iranian revolution was also a direct threat to Iraq's secular, socialist, Sunni-based regime, which governed a country in which 60 percent of the population, largely in the south, were Shia and late neighbors of Ayatollah Khomeini.[47] Instead of the anticipated easy victory against a demoralized Iranian military, the Iraqis got themselves a bloody, eight-year war of attrition. Saddam Hussein wrapped himself in Arab nationalism, while the Ayatollah waged holy war for the Iranian nation and Islam.[48]

After Iraqi gains were reversed in 1982 but Iran refused calls for a cease-fire, the specter of Iran possibly winning the war (plus not a little greed) drove the West to aid Iraq. The French sold Baghdad 40 percent of their arms exports for 1983, the Americans provided intelligence, looked the other way while their Arab Gulf allies provided money and arms, and led an effort to embargo arms shipments to Iran. When Iraq began to strike Iranian oil tankers and facilities in the Gulf, the outside world took the attacks on Iranian shipping in stride. When Iran retaliated against the Gulf shipping of Iraq's supporters in 1987 (Iraq itself exported oil via pipeline, not tanker), Moscow and Washington both agreed to escort tankers in the Gulf and to be on the lookout for mines and Iranian naval forces. One year later, an American cruiser on escort duty downed an Iranian airliner, mistaking it on radar for a military jet. After that incident, and with the tide of battle running heavily against it, Iran agreed to United Nations terms for a cease-fire.[49]

One year after the war ended, the Ayatollah Khomeini died, but before he died Khomeini issued a *fatwah* (or binding proclamation) defining the powers of the Islamic state, which derived, he declared, from divine sanction over all matters religious and secular. Having put clerics into government for the first time in Iranian history, Khomeini now gave government supreme power over Islam. That government, with an economy greatly weakened by war, has since seemed "incapable of coping with the challenges of a rapidly changing and increasingly interdependent world."[50]

The same passionate single-mindedness that helped Ayatollah Khomeini survive exile and grow in stature among millions of his countrymen ill served Iran's interests when applied to its relations with the rest of the world. Khomeini cared little for national, as opposed to "Islamic," interests as he saw them, but his vision of an Iranian-led Islamic commonwealth stretching from Morocco to Pakistan was always fatally flawed: since Iran represents both a singular culture and a minority, "heretical" sect within Islam, it could not "lead a unified movement of militant Islam against the West even if it wanted to."[51]

With a disdain for compromise typical of fanatics throughout history who have seen the "one best way" and insist that others take it, Khomeini turned a two-year war in which Iran was clearly the victim into an eight-year war in which it came be seen as the aggressor. No other man before or since did so much to raise the stature of Saddam Hussein's murderous regime in the eyes of the world, and no man since the Mongol chieftain Tamerlane was so instrumental in causing so many Iranian deaths. The net result of his uncompromising zeal was battlefield defeat, political isolation, a wrecked economy, a clergy seduced, corrupted, and discredited by their temporal powers, and increased American military presence and political influence in the Gulf.

Post-Khomeini Iran

Khomeini's successors have been somewhat less single-minded. They remained neutral in the Gulf War, despite invitations from Saddam Hussein to join forces against the outsiders, and they did not visibly react even when Saddam bombed Shia holy sites in Najaf and Karbala in his war against southern Iraq's Shia population.[52] Thus far, however, they have been unable or unwilling to undo those parts of Khomeini's legacy most distasteful to American policymakers, partly because the pragmatic and extremist wings of Iranian politics do not so much vie for control of policy as conduct parallel versions of it. Thus the clerical foundations known as *bonyads,* and not the Iranian foreign ministry, offered the $2.5 million reward for carrying out the *fatwah* threatening death to author Salman Rushdie. In 1998, the foreign ministry, not the *bonyads,* promised no longer to enforce it.[53]

The U.S. State Department's 1997 report on terrorism did not say whether mosque or ministry sponsored the eight assassinations of Iranian dissidents abroad in 1996, but Iran still officially opposed the Arab-Israeli peace process and supported such terrorist organizations as Hezbollah in Lebanon, Hamas and Islamic Jihad, which violently oppose a Palestinian deal with Israel, and the Islamist regime in Sudan. Even President Khatami, viewed as a political moderate (see below), made a point of meeting with Hezbollah leaders on his May 1999 visit to Syria. On the other hand, while Iran remained on the State Department's list of states supporting terrorism in its 1998 report, Iran was no longer singled out as terrorism's "most active state sponsor."[54]

The revolutionary regime still has not managed to revive the Iranian economy, which depends heavily on petroleum exports. Adjusted for inflation, oil prices by 1994 were little more than one-third of what they were a decade earlier, and about the same as in 1973, before the October war, the Arab oil embargo, and the price jumps they triggered. Meanwhile, Iran's international purchasing power was being decimated: the

Iranian rial was devalued by 97 percent against the U.S. dollar in March 1993.[55] Oil prices recovered in 1996, only to plunge again in 1998 in the wake of the Asian economic crisis to prices below the median for the entire period since 1945.[56]

Iranian voters, looking for economic betterment, have in recent elections voted for more pragmatic leadership in the Majlis (in 1996 and 2000) and the presidency (in May 1997). Dark horse presidential candidate Mohammad Khatami soundly beat the inside candidate for president whom most analysts expected to win, in an election that seemed to have been conducted fairly. Women, young people, urbanites, the educated, and those affiliated with larger businesses supported him in large numbers.[57] Most analysts expected policy changes under Khatami's leadership to be inwardly directed, with little change in Iranian policy either toward Israel and the Middle East peace process or toward American military presence in the Gulf region. Two years after his election, President Khatami stood by his domestic reform agenda, but it remained blocked by what he termed "monopolistic forces which seek to model society according to their interpretation of religion and law." Reportage from Iran indicated that popular opinion remained supportive of Khatami and was increasingly indifferent to the revolution (more than half of Iran's growing population has been born since the shah fell from power). The revolution's clerics retained their hold on key levers of power, however, including the military, the intelligence agencies, and the Republican Guard, which was created, as the name suggests, to protect the revolution. All of these institutions report ultimately to Iran's supreme religious leader, Ayatollah Ali Khamenei.[58]

Security and Foreign Policy

In most of the states surrounding the Persian Gulf, the most serious threats to ruling regimes derive from fundamental issues of social identity (Persian versus Arab, Shia versus Sunni versus Wahabi, guest workers versus citizens, rich versus poor, ancient versus modern), issues that are partly cross-border but also internal in nature. Certainly this is the case with Iran. Azeri losses in the long Armenia-Azerbaijan war over the enclave of Nagorno-Karabakh have threatened to arouse the nationalist feelings of the one-fourth of Iran's population in the northwest that is ethnically Azeri. Fighting between Iraq and its Kurdish population periodically spills across Iran's border. Iran's own Kurds, Sunni Muslims like their fellows in Iraq, are viewed with suspicion, as are other large ethno-religious minorities, the Baluchi in the southeast and the Sunni Turkomen in the northeast. The Baluchi's transborder nationalism (substantial numbers also live in neighboring Pakistan) and drug-running make southeastern Iran a difficult area to control. The Turkomen share Khorasan province in the east with a large population of refugees from Afghanistan. That country's combat anarchy worsened year by year through the 1990s, fueled by arms from surrounding states that supported their religious or ethnic brethren. The dominant fighting faction in 1998–1999, controlling most of the country although not widely recognized as a government, was the Taliban, ethnically Pushtun, denominationally Sunni, and politically retrograde. A major military clash between Iranian and Taliban forces was narrowly averted in late 1998.[59]

Iran would like to wean the Saudis and the other Gulf Arab states from their political and military dependence upon "global arrogance"—that is, the United States, its Fifth Fleet, and the sundry U.S. forces deployed on land to guard against Iraq, which Iran still has some reason to fear as well. Tehran would prefer a regional security regime that involves Gulf littoral states only; as the largest state in the region, it could resume the role of quasi-hegemon that the late shah vigorously claimed. A Gulf security regime without the United States is unlikely because the Saudis, with their much smaller native population, have no way to counter Iraq or Iran militarily without a heavy Western presence. They see the former coveting their resources and the latter seeking a "fundamental reordering of the Saudi political system."[60] Saudi Arabia's absolute monarchy is an anachronism to much of the world, but its governing premise is not so different from the strongman rule that prevails in much of the rest of the Arab world. All of these brittle regimes face a rapidly changing world in which adaptability is the key to prosperity and, ironically, to stability.

President Khatami worked assiduously to build bridges to Saudi Arabia and the other Gulf Arab states and to Western Europe. He was the first Iranian leader to visit Europe since the revolution. Paying a call on Pope John Paul II in Rome, Khatami stressed commonalities between Christianity and Islam. He also met with executives of Italian and French energy companies to promote investment in Iran. (Continuing U.S. sanctions against American business deals with Iran, embodied in the 1996 Iran-Libya Sanctions Act, also penalize foreign firms investing more than $40 million per year toward Iran's "ability to develop its petroleum resources."[61])

In mid-May 1999, Khatami was the first Iranian leader since the revolution to visit Syria, Saudi Arabia, and Qatar, having hosted the Saudi defense minister and the foreign minister of Bahrain in Tehran—also firsts for the Islamic Republic—a few weeks earlier.[62] Some analysts suggest that these new rounds of diplomatic activity reflect not only new openness on the part of Tehran but also renewed and shared concerns about the power of Iraq, which over the course of 1998 managed to limit and then terminate the work of the UN Special Commission that was charged, after the Gulf War, with dismantling its ballistic missile, chemical, and biological weapon programs.[63]

Military Forces and Priorities

Iran lost large quantities of equipment in its eight-year war with Iraq. Half of its ground forces equipment was destroyed or captured in the last six months of fighting.[64] Blocked from obtaining direct replacements for its American ground forces equipment and unable to afford new Western models, in the 1980s Iran turned to Russia and China. Iran sought to rebuild after the peace, and while its defense budget has been climbing, in 1997 Iran was estimated to have spent only about one-third as much on defense as Saudi Arabia, and only half as much as Israel.[65] It will be "well beyond the year 2000 before Iran's land forces can acquire anything like the full mix of modern equipment they need. It will be even longer before Iran can use such equipment effectively," due to shortages in training and discipline, divided command, and compartmented and ineffective logistics.[66] Iran's current ground force structure is split between the regular military and the Revolutionary Guards. Both lay claim to scarce equipment and they have refused to merge. While the Guards are nominally responsible for internal security, their strong ties to hardliners in Iran's clerical leadership make them difficult to control and to confine to internal security roles.[67]

Air Force, Air Defense, and Navy

Air defense is a long-standing Iranian weakness, and its current hodgepodge of suppliers, missiles, aircraft, and politics does not help matters. Aging American models acquired two decades ago by the shah are supported in the face of a U.S. embargo with secondhand parts scrounged on the international market and through the black market. Aircraft electronic systems are often non-operational, and deaths in aviation accidents are frequent. The air force can probably conduct operations with only half of its operational aircraft and sustain such operations for only a few days.[68] Iranian air force Sukhoi–24 strike aircraft nonetheless can and have attacked Iranian opposition forces based inside Iraq and could be thought to pose some threat to neighboring states. Russia's President Boris Yeltsin promised Washington that no more arms contracts with Iran would be written, but this may have coincided with Russian annoyance at Iranian failure to pay for its arms rather than reflecting a fundamental change of policy.[69]

Iranian naval forces are of special interest to the United States because of the West's reliance on oil flow out of the Persian Gulf and the U.S. Navy's forward presence in the region. Iran has been buying more sophisticated C-802 cruise missiles from China and installing them on launchers overlooking the lower Gulf and aboard missile boats. Iran has also taken delivery of its third and probably last Kilo-class diesel-powered submarine from Russia. The Kilos operate from a single base, Bandar Abbas, on the funny bone of the elbow formed by the Strait of Hormuz at the base of the Gulf. As Middle East military analyst Michael Eisenstadt notes, the handy drubbing given Iranian surface vessels by the U.S. Navy during its late-1980s tanker escort operations probably encouraged Iran to acquire less targetable assets. Yet the Kilos' operating environment is not particularly auspicious. The Gulf is shallow, the straits are narrow, and access to and from Bandar Abbas could readily be blocked with mines. In the open ocean the Kilos, although potentially quiet when running on batteries, would be vulnerable to concerted U.S. antisubmarine warfare operations and unlikely to survive more than a few days under combat conditions.[70]

Because its own oil exports must use the Gulf, Iran has no peacetime interest in obstructing passage through it. However, submarines, mines, and cruise missiles might allow Iran, in a future crisis, to make passage for others' vessels more difficult if its own access were to be obstructed. A similar purpose may lie behind Tehran's otherwise hard to fathom actions regarding the small islands (Abu Musa and the Tunbs) that lie near the shipping lanes of the Strait of Hormuz. They had been jointly held by Iran and the United Arab Emirates since 1971, but non-Iranians were ordered off in early 1992, a move that played badly in Gulf Arab capitals.[71]

Arms Industries

As of 1993, Iran claimed to have more than 240 state-owned arms plants controlled by the Ministry of Defense and Armed Forces Logistics, by the Defense Industries Organization, and by the Revolutionary Guards and the Reconstruction Jihad Ministry, in addition to 12,000 privately owned armaments "workshops." Iran can manufacture some parts for its aircraft, armored vehicles, and artillery, and it can build basic wheeled armored vehicles, unguided rockets, and mortars. But its leaders make exaggerated claims about indigenous production capabilities (a subject tailor-made for *taqiyeh*).[72] Although Iran is trying to build an independent technical/maintenance/support infrastructure, the revolution cost it much of a whole generation of technically trained people. Moreover, argues Ahmed Hashim, the country lacks the "culture of research" needed to stimulate advances in science and technology, military or otherwise.[73] In short, while Iranian leaders may claim self-sufficiency in arms production, the country is far from that point. Anthony Cordesman has argued that Iran "cannot currently mass-produce a single sophisticated guided missile system or advanced conventional weapon unless it is given the major parts needed to assemble one."[74] Iran did test-fire a 1,000-km-range missile in July 1998, but that is thought to have been a *Rodong I* acquired from North Korea. Iran also has sought missile-related technologies from China and Russia.[75]

Weapons of Mass Destruction

Under the shah, Iran had an ambitious nuclear energy development program, and the shah averred that Iran would "have to acquire nuclear weapons if some upstart in the region gets them."[76] Today, Israel, India, and Pakistan have nuclear weapons on hand or able to be assembled quickly. Circumstantial evidence and reports of Iranian efforts to acquire dual-purpose technology (that is, technology applicable to both nuclear weapon and nuclear energy programs) strongly suggest that Iran has embarked upon a program to build nuclear weapons. Unclassified U.S. Central Intelligence Agency reports state that Iran's efforts have focused on "production technology . . . for all types of WMD" although Iranian nuclear-related purchases in 1996 "were only indirectly related to nuclear weapons production."[77] Iranian recruiters have reportedly made concerted efforts to get Russian scientists with knowledge and experience in the Soviet/Russian biological weapons program to work in Iran.[78] The U.S. Defense Department's 1997 proliferation report noted that Iranian efforts to acquire fissile materials, suspended during the Iran-Iraq War, had been restarted. Although "Iran does not yet have the necessary infrastructure to support a nuclear weapons program," it is shopping for the technologies that would give it a "complete nuclear fuel cycle" (uranium processing and enrichment facilities, reactor fuel fabrication plants, and spent reactor fuel reprocessing plants) that could be directed to civilian or to military purposes.[79]

Russian technicians are busy installing a 1,000-megawatt nuclear power reactor at Bushehr, a coastal site originally intended for reactors of German design. Fuel from these pressurized-water reactors cannot be removed without shutting the reactor down, which will facilitate international monitoring of spent fuel removed from the reactor core. Iran is a signatory of the 1970 Nuclear Non-Proliferation Treaty (NPT), and its declared nuclear facilities are subject to inspection by the International Atomic Energy Agency.[80]

Iran argues that it needs nuclear energy for electricity production. Critics point out that its large reserves of natural gas offer a much cheaper alternative. Iran argues in public that nuclear weapons are immoral and notes that it has signed both the NPT and the 1996 Comprehensive Nuclear Test Ban Treaty (CTBT). Critics note that Iraq and North Korea were NPT signatories, too, and yet both had clandestine weapon development programs. Thus Iran's protestations may simply be atomic *taqiyeh*. Critics note patterns of covert, dual-purpose technology acquisitions much like Pakistan's when it was seeking to build the bomb in the 1970s.[81] There are, critics argue, several plausible Iranian motives for building a bomb, including regional prestige, a relatively cheap deterrent to American power (although the United States would probably vaporize significant parts of Iran should such weapons be used against its forces or territory); and a deterrent to neighbors with nuclear bombs in their own basements. However, Israel would probably contemplate attacking Iran *only* to cripple a budding Iranian nuclear weapon program or to retaliate in kind for nuclear use. A nuclear-armed Iraq, on the other hand, with a track record of using WMD against Iran, might show less forbearance. Reducing Iran's incentive to acquire nuclear weapons may therefore depend on keeping Iraq from acquiring them.

The weakness of Iran's industrial infrastructure and technological research and development base presents problems for a nuclear weapons program. While it may aspire to have uranium enrichment facilities (used to extract the tiny amounts of fissionable uranium isotopes present in natural uranium ore) or plutonium reprocessing facilities (used to chemically separate weapons-usable plutonium from other elements in used reactor fuel rods), thus far it has neither. Nor does Iran possess reactors optimized for plutonium production. Since production of fissile materials is the hardest part of making a fission bomb, Iran's best bet would be direct purchase of such materials on the black markets that appear to have opened up since the demise of the Soviet Union. Partly to prevent such diversion, the United States arranged to purchase 600 kilograms of highly enriched uranium from the republic of Kazakhstan, pressed Ukraine to send all of the ex-Soviet nuclear weapons on its soil back to Russia for dismantling, and set up the Cooperative Threat Reduction (CTR) program to help Russia securely transport and store its fissile materials and the warheads removed from its missiles and bombers. But highly enriched uranium destined for any of a number of research reactors around the world could in principle be targeted for theft. Given the rise of organized crime, and not just in the former USSR, and the extensive communications and transport networks developed for the international drug trade, a would-be proliferator's best bet might be to contract with the underworld for materials needed.

Iran: U.S. Policy Options

Iran's best options while it slowly rebuilds conventional defense capacity tend toward the low and high ends of the spectrum of violence: irregular warfare, subversion, or terror, and weapons of mass destruction. Neither track is in America's interest. Washington has taken a direct approach intended to smother the Iranian revolution economically and to sanction unilaterally anyone who doesn't agree with that policy. But the rough consensus of students of Iranian politics and culture, other than those who specialize in international security issues, is that Washington's policies have been counterproductive. "[T]here are only two viable options on the Iranian political spectrum—the pragmatists . . . and the hardline disciples of Ayatollah Khomeini's rigid ideology of militant Islam," and the pragmatists can win only by improving the economy; by breaking out of diplomatic and economic isolation. "Without that breakthrough, Iran falls to the hard-liners. Under their control, the Islamic Republic either shifts back to the wrath and zeal of the early 1980s or sinks into anarchy produced by a government that can claim little authority."[82]

James Bill has argued that U.S. efforts "to squeeze, isolate, and weaken the Islamic Republic," could cause it "to pursue increasingly desperate and aggressive political activities," and that the alternative to a future explosion of violence in the region is a lower U.S. military profile and "some kind of rapprochement" between the United States and the Islamic Republic.[83] To the extent the mullahs fear the United States as the origin of everything threatening to the revolution, they reflexively shun open trade, tourism, and cultural exchange. This provides all the more reason to try to open up relations.

Dual containment—the U.S. policy of putting pressure on both Iraq and Iran—is also a drag on the rest of the Gulf states, who increasingly seem to reject it. As Mohsen

Milani noted in the mid-1990s, the countries of the Gulf face a crisis brought on by economic globalization unless they reverse course and move toward regional economic integration, else "every single one of them, large and small, powerful and weak, risk becoming marginalized" in the global economy.[84] Iran's own road to economic recovery requires a painful shift to a more open, free-market system, which may engender civil unrest as subsidies are removed—especially public subsidies for the *bonyads*. But deregulation and other reforms are sorely needed to attract foreign investment, and without such investment, the Iranian economy will continue to languish.

The irony of American policy is that because the European Union, Japan, and other major trading states honor it mostly in the breach, Washington's containment policy is extremely leaky. And yet, while the American embargo is maintained, Iran's hardliners can blame America for economic ills originating in their own mistaken economic models and the corrupting influence of too much power. So while U.S. sanctions may have little other direct impact, they do help to keep the wrong people in control.[85]

Southern Asia:
Collisions of Religion and Nationalism

The South Asia subcontinent has cradled human civilization for thousands of years. The land itself has been pushing its way into Asia for even longer—about 50 million years—and the Himalayas, which separate Pakistan and India from Afghanistan and China, are the physical result.

Riding this slow geological train wreck are a melange of peoples, cultures, languages, and religions that are still in the process of sorting themselves into workable political arrangements 50 years after British colonial rule was removed. All of the smaller states in the region feel the influence of much larger India, the world's second most populous country. One of them, Pakistan, has built its national identity on resistance to that influence, and the resulting half-century of struggle between the two states has produced a great deal of fighting, suffering, and arms buying. Like Turkey, or Iran, what the region needs but has yet to produce consistently is political stability, honest governance, and healthy, educated peoples able to feed themselves, plan their families, and compete successfully in both the local and the global economies. What it gets instead is political infighting and competition, falling government legitimacy, rising religious militancy, mutual chest pounding, and nuclear bombs.

Domestic Politics and Regional Insecurity

On August 14, 1947, after 144 years of British suzerainty, the Indian subcontinent was partitioned into Hindu-majority India and Muslim-majority Pakistan. While more than 80 percent Hindu, India has a substantial admixture of Muslims, Sikhs, and Christians. Home to a dozen major linguistic groups, its "national" language, Hindi, is the primary tongue of just 30 percent of its people, so English is the language of politics and commerce. India had roughly 350 million people at independence, had over 980 million in 1997, and is projected to have 1.3 billion in 2025, potentially surpassing China as the world's most heavily-populated country.[86]

Pakistan was originally created with western and eastern parts separated by 1,500 km of Indian territory. Both were Muslim-majority areas but of quite different ethnicity. The smaller but more densely populated eastern half decided to secede in 1971. After Pakistani troops sent to quell the separatist movement caused substantial blood-shed and an explosion of 10 million refugees into Indian territory, the Indian army intervened and India supported the creation of the new state of Bangladesh. The net loss to Pakistan was 15 percent of its territory and over half of its population.

The Origins of Enmity

The traumatic birth of Bangladesh was just one of the later consequences of a flawed process of partition that began in violence and to this day remains both bloody and incomplete. Intercommunal violence broke out immediately after independence, ul-timately causing a million deaths and the frantic emigration of 12 million people—Hindus to India and Muslims to Pakistan—with many fatalities occurring in transit. Britain's Indian Independence Act of 1947 also liberated nearly 600 mostly minor princely states within the subcontinent, adding further complexity and inviting fur-ther violence. Most of these principalities opted to join either India or Pakistan, but three did not. Indian forces invaded and annexed the two that were Muslim-ruled but had Hindu majorities (Hyderabad and Junagadh). The third state, Kashmir, in the northernmost reaches of the subcontinent, had a Muslim majority but a Hindu ruler. Shortly after independence, Kashmir's maharaja faced a mutiny by his mostly Muslim troops and an invasion from Pakistan of several thousand well-armed Push-tun tribesmen roundly considered to be proxies for Pakistan. He sought Indian mil-itary help. India would help only if Kashmir joined India first, and the maharaja agreed to do so. By his signature of the Instrument of Accession in late 1947, Kash-mir became part of India. Although the Instrument referred to a plebiscite to de-termine Kashmir's ultimate political disposition once peace was restored, no plebiscite has ever been held, and for the past three decades India has been publicly opposed to one.

A UN-brokered cease-fire in 1948 left Kashmir divided into Pakistani- and Indian-controlled sectors, with the "line of control" that separated them monitored by a small UN observer group. The 1965 war between the two countries started in Kashmir. The Simla Agreement that followed the 1971 war pledged the two countries to seek a peaceful resolution of their differences and not to attempt a unilateral solution to any of their mutual problems.[87] Through 1998 there was no more open warfare between them, although periodic shelling occurred frequently across what is now called the "line of actual control" (LAC) in Kashmir, as well as in the northern mountain fast-ness of the Siachen Glacier and other areas where mutual borders are murkily de-fined.[88] In 1989, however, Kashmir erupted in insurrection that continued to tie down several hundred thousand Indian troops and police a decade later, with an ongoing toll in lives lost and human rights violated. Pakistan covertly supported the insurrection-ists, some of whom want outright independence, not association with Pakistan. It was widely assumed to have given substantial material and logistical support to 500 to 700 fighters who moved into a 50 km stretch of Himalayan ridgeline on the Indian side of the LAC in April 1999, occupying at least 24 three-season positions from which In-dian troops normally withdraw during winter months. After three weeks of Indian ground assaults failed to dislodge the fighters, Indian commanders called in the first air strikes to be undertaken in Kashmir since the 1971 war. Artillery duels across the cease-fire line continued for more than a month; by mid-July, all of the fighters had withdrawn back to the Pakistani side of the line.[89]

Politics and Economics

While the troubles in Kashmir are the most visible problem in India–Pakistan relations, other sources of instability are worrisome as well. Both countries have severe population problems and, as a result, looming economic and resource problems. Resulting socioeconomic stresses help give rise to ethnic and religious-based self-determination movements. But sustained Indo-Pakistani enmity has helped to divert the energies and resources of both countries from the business of creating modern states, and corruption endemic to both societies has eaten away both at efforts to modernize and at people's faith in government.[90]

India worked for three decades to build an advanced yet self-reliant economy, but its centralized planning models, discarded only in 1991, worked little better than those discarded by the Soviet Union and China. India has maintained a secular, democratic form of government throughout, with orderly changes in power excepting a three-year period of emergency (1975–1977) declared by Prime Minister Indira Gandhi, at the end of which she and the long-ruling Congress Party were (temporarily) booted from office by the voters.

Although elections take place with regularity, India's political culture is nonetheless "intensely hierarchical," and rising religious chauvinism poses some threat to its commitment to secular governance.[91] Hindu chauvinism triggered communal killings and riots in 1992–1993. India also has battled Sikh militants who want independence for "Khalistan" (the present state of Punjab) and tribal insurgents in its far northeastern provinces who seek the same for their groups. Hindu militants' political vehicle, the Bharatiya Janata Party (BJP), gained ground throughout the 1990s, heading a coalition government for a little over a year. It lost a parliamentary vote of confidence in April 1999, but emerged from elections in October 1999 with a stronger governmental coalition, propelled in part by the victory over Islamic militants in the mountains of Kashmir a few months earlier.[92]

Pakistan has enjoyed less political stability and less democracy than India. The military has intervened on three occasions (1958, 1977, and 1999) to take the reins of government. The coup leaders held power for 11 years in the first two cases.[93] During the second period of military rule, 1977–1988, Gen. Zia ul-Haq implemented Islamic law, and religion and politics became more closely intermixed. Pakistan's latest military ruler, Gen. Pervez Musharraf, seized power in a bloodless coup shortly after Indian voters returned the BJP to power but targeted Pakistan's own corrupt politicians and in particular Prime Minister Nawaz Sharif, in explaining his action.[94]

Aside from the military's praetorian tendencies, Pakistan suffers from a near-feudal socioeconomic system. Although its economy is more open to imports and investment than is India's, its land holdings remain concentrated in the hands of fewer than 50 wealthy families who control 70 percent of the country's wealth.[95] Lawlessness and violence between the Sunni and Shia sects of Islam has been growing, especially in the southern province of Sind and its major city, Karachi. The country's government budget deficit consistently runs from 6 to 8 percent of GDP (roughly equivalent to an annual U.S. government budget deficit of $600 billion). Two-thirds of government expenditures go to debt service and the military. Fewer than 1 million out of 140 million Pakistanis pay income taxes, and efforts to impose sales taxes led to a general strike. After its nuclear weapon tests in May 1998, Pakistan lost most of the foreign aid and foreign investment that helped it keep the economic effects of that deficit spending under control.[96]

Literacy in the subcontinent is limited: only half of the male population of Pakistan and fewer than one in four females can read and write. The equivalent numbers for India are about two-thirds and one-third, reflecting both a failure of education in general and the region's deeply rooted gender prejudices. The former contributes to the region's high birth rates. The latter produces a skewed ratio of males to females in the population at large, reflecting cultural preferences for male children and a general "devaluation of females" that hurts development.[97]

Threat Perceptions and National Self-Images

Asymmetric self-images, threat perceptions, and security policies help to keep defense dialogues in South Asia from coming to closure. India sees itself as deserving a regional, even global security role befitting its size. Its air force and naval programs, in particular, have long been geared toward greater power projection capabilities than needed to deal with Pakistan, whose territory is relatively shallow, whose coastline is short, and whose significant infrastructure, military and otherwise, lies relatively near the border with India. The navy's aircraft carriers and submarines, in particular, betoken a blue-water strategy that has nothing to do with the status of Kashmir and only a bit more to do with protecting offshore territories like the Andaman Islands, even though some officials are concerned with increasing Chinese naval activity in Burma. India's major fears about Chinese encroachment are directed toward its land borders, growing in part out of its forces' defeat by the Chinese Peoples' Liberation Army in the short, sharp border war of 1962. The border issues have never been fully resolved. China's successful tests of fission and fusion bombs in 1964 and 1967, and subsequent deployment of missiles able to reach India further encouraged Indian programs of nuclear and missile development and reinforced Indian views of China as India's principal "peer competitor."[98]

Pakistan's dominant problem is its insecurity, not just about its borders but about its national identity. Created as a Muslim homeland in the Subcontinent, it is hardly that today. Pakistan's present population is roughly the same as that of largely Muslim Bangladesh and no larger than the Muslim population of India itself. The identity that remains is "not India": the Pakistani state and military exist to resist the policies and ambitions of India, whose founders did not really, in their heart of hearts, believe that Pakistan had much business as an independent state. Similarly Pakistan's founders did not believe that an avowedly secular state like India could long survive.[99] Hoping to realize that belief, Indian analysts argue, Pakistan has supported insurrectionists not only in Kashmir but also in Punjab and elsewhere.[100]

India's aspirations to be a regional great power and the forces it has assembled in support of those aspirations contribute to Pakistan's fundamental insecurity. With one-seventh the population and economy of India, Pakistan cannot match Indian conventional military power. Positioned essentially "broadside" to India, Pakistan lacks strategic depth: aircraft flying from Indian bases can reach any significant part of the country within minutes. Feeling nonetheless compelled to match it, Pakistan turned to nuclear weapons.

Defense Policies and Politics

Despite a "discernible warrior culture" in Indian history, the armed forces of independent India do not enjoy high social status, argues British defense analyst Chris Smith. The military's political role is minimal. The Ministry of Defense (MoD) is civilian-staffed and the Defense Research and Development Organization (DRDO) is civilian-run. There has been no military Chief of Defense Staff (CDS) to "manage rivalry between the services," and the civilians in the MoD historically have been "technically incapable of challenging the [service] chiefs of staffs' interpretation of the relationship between defense and security." Military frustration with this state of affairs helped convince India's defense minister to ask the service chiefs in late 1998 to create a proposal to restructure the MoD and create a CDS position. Interservice rivalries and differing strategic visions delayed their recommendation.[101]

Hewing to Nehru's dictum of neutrality and self-reliance, India built upon an industrial base established by the British to develop a substantial military industry. By the late 1980s, the Indian defense establishment employed over 100,000 people.[102] Emphasizing licensed production of Western and Soviet weapon designs from the early 1970s onward, India built combat aircraft (for example, the Anglo-French Jaguar strike fighter and two generations of Soviet MiGs), armored vehicles (including a version of the British Chieftain main battle tank), British- and Soviet-designed surface ships, and German-designed submarines.

DRDO has been stymied in design and production of many indigenous weapon systems, however. A 1995 goal of 70 percent self-sufficiency in defense by 2005 is unlikely to be met; by the end of the 1990s the reality was closer to 30 percent. Prototype versions of the Arjun main battle tank, under development for 25 years, had dropped to 40 percent indigenous content. The Arjun still had severe problems with its fire control system, its engine overheated in the desert conditions in which it was likely to operate, and at 58 tons it had poor mobility and could prove too big to move by rail. The Light Combat Aircraft (LCA) program, begun in 1983 to replace India's MiG-21 fleet, had yet to fly even a prototype aircraft 16 years later and was not expected to enter squadron service for another 14 years. The U.S.-built engines it was to have used were rendered unavailable by sanctions imposed after India's 1998 nuclear tests. U.S. sanctions also affected the engines that were to power the Advanced Light Helicopter, a program under development since 1976 that Indian air force officers say is underpowered and overweight. Also begun in 1976 but not revealed until late 1998 is India's nuclear-powered submarine program, which experienced design problems with the nuclear reactor. Powered by 20 percent enriched uranium, the first boat is not expected to be commissioned before 2009.[103]

The delays and costs associated with local production, let alone indigenous development, of advanced weapon systems have led the Indian military to favor imports. In general, the market has provided "better, cheaper and more acceptable weapon systems" than Indian industry could produce on its own. Frequent specification changes and "shifting demands" from the armed forces helped to induce some production delays (and raise unit costs). The delays in turn facilitate service lobbying for "interim" imported systems in the interests of national security. For example, the Indian army will buy 200 Russian T-90S main battle tanks to substitute for the long-delayed Arjun and to match Ukrainian T-80UD tanks recently bought by Pakistan. The T-90S is 12 tons lighter than the Arjun, more maneuverable, and can fire a laser-guided antitank missile to a distance of five kilometers. Thus the existence of a substantial Indian defense industry "actually *increases* the demand for imported weapons," which suggests in turn that the hundreds of billions of rupees shoveled into the industry over five decades have bought more status than security.[104]

Pakistan inherited virtually none of the defense industrial base built by the British and has had few mobilizable resources with which to build a new one from scratch.

Historically, it relied heavily on concessional arms imports, initially from the United States, and facilitated such supplies by joining the Western alliance network that India shunned. As a result, Pakistan enjoyed relatively early access to advanced American aircraft and air-delivered missiles, at least until the 1965 Indo-Pak war, when the United States imposed what would be a 16-year embargo on arms sales to Pakistan. Thereafter, Islamabad turned to China for sales and assistance, but the Chinese could not provide cutting-edge technology. A U.S. delivery window opened during the Reagan administration while the United States needed Pakistan as a conduit for support of Afghani guerrillas fighting Soviet forces in Afghanistan. The window closed again at the end of the 1980s when the Soviets withdrew and the White House would no longer certify to the Congress that Pakistan was not building nuclear weapons. Caught in that policy change were 28 U.S. F-16 fighter aircraft paid for by Pakistan but not delivered. In late 1998, Washington finally reimbursed Islamabad $324 million in cash and $140 million worth of commodities after the Congress agreed to allow the president to waive U.S. sanctions.[105]

The greater political role of the military in Pakistan's largely nondemocratic system is a key determinant of the country's defense spending. Historically, when in power the army was "omnipotent" and even when it relinquished power the country was ruled, in effect, by a triumvirate of president, chief of the army staff, and prime minister. However, constitutional changes pushed through by Prime Minister Nawaz Sharif in 1997 ended the president's power to remove the prime minister and to appoint the army chief, and allowed the prime minister to dismiss any member of parliament from his own party who votes against him. In October 1998, Sharif sacked the army's chief of staff for publicly criticizing his government, without immediate counteraction by the military. However, a "generation of officers with Islamist leanings is rising through the ranks in the army," as a result of the Islamisation policies instituted under General Zia's rule. Thus, the army is thought to be increasingly sympathetic to—or unwilling actively to oppose—the rise of Islamic militancy or its goals of fundamental political change.[106] A year after Sharif's action, the prime minister himself was, as noted earlier, sacked by the army.

More highly regarded by the Pakistani public than either political leaders or the government's civil bureaucracy, the military devotes a large proportion of its spending "towards maintaining the military as an institution rather than as a fighting unit." Unable to match India quantitatively, Pakistan has "concentrated, where possible, on acquiring state-of-the-art defense technology," but that, too, is a costly strategy.[107] In the arms market of the 1990s, concessional prices for cutting edge technology are hard to find, and financial aid providers are leaning on the government to reduce defense spending, which leaves Pakistan in something of a box as far as conventional capabilities are concerned and, as long as India and Pakistan remain at loggerheads, points to WMD as the great equalizer.

Nuclear and Missile Programs

India detonated a "peaceful nuclear device" underground in 1974 but would not admit that it had "weaponized" its nuclear know-how or acknowledge possession of other "devices." It was roundly assumed, however, to have built several weapons. A subdiscipline of regional security studies developed around what came to be called "recessed deterrence," involving the strategic utility of a rumored but undeclared nuclear arsenal. Israel has long been assumed to own a recessed deterrent. India may or may not have turned its technology into weapons in the 1970s. Pakistan is thought to have done so by 1989.

Recess ended in South Asia in May 1998, when first India and then Pakistan detonated a series of nuclear bombs underground in quick succession. Pakistani leaders spoke of restoring Pakistan's credibility and the strategic balance in the region, and early public jubilation at the blasts led the *Pakistan Observer* newspaper to argue that the blasts had "instantly transformed an extremely demoralized nation into a self-respecting, proud nation . . . having full faith in their destiny."[108]

Shekhar Gupta has observed that in addition to deterring major warfare, or perhaps *because* they deter major warfare, nuclear capabilities may make low intensity conflict more likely:

> A thought did emerge around 1988–89 that the nuclear capability had provided Pakistan an unassailable deterrent against India. It was from this belief that the new Pakistani adventurism in Indian-held Kashmir emerged. India's classical doctrine for the defense of Kashmir consists of a riposte in the plains if Pakistani pressure became too much in the [Kashmir]Valley. Many within the Pakistani military establishment now conclude that the emergence of the nuclear deterrent has made this Indian doctrine obsolete.[109]

The nuclear option may have been made even more attractive, ironically, by the U.S. policy of penalizing Pakistan for seeking such capability. The more the United States squeezed the conventional arms pipeline, the more attractive nuclear weapons may have appeared as a substitute for unreliable foreign suppliers. But this is not just a "dove's dilemma." It is also a realist's nightmare: selling Pakistan first-line F-16 fighters in the mid-1980s did not lead to curtailment of its nuclear program but did give it excellent, long-range platforms to deliver the bombs that it continued to build.

India, meanwhile, stressed that its nuclear forces and policies were "a sovereign function and not subject for negotiation," and that India would press ahead with ballistic missile development and testing.[110]

Both countries are pressing the acquisition or development of medium-range ballistic missiles. India had tested the 1,500-km-range Agni (named after the Vedic god of fire) once in 1994. Pakistan tested a 1,000-km-range missile that it called Ghauri (after a twelfth-century Muslim conqueror of India) in April 1998. This was probably a North Korean Rodong I in Pakistani colors. In April 1999, 11 months after their nuclear tests, both countries tested even longer range missiles. India tested its 2,500-km-range Agni II on April 11 and Pakistan tested the 2,000-km-range Ghauri II on April 14. Pakistan also tested the 750-km-range Shaheen. The Agni II is reportedly indigenously developed, with a solid fuel motor more suitable to military deployment and rapid use than the previous liquid-fueled design. The liquid-fueled Ghauri series relies on North Korean missile technology while the Shaheen, which is solid-fueled, uses Chinese technology.[111]

Although another Indo-Pakistani war is thought to be unlikely, should war break out it could easily involve nuclear arms unless both countries take steps to harden and/or conceal both the weapons themselves and their command and control systems, so as to make preemptive attacks in crisis unproductive and less attractive. Less predictable, perhaps, would be the impact of growing domestic disorder or political crisis on either state's nuclear arsenal. Seizing control of nuclear facilities and devices in both countries could become an urgent objective of various fighting factions—and an urgent concern of outside observers as well.

East Asia: China and Its Neighbors

For most of recorded history, a succession of Chinese empires dominated the mainland of eastern Asia, treating areas on their eastern and southern periphery as tributary states. In the closing moments of the twentieth century, China's neighbors wondered whether the historical pattern, interrupted first by Western intervention and then by six decades of anarchy, civil war, and revolutionary excess, might not be reasserting itself as China strives to transform itself into a twenty-first-century great power.

Western analysts worry in turn that unless China's politics liberalize and the country begins to follow the rule of law as its economic and military power grow, old habits of regional dominance will reassert themselves and old territorial claims will be actively pursued. Whether or not one believes that the Middle Kingdom will rise again, China has a major role to play in the resolution (or the worsening) of a number of long-running regional disputes, and its military policies and territorial claims have an impact on the military purchases and policies of its neighbors. The key disputes include the status of Taiwan, the stalemate in Korea, and disputed jurisdictional claims in the South China Sea.

Contemporary China

Overrun periodically by northern invaders, China's stability-oriented culture and institutions historically co-opted the invaders and resumed their previous dominant role in the region. The long era of Chinese dominance ended, however, when Western colonial powers with superior military technology could neither be driven off nor absorbed. For more than a century, from the early 1840s to the late 1940s, China suffered continuing indignities at the hands of Western powers and Westernizing Japan, which adopted the new forms of military technology and organization and used them to win successive territorial concessions in China.

In 1911, a republican revolution overthrew the Ch'ing Dynasty and China shattered within a few years into warlordism.[112] In 1915, Japan presented to what remained of the Chinese government its "Twenty-one Demands," which insisted that China, in effect, be turned over to Japan for safekeeping. After World War I ended, Japan occupied China's large northeastern province of Manchuria, parlaying staged assassinations and other events between 1928 and 1932 into excuses for further intervention. Finally, in 1937 Japan invaded China outright, launching the Pacific War (1937–1945).

Meanwhile, Chinese revolutionaries struggled to take the country back from the warlords. In 1922, at the height of the anarchy, the ten-year-old Kuomintang (KMT) or Nationalist Party formed a "united front" with the one-year-old and much smaller Chinese Communist Party. In 1924, the parties set about building a military wing. In July 1926, the joint National Revolutionary Army headed by Chiang Kai-shek launched the Northern Expedition. Marching from Canton (Kwangtung) and growing rapidly in size as it moved north, part of the army reached the Yangtze River by the fall and liberated Shanghai with the help of a communist-organized general strike the following March. Chiang shortly thereafter violently purged the communists from the KMT. By late 1928, he and the KMT, doing business as the Republic of China (ROC), controlled most of the country from their new capital Nanking, on the Yangtze about 300 km upriver from Shanghai. By 1930, the KMT regime was strong enough to go after its regrouped communist rivals once again and in late 1934 forced them to undertake the strategic retreat that came to be known as the Long March, to find haven in a desolate part of northern China. The KMT, whose military advisers early on had been sent by the Soviet-led Comintern, now got its advice from generals on detail from Nazi Germany. The 1937 Japanese invasion halted the campaign against the communists and a new nationalist-communist united front held for the duration of the war with Japan. Communist guerrillas pursued Japanese forces rather more energetically than did the KMT, which hoped that the People's Liberation Army (PLA) would wear itself out fighting the Japanese. In the event, the PLA's fight earned it popular respect that served it well when China's civil war resumed in earnest after Japan's August 1945 surrender to the United States.

KMT armies swept through communist-held parts of northern China by spring 1947 and onward into Manchuria—which proved to be a fatal overextension. Eighteen months later, the nationalist armies were reeling southward, back across the Yangtze in a reversed replay of the Northern Expedition of 22 years before. Remnants of nationalist forces fled, along with Chiang Kai-shek, his family, and many other leaders of the KMT, to the large island of Taiwan off the coast of Fukien, where they planted the flag of the ROC. As the PLA prepared to invade the island in May 1950, North Korea invaded South Korea and the United States not only intervened to stop that invasion but also interposed the Seventh Fleet between Taiwan and the Chinese mainland, interrupting the final act in the Chinese Civil War. When U.S. forces outflanked the North Korean army in October 1950 and headed north toward the border with China, Chinese forces poured into Korea, inflicted punishing casualties, and turned what might have been a six-month conflict into a bloody, three-year stalemate.

The United States signed a mutual defense treaty with South Korea in 1953, and a year later signed a similar treaty with the ROC. In 1960, Washington and Tokyo inked a Mutual Security Treaty. By the mid-1960s, Washington was hip deep in the Indochina War, a struggle that brought U.S. war planes relatively close to China's southern frontier on a daily basis. Although the United States ultimately lost that war, it and other U.S. political and military engagements around China's eastern periphery shaped the Asia-Pacific security environment for a half-century.

While the United States built its security relationships, the People's Republic of China (PRC) under the leadership of Mao Zedong lurched from one disastrous socioeconomic experiment to the next, from the Great Leap Forward in the latter 1950s to the Cultural Revolution in the latter 1960s, breaking ties with the Soviet Union in the interim and losing what access it had to outside sources of military technology. Still, by 1967, the PRC had developed and tested in very quick succession both principal types of nuclear weapon. In 1971, the PRC supplanted the ROC as the acknowledged representative of China in the United Nations. Occupying China's seat on the UN Security Council gave Beijing a veto over council activities, although it rarely exercised that power until the late 1990s. In 1979, the United States finally recognized the PRC as the sole government of China and agreed that Taiwan was a part of China. Nearly two decades later, in his mid-1998 visit to the PRC, President Bill Clinton affirmed Washington's "one China" policy.[113]

In those intervening decades, China had undergone a significant internal transformation. Following the death of Mao Zedong in 1976 and his replacement as the PRC's de facto paramount leader by sometime-protégé, sometime-rival Deng Xiaoping, China embarked upon a course of radical change that left the Communist Party firmly in charge of politics but opened up the economy to domestic and foreign entrepreneurship. The Chinese economy, still largely peasant-based in 1979, took off.

Even the PLA took to heart the new slogan, "To get rich is glorious." In 1985, a new arm of the PLA's General Logistics Department was created to oversee PLA commercial ventures. A decade later, the military controlled 10,000 to 15,000 profit-making businesses under roughly 50 umbrella groups in the areas of real estate, finance, pharmaceuticals, engineering, audio-video, and automotive products. These generated an estimated annual income of $5 to 10 billion, a good fraction of which may have gone unreported for tax purposes. PLA businesses were eligible for special tax breaks, had access to military transport, and enjoyed "effective immunity from civilian law enforcement." The government estimated that China was losing up to $12 billion in customs revenues annually in the mid-1990s, due to smuggling of all types, and the military's special economic status, combined with its "privileged access to national borders," may have contributed to that total and to the corruption of the institution.[114]

In July 1998, President Jiang Zemin ordered the PLA to withdraw from management of its nondefense businesses. That order was followed by a reorganization of PLA weapons development and procurement in September. A new General Armaments Department would henceforth be responsible for development (and for overseas arms sales), while the State Commission of Science, Technology and Industry for National Defense (SCOSTIND)—a body with no military representatives created in April 1998—would be responsible for managing China's large defense industry. Perhaps the most significant wrinkle in the new arrangement, however, was that the PLA henceforward would have to "pay market prices for weapons and equipment . . . instead of receiving them virtually free of charge." Analysts viewed this as a further move by Jiang to assert control over a military institution that had become too financially and operationally independent. In exchange for relinquishing control over external sources of income and paying market prices for weapons, the PLA was promised increasing regular budgets. (Similar divestiture rules were applied to the Communist Party itself in November, to end a corrupting situation in which "Communist Party and government officials routinely run companies in the industries they regulate, granting themselves the best government contracts and writing rules that benefit their firms."[115])

The PLA, its doctrine no longer based on Maoist "people's war," prepares instead for "regional active defensive warfare in support of Chinese economic interests and sovereignty claims."[116] It has been shrinking in size and attempting to implement the lessons of the 1991 Gulf War, where the smart technologies and systems of the "revolution in military affairs" (RMA) helped turn the liberation of Kuwait into a rout of Iraqi forces. The PLA has acquired some front-line weapon systems from Russia (for example, Su-27 fighter aircraft, Kilo-class diesel submarines, and Sovremennyy-class guided missile destroyers) but neither deploys such systems in quantity, nor possesses some of the other necessary piece-parts of the RMA needed to use them effectively, such as joint doctrine, command initiative, operational flexibility, intensive training, and high-grade maintenance practices.[117]

China's defense industry has been trying for some time to produce high-end weaponry, with technical assistance from Russia and Israel. But even with foreign assistance, China's ability to design and produce cutting-edge weaponry has been marginal to date. It is only now updating its diesel submarine fleet from a 1950s Russian design to one from the 1970s. Its current nuclear submarines seem comparable to the first generation of Soviet nuclear subs of the late 1950s, and both suffered radiation leakage problems. China has so far been unable to mass-produce a long-range fighter-attack aircraft on its own. Its current models, even those infused with Western sensors and engines, appear to have the performance of 30-year-old designs.[118] Its J-10 lightweight fighter, which draws technology from the cancelled Israeli Lavi program, began development as the delta-wing J-9 in the late 1960s before being revamped with Israeli help starting about 1984. First flight occurred in 1998, but operational deployment is not expected before 2005, for a total development cycle of nearly four decades.[119]

In mid-1999, China loomed large as ever over eastern Asia after a sometimes turbulent decade that began in June 1989 with the violent suppression of democracy protesters by the army in Beijing's Tiananmen Square. China's expansive (and regionally contested) claims to sovereignty over islet archipelagoes in the South China Sea led to periodic mini-crises over fishing, oil and gas concessions, and military occupation of islets also claimed by other states. In March 1996, Chinese military missile tests punctuated the continuing standoff with Taiwan, splashing down near the island's two major ports as the general populace streamed to the polls in Taiwan's first direct election of a president since KMT forces came ashore in 1949. In 1997, Great Britain's 99-year lease on Hong Kong expired, and China resumed control under the slogan "One China, two systems," hoping to demonstrate by example that Taiwan had nothing to fear by rejoining the mainland's political fold.

Northeast Asia: Old Divisions, Heavy Armaments

The northeast Asian security complex is tied together by the interests and activities of the major powers engaged in its affairs: China, Japan, and the United States. (Were Russia to resume an active political-military role in the Pacific, it would be included in this list as well.) We have already noted the role of China and, more briefly, the United States in this security complex and want to stress here that Japan has quietly but steadily built up its military capability in recent years. Although Article 9 of Japan's postwar constitution prohibits the acquisition of offensive military capacity, it has been interpreted over the years to allow the development of substantial forces for national defense. With appropriate logistical support, its naval forces, in particular, could range far afield. There are those in the region who worry that Tokyo would be tempted to create that support, especially should Chinese power grow substantially and the American security presence diminish. Tokyo's reaffirmation of the U.S.-Japan Security Treaty in April 1996 and its agreement to new defense guidelines that would deepen U.S.-Japan cooperation in the event of conflict in the "areas surrounding Japan" have tended to reinforce America's role in the region. China worries, however, that such Japanese combat support may extend to U.S. efforts to intervene militarily in some future conflict between China and Taiwan.[120] Such concerns form the backdrop to the assessment, in this section, of the security situations and armaments policies of the smaller members of the Northeast Asian security complex—Taiwan and North and South Korea.

The Taiwan Factor

China is doing its level best to intimidate Taiwan militarily and to isolate it diplomatically, even as cross-strait trade and investment continue to boom.[121] Martial law on Taiwan was lifted in 1986 and the island has since transformed itself into not just an economic powerhouse but also a budding multiparty democracy. Its government has long since abandoned the fiction that it is the true, legitimate government of the whole of China, as well as the policy of "recovering" actual control of the mainland by military force. Only 14 percent of Taiwan's population is of recent mainland Chinese ancestry and as the majority of native-born Taiwanese has gained political clout, rejoining the mainland politically has begun to look less attractive than does outright independence. Beijing's rulers made their objections to such sentiments clear in March 1996 as Taiwanese went to vote.[122]

China's military posturing may have had the desired political impact at the time. After giving independence advocates like the Democratic Progressive Party (DPP) encouragement in December 1997 municipal elections, Taiwan's voters did not cast many ballots in favor of the DPP in parliamentary elections one year later, deciding instead to stick with the KMT. High-level talks between the PRC and ROC resumed in October 1998 after a three-year hiatus. The meeting between Taiwan's envoy, Koo Chen-fu, and PRC president Jiang was described as "the highest-level contact between the two sides in nearly 50 years." During those meetings and in a press briefing afterward, Koo emphasized the need for democracy and "competitive party politics" if China was to have a stable society and healthy economy in the long term—and, by implication, achieve reunification.[123] But in the 2000 presidential election, as the independence-minded opposition candidate led in the polls, China issued a white paper warning that a declaration of independence by Taiwan, foreign occupation, or "indefinite postponement" of reunification would cause China to seek reunification by force.[124]

Taiwan is using its economic success and its accumulated reserves of foreign currency to make itself a tougher military nut for China to crack. According to University of British Columbia analyst Shannon Selin, 67 ships were ordered or planned to be built for Taiwan's navy from mid-1994 onward. Six French La Fayette-class frigates and seven modified U.S. Perry-class frigates built in Taiwan as the Cheng Kung class were in service by the end of 1998. Two of five Knox-class frigates leased from the United States arrived in 1998 to supplement ten locally built, missile-equipped corvettes and updated World War II–era destroyers.[125] Taiwan's heavy armored forces will have nearly doubled when deliveries of 340 ex-U.S. M-60A3 tanks ordered in 1995 are completed. Its air force took delivery of 60 French-built Mirage 2000–5 and 150 ex-U.S. F-16A/B fighter-bombers in 1997 and 1998. Air defenses are being augmented by six Taiwan-built, U.S.-designed Patriot missile batteries that have been upgraded to defend against short-range ballistic missiles, and by the U.S. sale of 400 to 500 infrared-homing Stinger antiaircraft missiles.[126] Taiwan would have liked to acquire a dozen more submarines by 2010 to augment the two now in active service, but had difficulty in finding a supplier willing to ignore Beijing's opposition.[127]

The main non-nuclear threats that Taiwan faces from the PRC are not invasion but blockade and missile brinkmanship. China wants to absorb rather than destroy Taiwan, and using a blockade or selective use of conventionally armed ballistic or cruise missiles to collapse its stock markets and disrupt its international trade would be a much smarter strategy than collapsing its buildings and killing its people. Once Taiwan builds up its own fleet, it might be able to keep Chinese ships and submarines away from its ports, but it would still face the problem of escorting ships into and out of port safely in a crisis. With China's handful of nuclear-powered submarines and newer diesel submarines theoretically able to attack convoy formation points and harass their escorts at some distance from Taiwan, running a Chinese blockade could be difficult.[128] Combined with minefields that Chinese ships and submarines would likely lay in such a scenario, the risk to shipping might prove much greater than most shippers and their insurers would be willing to assume or that securities markets would be willing to brush off.

Dealing with such a blockade would require outside help in the form of political or economic pressure on China by China's (and Taiwan's) major trading partners, and the willingness of third countries to escort cargo ships to and from Taiwanese waters, as was done for oil tankers in the Persian Gulf in 1987 and 1988. Together, the United States, Japan, and (by the end of this decade) South Korea could bring to bear much more naval power in the vicinity of Taiwan than could China, on a sustained basis. Whether they would be willing to take the political or military risk of doing so is of course another matter.

The military risk derives from China's investment in cruise and ballistic missile technology. Some Chinese bombers are equipped with C-601 antiship cruise missiles having a range of 150 km. Newer ships carry the Exocet-like C-802 (roughly 50 km range), while the Russian-design Sovremennyy-class destroyers delivered beginning in 2000 are equipped with supersonic SS-N-22 Sunburn cruise missiles. Development by China of a longer-range land-attack cruise missile, probably based on Russian designs, has high priority. While Taiwan's forces may be able to defend themselves against low numbers of cruise missiles, a barrage would be more difficult to defeat.[129] Since China's navy to date is said almost completely to lack coordinated air and missile defenses, Taiwan's best defense in such circumstances may be a good offense: cruise missile attacks of its own against Chinese forces. A cruise missile exchange, in other words, could be a costly affair for both parties.

The PLA has deployed a regiment of DF-15/M-9 short-range ballistic missiles (capable of delivering a 500 kilogram payload 600 km) opposite Taiwan, and in future it is expected to deploy shorter-range M-11 missiles (estimated range 300 km) as well. Although Taiwan is deploying a version of the U.S. Patriot air defense system, its ability to intercept missiles like the M-9 is at present essentially zero. Taiwan's forces could attempt to attack the M-9's mobile launchers, but a direct attack on PRC territory would be very dangerous and likely to provoke an unwanted response. Taiwan is thus weighing the costs and benefits of "theater missile defense" (TMD) systems, currently under development in the United States but years from being operational with U.S. forces. Although TMD had not been offered to Taiwan as an option by the United States, in the wake of growing Chinese missile capability Taiwan sought to acquire four U.S.-built Burke-class destroyers with the potential to serve as floating TMD platforms. Meanwhile, the United States and Japan have begun to talk about cooperation in TMD development, and U.S. leaders are committed to providing TMD coverage to forward-deployed U.S. forces in Asia.[130]

A spring 1999 U.S. Department of Defense (DoD) report to the U.S. Congress on requirements for TMD in East Asia laid out three TMD architectures for the defense of Japan, South Korea, and Taiwan respectively. The Japanese scenario explicitly focused on the North Korean missile threat and not on Chinese missile capabilities, and it involved three or four land-based "upper tier" defense sites or one to four TMD-capable ships positioned in the Sea of Japan. ("Upper tier" refers to missile defenses capable of engaging targets as high as 100 km altitude.) The South Korean TMD scenario, involving many short-range ballistic missiles rising from a rather narrow azimuth in North Korea, would require land-based defenses with perhaps four upper-tier and seven lower-tier TMD installations. Taiwan's defense against the M-9 would require the ability to detect and intercept ballistic missiles rising anywhere along a 1,000-km arc of China's coastline (roughly from Hangchow to Swatow). DoD's study results indicated that either a land-based upper-tier system or a single sea-based system could protect Taiwan from such longer-range threats.[131]

The DoD study emphasized that "the architecture options analyzed for each entity in no way suggest or imply a region-wide architecture network," nor do they include "US TBMD forces that may be deployed in the region. . . ."[132] However, the single sea-based system indicated as needed to protect Taiwan could just as well belong to the U.S. Navy as to Taiwan's navy should the United States choose to intervene once again in a future missile standoff with the PRC. Indeed, any TMD-capable forces afloat in the region would be flexibly deployable to the limits of political interest, tolerance, and courage.

China is well aware of this potential, and it may have contributed to the white paper that set the clock ticking on its policy of reunification by force if necessary. After 2010 or thereabouts, that policy may lose its remaining military clout as TMD systems are deployed in the U.S. fleet, unless Beijing chooses to escalate the threat against Taiwan to include nuclear weapons. Thus, from perhaps 2001, when new generations of Chinese missiles, ships, and aircraft may begin to be deployed, a military window of opportunity may open for China that could slam shut again at decade's end. Let us hope that Taiwan's stumping for democracy in China and Chinese leaders' ability to see its necessity open new doors that make the window irrelevant.

The Korean Stalemate

On the Korean peninsula, the South still faces the heavily armed, communist North. Many of the 36,000 U.S. troops still in South Korea are based between its capital, Seoul, and the four-km-deep, 243-km-wide Demilitarized Zone (DMZ) that bisects the peninsula, roughly along the thirty-eighth parallel of latitude.

After a long and painful period of military and quasi-military rule, South Korea experienced its first democratic handover of governance from one elected civilian president to another in February 1998, when long-time dissident Kim Dae Jung assumed office. Elected in December 1997, several months after the Asian economic crisis had begun to sweep through the region, the new president cooperated with the International Monetary Fund, reassured international investors, and steered the country toward recovery. Korea's gross domestic product (GDP), which shrank at an annualized rate of about 7 percent in the second and third quarters of 1998, reversed course in the fourth quarter and actually began to grow again in 1999. The financial system remained burdened by more than $100 billion in bad debts, however, and the government's annual budget deficit had grown to 5 percent of GDP.[133] One impact of the economic crisis was the delay, in early 1998, of a number of expensive weapon acquisition programs such as airborne early warning aircraft.[134]

South Korea has attempted to achieve as much self-sufficiency in defense as possible. Its defense industries build modified U.S. M1 Abrams tanks and indigenous-design infantry fighting vehicles, American F-16 jet fighters, indigenous-design naval destroyers and frigates, and German-design submarines. The submarine program, begun quietly in 1987, became public knowledge in 1991; the first boat was launched in late 1992, six were in service by late 1998, and at least ten were planned to be built.[135] Unwilling to remain wholly dependent on the United States for high-end military equipment, South Korea has been increasing European manufacturers' shares in its market, both for major weapons and for the subsystems that go into its ships. Most of its weapons are produced by the integrated conglomerates known as *chaebols,* for which military products are a small percentage of total output, and are subsidized by commercial profits.[136]

North Korea's armed forces operate equipment—tanks, aircraft, submarines—that was designed mostly in the 1950s. Being largely self-sufficient in defense, partly because it has very little outside support and partly because self-reliance is the motto of the Korean Workers' Party, North Korea does not have ready access to cutting-edge military technology and could not afford to acquire it in quantity should the opportunity present itself.[137] North Korean pilots also may receive as few as 30 hours of training a year, making the North's air forces even less of a match for those of the United States and South Korea. The majority of the North's army is, however, deployed close to the DMZ, within artillery and rocket range of Seoul.

In the event of conflict, South Korean forces would provide the bulk of the manpower, with a half-million-man army and substantial naval forces. They would be backed up by the U.S. Army brigades based in-country, the U.S. First Marine Expeditionary Force (MEF) based on Okinawa, and the Army's Twenty-fifth Infantry Division, based in Hawaii. Revised war plans reportedly call for South Korean and U.S. forces to block, then repel a North Korean invasion, pushing into the North to capture Pyongyang and destroy the Korean People's Army and the ruling regime. The full weight of the U.S. Marine Corps (that is, all three MEFs) could be used in landings on both coasts of North Korea to cut off the North's forces near the DMZ, replicating in stereo Gen. Douglas MacArthur's famed landing at Inchon in October 1950 that reversed the early course of the Korean War.[138]

Elements of this plan were briefed anonymously to reporters in late 1998, when North Korea appeared on the verge of reneging on the 1994 Agreed Framework, a package of incentives designed to halt its efforts to build a nuclear bomb.[139] (The Framework will be discussed in greater detail in chapter 5.) In August 1998, press reports revealed U.S. intelligence officials' concerns about what appeared to be underground tunnels suitable for housing a nuclear weapons facility, located 40 km north of the existing North Korean nuclear facilities at Yongbyon. At about the same time, the North launched a new, three-stage missile, the *Taepodong I,* whose flight path took it over northern Japan. The third stage, apparently intended to place a payload in Earth orbit, failed to do so, splashing down instead in the north Pacific. Any missile capable of accelerating a payload to orbital velocity (approximately 8 km/second) can accelerate it just a little bit *less* and have that payload fall back to Earth instead, at intercontinental range. The test came only weeks after a U.S. blue-ribbon panel warned that countries like North Korea could be developing long-range missiles much faster than projected by U.S. intelligence agencies.[140] Japan suspended payments under the Agreed Framework, and the U.S. Congress refused to appropriate funds in support of the accord.

North Korea demanded a substantial payoff ($300 million) from the United States for the privilege of examining the suspect tunnel complex close-up. After six months and several rounds of haggling, it settled for additional food shipments and assistance from U.S. NGOs in establishing new crops to ease the country's chronic food shortages. When U.S. inspectors finally visited the construction site in late May 1999, they found "a huge empty tunnel." The site was to be reinspected after one year.[141]

Systemic mismanagement of agriculture produced widespread malnutrition among North Korea's civilian population by the mid-1990s. Two years of floods (1995–1996) followed by a summer drought in 1997 pushed large segments of this weakened population toward starvation. Relief agency personnel surveying the scene in July 1997 reported a situation as grave as that found in Ethiopia in the mid-1980s, and perhaps more widespread. Reports of total mortality ranged from several hundred thousand to several million, all of them estimates, because no outside observers were ever allowed to see the entire country.[112] In September 1998, angered that food and medical supplies were being distributed on the basis of party loyalty rather than human need, the international NGO Medcins Sans Frontieres (Doctors Without Borders) pulled out of North Korea. The World Food Program also threatened to cut food deliveries if it was not allowed to monitor their disposition more thoroughly. Most observers agreed that those who survived the worst of the famine, especially children, would probably suffer severe long-term physical and mental developmental problems.[143]

In upgrading its military and building a modern defense industrial base, Seoul is both looking to bolster its security vis-à-vis the 1.1-million-strong North Korean army and looking toward the sort of military face that it wishes to present to the rest of the region. Sandwiched between China and Japan, Korea has historical reason to make itself a tough target. Whether or not the peninsula is eventually unified—and an estimated price tag of $1 trillion to rehabilitate and reintegrate the North has given Seoul pause[144]—South Korea is building forces that befit an industrial power, probably also with an eye toward defending its seaward interests in future circumstances of much-reduced American presence.

Southeast Asia: Watery Claims, Economic Meltdowns, and Political Change

Southeast Asia's security complex is not so riven by interstate disputes as Northeast Asia, and those that exist mostly involve claims to seabeds and atolls in the South China Sea rather than contests for dry land. China claims as its own virtually all of the South China Sea outside the territorial seas of the other riparian states, which in turn have overlapping claims to parts of it. The PRC's adamance appears fueled in part by Chinese nationalism (Taiwan makes the same claims as the PRC), in part by desires to exploit potential seabed oil and gas, and in part by domestic politics and the implications of Chinese backpedaling here for other issues of sovereignty such as Taiwan and Tibet.[145]

Vietnam and Cambodia

Southeast Asia used to be much more violent, of course. The Vietnamese communists' long struggle to end French colonialism and North Vietnam's subsequent fight to wrest control of the southern part of the country from its U.S.-backed regime kept the region boiling through 1975.[146] The murderous Khmer Rouge forces that seized power in Cambodia that same year instituted a four-year reign of terror that killed one Cambodian in seven. It was ended only when Vietnamese forces invaded Cambodia in 1978, driving the Khmer Rouge into the jungle. China fought a short, sharp border war with Vietnam in 1979 to teach its government a lesson in frontier manners and remind it not to push the Khmer Rouge—China's clients—too hard. Cambodia then plunged into another decade of civil war between Vietnam's hand-picked regime, led by Khmer Rouge defector Hun Sen, on one side and an awkward, red-gold coalition of Khmer Rouge and Cambodian royalists on the other. An international peace process chaired by France and Indonesia and backed by the UN Security Council led to a peace accord, a large UN peacekeeping force, and 1993 elections. The civil war resumed on a reduced scale after the UN left.[147] Six years, one coup, and another election later, Cambodia, with the ruthless and durable Hun Sen still at the helm, was finally welcomed into the Association of Southeast Asian Nations (ASEAN), the region's high-level political and economic discussion forum (about which more in chapter 5), and treated as a normal state.

Vietnam was admitted to ASEAN in 1995, and while benefiting to a degree from the region's economic uplift, its per capita income remained under $1,000 (U.S.) in 1997, the second-lowest in this security complex (only Cambodia is poorer). Its nonconvertible currency helped to insulate Vietnam from the shocks felt elsewhere in the region in 1997–1998, but the government's failure to open up the economy has left the country well behind other Southeast Asian states. State-owned enterprises still account for nearly half of the country's GDP and two-thirds of its industrial output. Corruption is endemic. Vietnam cannot afford many new weapon systems but did order a squadron of Russian Su–27 fighters in the mid-1990s and a small flotilla of Russian missile boats in 1998.[148]

Burma (Myanmar)

Burma (Myanmar) was invited into ASEAN in 1997. Ruled by a succession of military governments since its independence from Britain in 1947, Burma has long contended with a dozen or more ethnically based, drug-running insurgencies. The current military regime, which calls itself the State Peace and Development Council (SPDC), took power in 1988, killing as many pro-democracy student protesters in the process as died in the Tiananmen Square massacre in China one year later. In September 1990, the regime voided the results of Burma's first democratic election in 30 years and continues to hold at bay the country's democratic forces and their quietly charismatic leader, Daw Aung San Suu Kyi. The closest analogue in Southeast Asia to North Korea's hermit kingdom, the country has been further isolated since 1997 by U.S. and European Union rules against new investments there by their nationals.[149]

In the first visit to Thailand by a leader of the SPDC, Gen. Than Shwe met with Thailand's Prime Minister Chuan Leekpai in March 1999, not in Thailand's capital but in the far northern city of Chiang Rai, in the so-called Golden Triangle, to discuss measures for combating the drug trade.[150] The Triangle, overlapping Thailand's northernmost provinces, northern Laos, and northeastern Burma, has for decades been the source of much of the world's illegal opium and heroin, and 90 percent of the Triangle's opium originates in Burma. After the current Burmese junta signed cease-fire agreements with many of the country's major drug trafficking insurgent groups in 1989, Burmese opium production doubled. As a by-product of the 1989 cease-fires, the government also gained full access to Burma's borders and border checkpoints for the first time, giving field commanders dual leverage over drug exporters. While the U.S. State Department's annual narcotics control report says carefully that there is no government involvement in the drug trade "on an institutional level," it also notes "persistent and reliable reports that officials, particularly corrupt army personnel posted in outlying areas, are either involved in the drug business or are paid to allow the drug business to be conducted by others." It also notes "lack of enforcement against money laundering."[151] The latter point means that Burma's banking system can be used to launder drug money with impunity.

Private analysts are more blunt, contending that the SPDC lives off the drug trade, with profits from extensive drug money laundering used to finance over $1 billion worth of arms from China and, in particular, from PLA-run Polytechnologies Corp. Indian researchers reported work by Chinese technicians at several secondary Burmese ports in the early- to mid-1990s, along with construction of a radar installation on Burma's Great Coco Island, just north of India's Andaman Islands, with the implication that access would be given to the PLA navy. Finally, reports suggest that China is upgrading the old "Burma Road" from Kunming to the Burmese town of Bhamo on the Irrawaddy River and improving the river's navigability to create an integrated transportation channel from southeastern China to the Indian Ocean.[152]

Thailand

Historically prone to military rule and coups d'etat, Thailand has enjoyed a measure of political stability and democracy under civilian rule since 1993. However, it also had the misfortune of being the first country laid low by the Asian economic crisis that began in July 1997.

In the decade leading up to that crisis, Thailand and the other nonsocialist countries in this security complex—Malaysia, Singapore, the Philippines, and Indonesia—experienced high rates of economic growth and dramatic gains in personal prosperity, as much as doubling per capita GDP.[153] Their governments were able, as a result, to devote substantial resources to military modernization, which they focused mostly on air and naval forces, the better to protect regional sea lanes, fisheries, and territorial claims. Highly capable U.S. F-16 fighters in several instances replaced larger numbers of lightweight, short-range F-5 fighters built by the U.S. firm Northrop for the export market in the 1970s. Missile frigates and corvettes augmented or replaced smaller fast attack boats of more limited range. Thailand even took delivery in 1997 of a relatively small (12,000-ton) aircraft carrier built in Spain and nine Harrier jump-jets to go with it. Some analysts declared an arms race.[154] If there was a regional competition, however, it was as much a competition to look sharp as to counter threats, with the possible exception of China's maritime claims and the activities of pirate gangs who prey on commercial shipping in the region.

When the hard times hit, many arms acquisition programs in the region were cut back or canceled. Thailand, for example, terminated a $392 million agreement with the U.S. government to buy eight F/A-18 strike fighters.[155] Its economy shrank by double digits, and even a year after the onset of the crisis, a liquidity crunch left Thai businesses unable to secure loans needed for new production. Without revenue, firms were increasingly unable to pay existing debt, and the banking industry's fraction of bad loans continued to climb. Private investment in the country in 1998 dropped by nearly half in comparison with 1997.[156]

Although economists continue to argue about the causes of the crisis, Steven Radelet and Jeffrey Sachs of Harvard's Institute for International Development make a plausible case for a convergence of conditions that triggered an investor panic. Thailand is used here to illustrate their argument. First, government policy kept the exchange value of the national currency (the baht) tied to the value of the U.S. dollar. The predictability of its value "helped encourage short-term capital inflows, since investors perceived little likelihood of a loss from exchange rate movements." Second, in the late 1980s and early 1990s, banking regulations were loosened to allow private banks to open and to raise funds offshore, including short-term funds. Largely unregulated international capital flowed into Thai (and Malaysian and South Korean) banks to buy a piece of these well-performing economies, and the banks in turn made loans to industry, the sum total of which "increased by more than 50 percent relative to GDP in just seven years."[157] That is, bank loans were rising much faster than the overall output of the economy, and the banks were dependent on foreign capital inflows to fund those loans. As other prices and exchange rates fluctuated, however, and especially when the Japanese yen lost 35 percent of its value versus the U.S. dollar in early 1997, the dollar-linked Thai baht gained in value right along with the dollar, making Thai exports less competitive and leading currency traders to sell baht.[158] The government depleted its reserves of foreign exchange buying baht in an effort to maintain the pegged exchange rate vis-à-vis the dollar but could not compete with the global currency markets and ultimately gave up and devalued the currency. The markets, argue Radelet and Sachs, saw this as a "broken promise" on the government's part and sold baht even faster, devaluing it further. Foreign owners of short-term capital investments in Thailand, meanwhile, began to pull their money out of the country as those investments reached maturity, in what amounted, argue Radelet and Sachs, to an investor panic. To buttress their case, they note:

- that Thailand (and other hard-hit states) enjoyed favorable economic "fundamentals," indicating no economic downturn on the immediate horizon;
- that hard-hit countries had "high levels of short-term foreign debt relative to short-term foreign assets" (such as hard currency reserves) and were thus particularly vulnerable to sudden shifts in international investment patterns;
- that the crisis hit countries with markedly different economic structures more or less simultaneously; and
- that it "eased up after about one year, even though several fundamental conditions . . . were not significantly improved."

The Asian crisis, they concluded, was first and foremost "a cautionary tale about rapid financial liberalization in emerging markets," and the dangers of opening them to cross-border lending that is "much less regulated, and therefore more unstable, than domestic finance."[159] Such liberalization is especially dangerous when countries simultaneously try to maintain a fixed exchange rate, since a run on the currency will then *force* the treasury to spend out its hard currency reserves in a vain effort to compete with a much larger currency market whose instincts and interests send it pounding for the exits at the first rumor of smoke. Moreover, the International Monetary Fund (IMF) misperceived the crisis, seeing not an investor panic but bad governance (and there were indeed problems with "crony capitalism"). Initially, therefore, the IMF made things worse, pressing for major structural reforms that had little to do either with the immediate causes of the crisis or their solution.[160]

Malaysia, Singapore, and the Philippines

Until the economic crisis, Malaysia and Singapore were both examples of statist prosperity with growing economies open to foreign trade and investment. In both cases, the government has invested heavily in education and physical infrastructure and rules with a fairly heavy hand despite the nominal trappings of parliamentary democracy.

Malaysia's territory comprises the bottom third of the Malay peninsula that hangs down from Burma and Thailand, plus the northwestern third of Borneo, the two parts being separated by a 500 km stretch of the South China Sea at their closest point. Dr. Mahatir Mohamad has been elected prime minister successively since 1981. Mahatir is one of the region's most vocal proponents of "Asian values" and its most bitter critic of economic globalization, even though Malaysia's economic success in the 1990s, from oil and gas exports to electronic components, required and benefited from expanding global markets.

Malaysia's military grew out of a counterinsurgency tradition, fighting against communist guerrillas, first alongside the British from the late 1940s, and then from independence in 1957 to 1960. From 1963 to 1965, it faced the further challenge of "Confrontation," the effort of Indonesia under Sukarno to drive Malaysia out of Borneo with a campaign of infiltration and sabotage. Confrontation ended when Sukarno fell from power. With North Vietnam's triumph in 1975, Malaysia began to worry about conventional warfare and a new communist challenge, worries reinforced when Vietnam invaded Cambodia. A major armaments program in the early 1980s in response to that invasion was soon tempered by recession, however, and then by the realization that the Vietnamese were, in fact, stuck in Cambodia and would be going nowhere else soon.[161] In the 1990s, Malaysian acquisitions focused on the navy and air force, giving both services better capability to patrol lines of communication between the two halves of the country. Malaysia invested in British- and German-built missile frigates, medium transport aircraft, first-line interceptors (a squadron of Russian MiG-29s, delivered in 1995) and multi-role fighters (a half-squadron of U.S. F/A-18s, delivered in 1997). But even before the economic contraction, Malaysia had cut defense to about 10 percent of government spending, from 17 percent, in light of growing budget deficits.[162] By the time the economy slowed, most of the major acquisitions, frigates excepted, were already in hand.

Singapore, a small island nation at the tip of the Malay peninsula with a population that is three-quarters Chinese, was federated with Malaysia for two years but separated in 1965. The country had just one prime minister, Lee Kwan Yew, from independence in 1959 until 1990. Lee is assumed still to wield considerable influence behind the scenes. His People's Action Party retains a tight hold on power and on Singapore's three million people, but crime is low and the sidewalks are clean and gum-free (in Singapore, chewing gum is illegal). Singapore's defense spending absorbs about one-quarter of the national budget and 6 percent of GDP. Since 1995, the country has bought three squadrons of F-16C/Ds and four ex-U.S. jet tankers to extend their range, plus 40 surveillance drones from Israel and 18 CH-47D heavy troop transport helicopters, with most due for delivery by 2000. Moreover, in mid-1999, the United States approved Singapore's request for eight Longbow Apache helicopters. The

Apache is the premier U.S. all-weather antitank helicopter and the Longbow is the latest version, able to attack multiple targets with radar-guided missiles.[163] Singapore clearly will be prepared to reach out and touch someone at a distance, if necessary, in the new century but is typically close-mouthed about its forces and strategies.

The Philippines was the first state in this security complex to make the transition from the trappings to the reality of democracy. In 1986, "people power" (and a little air support from U.S. forces based in the country) prevented longtime dictator Ferdinand Marcos from stealing a contested election from Corazon Aquino, widow of Sen. Benigno Aquino, a democratic opposition leader assassinated in 1983 upon his return from the United States. Although it has yet to end long-running insurrections in parts of the country, the country enjoyed several peaceful, democratic transfers of power during the 1990s.

One of the major changes undertaken by the Philippines' new management was the termination of long-running military basing agreements with the United States. Subic Bay naval base and Clark air base had been the anchors of American military presence in Southeast Asia. Clark was rendered unusable in any case by the eruption of nearby Mount Pinatubo in 1991.

Having lived under the protective wings of the U.S. Navy and Air Force for four decades after World War II, the Philippines' own military capabilities were quite modest when U.S. forces departed. In early 1995, the government decided that it needed to augment those capabilities, which consisted of just one aging frigate and several dozen naval patrol craft and an air force limited to one squadron of F-5A light fighters. Parliament approved then-president Ramos's initiative to spend some $15 billion on defense acquisitions over the next 15 years.[164] Hurt by the economic crunch, the defense ministry sought to bring the procurement program back to life in April 1999 with a $157 million infusion of cash for systems to protect the country's 200-mile economic zone and to combat smuggling, piracy, and drug trafficking.[165]

Indonesia

Although near neighbors, Singapore and Indonesia are opposites in many ways. One is small, self-contained, ethnically homogeneous, rich, relatively uncorrupt, and politically single-minded. The other is huge, sprawling over 13,700 islands in a 3,000-mile arc, with a population of 200 million that is ethnically heterogeneous, much poorer since the economic crisis, and coping with extensive business and governmental corruption, separatist sentiment, and Muslim-Christian violence. The first year of the Asian crisis saw the value of the Indonesian rupiah drop by as much as 85 percent against the U.S. dollar (businesses holding dollar-denominated debt, in other words, would have had to raise six times more revenue in local currency to pay off that debt). Indonesia's GDP shrank an estimated 14 percent in 1998.[166]

Major elements of the economy were owned and operated by President Suharto and his extended family. In office from 1967 until his forced resignation in May 1998, Suharto had the close support of the armed forces, whose dual military and political role ("dual function" or *dwifungsi*) was written into law and whose economic activities, like those of the Chinese PLA, both kept military leaders satisfied and corrupted the institution. Until ordered to choose between their military and civilian roles, many high-ranking military officers also held governing posts. In 1999, nearly half of Indonesia's provincial governors and 40 percent of its district heads were serving military officers.[167]

Until Suharto's resignation, Indonesia was as politically single-minded as Singapore and no more tolerant of criticism. But leading up to Indonesia's first contested parliamentary elections since 1955, in late 1999, a freed press provided criticism aplenty as 48 parties vied for votes and voters gave a majority to opposition parties. The subsequent election of President Abdurrahman Wahid appeared to represent a sea change in Indonesian politics as Wahid quietly set about replacing military leaders and reducing the institution's political influence.[168]

Indonesia remains, however, a patchwork not only of islands but also of religions and ethnicities. Its Chinese minority (3 to 4 percent of the population) is the core of the business community and controls much of the wealth not managed by the former ruling family or the military, but suffered from what appeared to be orchestrated attacks in May 1998. Anti-Chinese riots in 1965 took several hundred thousand lives, and Indonesia's continuing capital flight in 1999 was partly a function of Chinese Indonesians taking their liquid assets and themselves out of the country in anticipation of further turmoil.

The Indonesian army, 235,000 strong, seems at first glance relatively small for so large a country. Thailand's is about the same size, for a country with one-third the population. Indonesia's navy, at 31,000, is a third smaller than Thailand's while the expanse of waters and coastline for which it is responsible is perhaps ten times as large. Moreover, many of the waters that it needs to patrol are effectively open ocean, calling for oceangoing vessels in which Indonesia is underinvested. But where the navy is under-strength, the army is probably over-strength, given Indonesia's paucity of external foes and the army's primarily internal security functions, most of which could be served by an expanded national police force or regional police forces more accountable to civilian authorities. Suppression of secessionist movements in Irian Jaya (western New Guinea), Aceh (at the western tip of Sumatra), and East Timor has been the army's main operational activity.

East Timor was a Portuguese colony summarily granted independence in 1975 but subsequently overrun by the Indonesian army, which instituted a reign of terror that killed an estimated 200,000 Timorese and substantially blotted Indonesia's human rights record.[169] Indonesia's assertion of sovereignty was not recognized internationally. While Suharto had been adamant about holding onto East Timor, his appointed successor, B. J. Habibie, announced suddenly in early 1999 that East Timorese would be offered autonomy within Indonesia and, if they rejected it, Indonesia's legislature would vote on granting them independence. Between this announcement and a UN-supervised referendum (or "consultation") on autonomy in late August 1999, however, substantial anti-independence paramilitary groups, supported by local military commanders, waged a violent campaign of intimidation in the territory. Suspended briefly for the referendum, violence resumed after vote tallies showed that East Timorese had overwhelmingly rejected the autonomy option. Militias supported and assisted by some of the 26,000 Indonesian army troops deployed in the territory systematically destroyed the infrastructure of East Timor, displaced half of its inhabitants, and forced half of those into refugee camps in West Timor. After a week of unabated destruction, the United States threatened to suspend all aid and international loans to Indonesia unless it agreed to remove its troops from East Timor and to accept in their stead a UN-authorized peacekeeping force. President Habibie and his military chief of staff agreed, and the initial 8,000-strong, Australian-led force known as INTERFET (International Force for East Timor) began to deploy in the third week of September. Indonesian forces withdrew and militia violence abated, but East Timor was left an economic wreck for UN civil administrators to attempt to rebuild as the territory became a de facto, if temporary, UN protectorate.[170]

Meanwhile, the economic crisis forced Indonesia to curtail or postpone a number of military programs. The dollar-value purchasing power of the defense budget dropped by half in 1998, even though its total in rupiahs was 40 percent larger than in 1997. The military postponed indefinitely the purchases of a dozen Russian-built Su–30 multi-role fighters, eight Russian troop-carrying helicopters for its special forces units, and an order for five German submarines. The Russian orders had been placed in lieu of an order for F-16s that the United states rejected due to Indonesia's human rights record, primarily on East Timor.[171]

Observations and Conclusions

Looking closely at the sources of demand for arms in a broad arc along the entire southern rim of Eurasia has emphasized a number of key points for any program of regional stabilization that includes an arms control component. First, the sheer asymmetries in size and capability among the states of these regions impel a basic sense of insecurity among the smaller states. Those big or rich enough to be less than fatalistic about their position take steps to secure their interests. Some rely on alliance structures, others on the presence of U.S. forces, still others on weapons of mass destruction. The United States, Japan, and South Korea are trying to buy off North Korea and end its nuclear weapons development program, but they have been less successful in keeping Pyongyang from distributing medium-range missiles to regions of tension and unstable states.

U.S. sanctions on India and Pakistan following their nuclear tests cut into both states' ability to acquire new weapon systems but may help to reinforce the conventional force imbalance that Pakistan says drove its nuclear program. The asymmetries of power, politics, and perception in the Subcontinent make a meeting of the minds on nuclear issues, let alone conventional forces, very difficult to arrange. Both countries invoke nuclear deterrence, but against different adversaries, with Pakistan turned toward India and India turned toward China. The assumption that nuclear weapons deter conventional warfare owing to the risk of nuclear escalation may have made the problem of low-intensity conflict worse; it is a major gamble to assume that such fighting cannot escalate to conventional warfare and not risk recourse to nuclear weapons, else Pakistan, in particular, need not have bothered making the painful investment it did in that capability. In the end, the tyranny of sunk costs may yet provoke leaders, under stress, to use the power that pain has purchased.

In the western part of this arc of countries, passionate religious interests are putting Turkey's secular experiment under stress, sustaining Iran's aging revolution, and giving Pakistan a national identity to stretch over serious ethnic cleavages. India has successfully maintained both secular and civilian rule to date, but secularism is threatened by the rise of Hindu nationalism and its avatar, the BJP. The rising influence of Islamic militancy in Pakistan, however, poses growing dangers not just for the Kashmir dispute but also for governance in Pakistan. A common thread in all cases is popular disenchantment with the corruption evident in everyday politics and governance, which leads to the search for alternative rule. Sometimes the alternative is religious; sometimes it is military; sometimes it is both.

In the eastern part of the arc, fundamentalism or fanaticism has historically worn the secular face of militarism or communism. Communism is a fading force in China and perhaps in Vietnam; in North Korea, communist ideology may not have lost its force, but the state itself is fading as its people slowly starve. Militarism seems to have given way in Taiwan, South Korea, and Thailand to civilian democratic rule. In Indonesia, the military shared political power for three decades and faces in the new decade a much-diminished political and internal security role. In Burma, on the other hand, the military is the state, which it runs as a criminal enterprise with arms and other material support from China. In China itself, the military is being weaned from its profitable encounters with commerce and markets and redirected toward China's primary military concerns, namely, territorial integrity. The stability of both the Northeast and Southeast Asian security complexes hinges on how China decides to deal with old territorial issues in the decades to come, which will hinge, in turn, on the rate of growth in China's military power versus its rate of political development.

Major arms imports remedy some of the security problems faced by the countries analyzed in this chapter and deepen others. It should be clear from this analysis, however, that conventional arms imports are neither the fount of instability within countries or regional security complexes nor its solution, and that demand for arms stems from a wide variety of sources, some of which are amenable to external influence and others of which are not. This implies that any policy aimed at bolstering peace and stability within a regional complex must be "multivariate" in approach, treating armaments as one element of a larger set of problems and using arms transfer restraint, where appropriate, as one part of a wider stabilization program.

CHAPTER 4

Assessing Supply-Side Arms Control

Campaigns to rein in major weapon systems date back a very long time, at least to Pope Innocent II's effort to ban the crossbow in 1139. Most such efforts have been only partly or temporarily successful. (Innocent's own campaign failed to take hold, and the crossbow was widely used until rendered obsolete by a deadlier weapon, the English longbow.[1]) Governments are instinctively averse to abandoning militarily useful weapons and to waiving their rights to acquire them. Unless done unilaterally, with no expectation of reciprocity from others, arms restraint requires a great deal of international cooperation and inevitably some degree of trust and risk, three things to which countries that function in a formal state of anarchy are at least semi-averse. Still, the twentieth century—perhaps because it was one of the bloodiest on record—witnessed a greater number of arms control and disarmament plans, programs, and agreements than any century previous.

The capsule history of these efforts presented in the next section categorizes them by major objective—controlling dangerous weapons, controlling dangerous governments, or altering dangerous configurations of forces—and sets the stage for analysis of five decades of intermittent efforts to implement supply-side controls on conventional arms transfers and recipient governments. Since proposals to alter configurations of military capabilities via arms transfer restraints have been directed historically at specific recipient states or security complexes, the chapter considers these two objectives together under the heading of "recipient-based" restraint.

The analysis leads to the conclusion that whatever the motivations behind supply-side restraint (self-interested or other-directed) and whatever the emphasis (weapons or recipients), pure supply-side mechanisms to control the proliferation of major conventional weapons are of very limited utility—not useless, but limited. The sources of limitation include the natural competitiveness of national arms suppliers and their incentives to defect from restraint regimes in the search for market share. But most of the difficulty stems from a basic difference between supply-side arms transfer restraint and other forms of arms control: supply-side accords attempt to extend the writ of restraint outside the circle of states that has crafted and agreed to it, applying rules to governments that have not been consulted about them and that would likely object if they were consulted. They would object in part because supply-side restraint measures lack reciprocity, the mutual adjustment that makes international agreements fair bargains. Perceptions of unfairness gnaw at any supply-side restraint regime not devised in cooperation with the countries to which it is supposed to apply. This does not mean that such restraint regimes are never warranted. It does mean that any state, group of states, or international organization that imposes them must be prepared to enforce them vigorously for the duration of the regime.

A Capsule History of Arms Control

Arms control regimes and agreements focused on dangerous weapons include one group of treaties limiting WMD per se, and another series limiting antipersonnel weapons deemed especially cruel and inhumane in their effects. The impetus to constrain WMD derives in part from the nature of the weapons, biological and chemical weapons in particular, which are designed to cause mass casualties—targeted, as it were, against life itself. The Geneva Protocol of 1925, crafted eight years after the carnage of World War I, banned the use of lethal chemical arms, while the 1995 Chemical Weapons Convention (CWC) banned all possession of such weapons, with strict international inspection protocols to enforce the ban. Biological and toxin weapons were banned by treaty in 1972 but without measures to verify or enforce compliance.[2]

Nuclear weapons have been subject to testing and proliferation constraints. Most explosive testing was confined underground by the 1963 Limited Test Ban Treaty (China and France did not abandon atmospheric testing until years later). Proliferation controls were established in 1970 with the entry into force of the Nuclear Non-Proliferation Treaty (NPT), augmented later by a number of regional "nuclear-weapon-free zones."

The NPT is often criticized as "unfair" by developing, non–nuclear-weapon states for its division of the world into nuclear haves and have-nots. The treaty offers partial reciprocity: in return for giving up the option to acquire nuclear weapons, non-nuclear-weapon states were to be given access to other nuclear technologies, while Article VI of the agreement committed nuclear weapon states to eventual "nuclear disarmament." Twenty-five years and several thousand more nuclear weapons later, when the treaty came up for review, the have-nots decided that the original bargain was not good enough and extracted a pledge from the haves to negotiate a comprehensive ban on nuclear explosive tests as a step toward fulfilling Article VI.[3] A Comprehensive Nuclear Test Ban Treaty was endorsed by the UN General Assembly and opened for signature in September 1996.

Nuclear weapons have not as yet been subject to the sort of use prohibitions that apply to both chemical and biological weapons, although a de facto norm of non-use seems to have evolved since the last nuclear weapon was detonated in anger, in 1945. Moreover, it was the rapid accumulation of nuclear arms during the Cold War that gave arms control in general a new lease on life.

Limits on inhumane weapons have banned expanding bullets and poisoned weapons (Hague Convention of 1907); the use of weapons that produce fragments undetectable by X-rays, such as plastic land mines, the use of incendiaries against civilians, and the use of blinding laser weapons (United Nations [UN] conventional weapons treaty of 1980 and its 1995 protocol); and possession or use of antipersonnel land mines (Ottawa Convention of 1997). None of these accords includes measures to verify compliance, nor do they specify sanctions for non-compliance, relying instead on mutual interest and mutual revulsion to foster compliance.[4] As to effect: expanding bullets may remain a rarity on the battlefield, but rough substitutes—tumbling bullets and flechette weapons (which scatter dart-shaped projectiles)—are common, as are cluster munitions (bombs containing hundreds of grenade-like bomblets), and napalm (jellied gasoline). Purpose-built blinding lasers may not yet be deployed, but powerful laser rangefinders and target designators are ubiquitous. The land mines convention entered into force in March 1999 as a step toward elimination of these weapons. Some signatories began to destroy stockpiles, but mines also continued to be deployed in conflicts in Angola, Kosovo, and elsewhere, millions remained in the ground from old wars, significant states remained outside the convention, and verification of the production ban and destruction of existing stockpiles remained matters for future negotiation.[5]

Dangerous governments include those guilty of aggression, genocide, or other war crimes; or governments that support international terrorism. Measures focused on aggressor governments and their ability to make war include the punitive Treaty of Versailles (1919) imposed on Germany at the end of World War I and the sanctions and disarmament measures imposed on Iraq after the 1991 Gulf War. Versailles largely disarmed Germany, which the victorious allies blamed for the war. Its regime of reparations and disarmament seemed to work, until vitiated by the nationalist sentiments that brought the Nazi Party to power and by weak responses to Nazi policies from the other treaty members. Both Germany and Japan, held responsible for World War II, were politically and militarily restructured following that war, with better results. The UN imposed an arms embargo on Iraq following its August 1990 invasion of Kuwait, plus a punitive regime of economic sanctions and programs to dismantle its WMD and longer-range missile programs. UN inspectors were withdrawn in late 1998 and by 1999 the arms embargo was growing increasingly leaky (further discussion below).

U.S. legislation imposes unilateral sanctions on countries deemed to harbor or sponsor international terrorists. The State Department's 1998 list included Cuba, Iran, Iraq, Libya, North Korea, Sudan, and Syria. Some legislation, such as the 1996 Iran-Libya Sanctions Act, attempts to control the behavior of foreign as well as American firms in their dealings with the sanctioned states, with mixed results.[6]

Dangerous force configurations include structures vulnerable to surprise attack and national strategies built around offensive military doctrine, as well as "destabilizing" or "excessive" arms acquisitions deemed either to make war more likely in the event of political crisis or to encourage aggression. Arms control measures aimed at preventing or changing dangerous force configurations include a series of U.S.-Soviet arms control treaties inked between 1972 and 1991 to manage the size and shape of their respective strategic nuclear arsenals and the defenses arrayed against them. The first agreement froze numbers of nuclear weapon launchers (missile silos, missile submarines, bombers), while the accompanying Anti-Ballistic Missile Treaty limited missile interceptors to 100 on either side, avoiding an offense-defense race and ensuring that nuclear aggression would be met with retaliation uninhibited by defenses, creating the likelihood of mutual assured destruction in the event of war. Mutual safety thus depended heavily on joint rational behavior and on exceedingly good command, control, and security arrangements to prevent accidental or unauthorized launch of nuclear forces.

The United States and USSR never went to war despite their accumulation of large nuclear and conventional war machines and despite several political-military crises that might have tipped into war. The mutual caution that "assured destruction" is thought to have induced in U.S. and Soviet decision-making has been taken as a general lesson for interstate relations by some U.S. scholars, who argue that nu-clear proliferation may bring with it more stable deterrent relationships in erstwhile regions of tension.[7] As noted in chapter 3, however, adding nuclear weapons to a tense political relationship may also encourage covert action and other forms of low intensity conflict on the assumption that conventional military responses to such provocations are blocked by the risk of nuclear escalation.

While the early U.S.-Soviet arms control treaties managed the configuration of forces, later accords reached as the Cold War began to thaw started to take the force structures apart. The 1987 Intermediate Nuclear Forces treaty eliminated roughly 2,700 nuclear-armed missiles with ranges between 500 and 5,500 km. The first Strategic Arms Reduction Treaty (START I), signed in 1991, reduced both numbers of launchers and numbers of nuclear warheads deployed, the latter by about 40 percent. START II, signed in 1993, would cut warhead numbers by another 50 percent. Although ratified by the U.S. Senate in January 1996, START II remained unratified by the Russian Duma, a hostage to domestic politics, until April 2000.

Conventional forces arrayed in Europe by the United States, the Soviet Union, and the members of their respective alliances were the subject of arms reduction talks for two decades but negotiations made headway only after the USSR decided to end the European standoff, whereupon a Conventional Forces in Europe (CFE) Treaty was concluded in about 18 months. It led to the removal of roughly 65,000 heavy weapons (tanks, artillery, infantry fighting vehicles, combat aircraft) from an area that stretches from Europe's Atlantic coast to the Ural Mountains in Russia. (CFE is discussed in more detail in chapter 5.)

This brief review of other arms control arenas suggests that arms limitations can be valuable adjuncts to political reconciliation and worthwhile tools to enforce the will of the international community against a dangerous member, provided the members of that community are willing to work at enforcing the sanctions imposed. Supply-side arms trade restraints reflect suppliers' interests and objectives, meaning that they can't expect much cooperation from the targets of restraint. The shortage of cooper-ation is just as acute for programs rooted in *realpolitik,* emphasizing supplier interests and prudent behavior, as it is for programs that are more other-directed and morally based, stressing the well-being of recipient states' populations and the impact of arms transfers on regional peace.[8]

Arms Transfers and *Realpolitik*

States have been the primary actors in global affairs for several hundred years. The idea of the sovereign state spread outward from Europe along with European em-pires. In the twentieth century, as each colonial territory in the developing world shed its European overseers, the goal of its leaders was to form a new state, a po-litically autonomous unit with sovereign (traditionally, non-second-guessable) au-thority over its territory and population. Lacking a higher political authority than the state itself to protect their interests, national governments have traditionally opted for self-protection via national armed forces, reinforced by alliances with other states and by the international laws that groups of states have devised and promised to live by. Article 51 of the UN Charter affirms UN member states' rights to "individual and collective self-defense," makes state borders inviolable (and thus makes aggression illegal), and makes other threats to international peace and security a cause for voluntary international action. If governments cannot build the tools they need for self-defense, they have the right to contract with others to buy them, much as any other commodity. From this perspective, military items are just as legitimate an object of international commerce as textiles or foodstuffs. Hence the arms trade.

Analysts and policymakers who emphasize the primacy of the state and its prerogatives in international affairs focus on states' interactions with one another rather than their internal politics or the character of their governance. Maintaining peace with the neighbors has been a principal focus of many states' foreign policies as well as the subject of numerous détentes, ententes, and alliances, as states "balance" or "bandwagon" through history.[9]

Although the nature of the international system argues in favor of arms acquisitions by states, there are three supplier arguments from the realpolitik perspective in favor of arms trade restraint (assuming the suppliers are major powers with interests that can be affected by events beyond their immediate borders). Each argument is rooted in political-military prudence, that is, the notion that it can be *unwise* to sell certain weapons to certain customers abroad, not that it is *morally wrong* to do so. In brief,

- Supplier states could find themselves confronting, in some future conflict, the arms that they or other suppliers have sold;
- Weapons technology can find its way into the wrong hands, including terrorist hands; and
- In many countries, political development is so rudimentary that ruling groups and their armies cannot be relied upon to use what they acquire responsibly, and may be subject to violent replacement.

A prudent arms transfer policy would weigh the odds that weapons sold overseas might be used to attack one's own forces or the forces of one's allies in future military interventions. The United States had to face modern Western weapons sold to Iraq, for example, when it escorted oil tankers through the Persian Gulf in 1987–1988, including a French-made Iraqi Exocet missile fired in error at the USS *Stark*. Exocets actually sank British warships during the Falklands/Malvinas War in 1982, and the fact that Iraq owned 100 French-built Mirage F-1 fighters complicated French contributions to the allied air effort in the Gulf War, because those Iraqi jets could not be readily distinguished visually at a distance from similar-model French air force aircraft.

End-use restraints on military systems and technology sold to developing states are difficult to enforce, and most difficult in unconventional or guerrilla warfare, where balance of power imperatives at one time may have argued in favor of the provision of arms. U.S.-funded arms flooded into Afghanistan in the 1980s, with minimal U.S. oversight to support the Afghan guerrilla armies that were fighting Soviet troops. The United States even provided Stinger shoulder-fired, infrared-homing, surface-to-air missiles, which played havoc with Soviet air power but also found their way into the inventories of neighboring countries, including Ayatollah Khomeini's Iran.[10]

The United States poured first-line military technology into Imperial Iran in the 1970s, which became the only foreign country to operate such U.S. weapon systems as the Grumman F-14 Tomcat naval interceptor and its long-range Phoenix missile. When the shah's rule collapsed, this equipment fell into the hands of the theocratic

Islamic regime that succeeded him, forcing a redesign of the missile's electronics on the assumption that the original technology had been compromised and could be countered more readily by the USSR.

More generally, it might be argued that weak or corrupt civilian rulers, praetorian militaries, and/or ethnic and religious rivalries that are endemic to many developing states make such states poor candidates for cutting edge military technologies, indeed, for any power projection capability whatsoever. During the Cold War, this third criterion was perhaps the most difficult one on which to base arms transfer restraint decisions because Western restraint could result in greater influence for the Soviet bloc. The political risk entailed by such forbearance has since dropped dramatically. The economic opportunity costs to American firms of sales foregone might still be substantial, but the net political-military costs of restraint to the U.S. government are now far less than during the Cold War.[11]

Arms Transfers, Human Security, and Regional Stability

Whereas *realpolitik* stresses the effects of arms sales and restraint on supplier interests, what we are calling here an other-directed or morally based philosophy of restraint emphasizes the character and behavior of recipient governments toward their own people and within their security complex, and the contribution of certain types and quantities of arms to that behavior, especially as it relates to peace and war. Applying a restraint policy based on this philosophy requires close monitoring of foreign governments and their policies. Moreover, it involves mental models of how transfers of weapons reverberate within a regional security complex, the basis and validity of which are much more easily assumed than demonstrated.

Promoting the Good

In the years following World War I, when international arms sales by munitions firms were not closely regulated by governments, the evil effects of armaments were laid at the feet of the industrial "merchants of death" who built and profited from the weapons used in that war.[12] That term's contemporary, somewhat less derogatory equivalents, the "military industrial complex" and the "iron triangle," refer to the mutually supportive relations that evolved during the Cold War between Western governments and their defense industries.[13] The image of the death merchant hawking destruction could be seen in U.S. president Jimmy Carter's decision to distance the U.S. government from defense firms' overseas marketing efforts by curtailing support by U.S. embassies and military representatives abroad for U.S. firms seeking contacts for arms sales.[14] Reversal of that decision was completed by President Clinton, whose arms transfer policy treated defense items much like any other commodity that Americans excel in making and trading, and made arms sales as much an element of trade policy as of national security strategy.[15]

Clinton administration policy has been criticized by non-governmental organizations (NGOs) who favor a "code of conduct" for arms transfers that would require potential recipients to meet specified norms of democratic governance (see further discussion, below). Behind the push for such codes is the notion that security for the recipient state itself or profits for the seller are insufficient reasons to make armaments available. Advocates argue that countries like the United States that are governed by democratic principles should supply arms only to states similarly based on concepts of popular sovereignty, whose governments respect the basic rights and further the well-being of their populations at large, and not just the well-being of governing elites and their institutions.

Looked at from such a normative viewpoint, a focus on recipient government intentions and behavior is self-interested. The United States has long said that it supports open, democratic governance and respect for human rights, as well as open markets. Doing so furthers U.S. interests because democratic political values and regulatory institutions are what will make the market system tolerable (and viable) over the long run. They counterbalance the market's self-destructive tendencies toward monopoly and concentration of wealth, and help forestall possibly violent popular rejection of such concentration. Democracy's empowerment of civil society makes government more responsive to popular demands, makes it more adaptable to changing circumstances, and provides nonviolent channels for social, political, and economic protest.

Respect for human rights keeps democracy viable, in turn, by deterring absolute majorities of the like-minded from ruling at the expense of a country's minorities. What democratic political institutions do for the market, institutionalized respect for human rights does for democracy itself. States whose governing systems are undemocratic and disdainful of human rights (such as the shah's Iran, South Korea until the 1980s, all of eastern Europe until the 1990s, and a good deal of Africa, Asia, and the Arab world today) run a greater risk of internal upheaval. Such governments may also succumb to distracting foreign adventures (as did the Greek junta in its fatal attempted coup in Cyprus in 1974, Somali dictator Siad Barre in his failed attempt to seize land from Ethiopia in 1977, and the Argentine junta, which failed to conquer the British-held Falkland Islands in 1982).

How could arms transfer restraint contribute to the evolution of good governance? If there were general supplier agreement to minimum criteria of governance to be met to justify arms transfers, then some or all transfers could be made contingent on a government's meeting some or all criteria. Or there might be an agreed surcharge attached to sales made to states that don't meet the governance standards, with the proceeds used to fund, for example, grass roots campaigns to support democracy and human rights by international NGOs. Unfortunately, NGOs are likely to view such funds as tainted, and arms recipients would object to the surcharge and the scrutiny, as do Latin American states that the U.S. Congress has demanded be "certified" by the U.S. government as cooperating in the fight against illegal drugs. The U.S. State Department already publishes annual human rights reports that are frank assessments of governments' performance, as are the periodic reports of the UN Human Rights Committee and Commission on Human Rights.[16] At issue is whether reports such as these would have greater impact or generate greater backlash if linked to a country's eligibility for arms. Suppliers that did not meet the good governance criteria themselves would have little incentive to participate in a rights-contingent restraint program, unless a second tier of sanctions penalized suppliers operating in violation of a rights-arms regime that they had not joined.

One might also attempt to promote good governance by arming the enemies of bad governance. Historically, however, prevailing political wisdom and power have played a large role in defining "good" and "bad." The Soviet Union and China armed "national liberation" movements in developing countries and remaining European colonies throughout the Cold War, while the United States funneled arms and advisers to the governments fighting such movements. Individual human rights took a back seat to the larger political objectives of both sides. In the 1980s, the Reagan administration's interest in the global rollback of communism produced the Reagan Doctrine, through which the United States made arms available to what might be labeled "restoration" movements, guerrillas who fought Soviet troops in Afghanistan, fought Cuban troops in Angola, and fought the Sandinista government in Nicaragua. The opposite notion of sending arms to local parties struggling *for* reform—a sort of left-shifted Reagan Doctrine, if you will—has never really enjoyed comparable U.S. government support. In 1994, Congress did terminate funds to enforce the UN arms embargo on Croatia and Bosnia, facilitating the smuggling of arms to Bosnia's Croatian and Bosniac (Muslim) communities, and in late 1998 it passed an "Iraq Liberation Act" to channel funds to Iraqi groups seeking to overthrow Saddam Hussein. These were, however, exceptional actions; the conservative lawmakers who dominated the Congress from 1995 through the end of the decade generally had little interest in foreign intervention. Early in the decade, political scientists and liberal legal scholars pointed to humanitarian military intervention as a tool for easing or ending disasters such as the war-worsened famine in Somalia or ethnic cleansing in Bosnia.[17] However, after U.S. commandos and special forces suffered about 100 casualties trying to capture one of Somalia's faction leaders in October 1993, even liberals lost interest in humanitarian intervention, except in those established zones of interest where U.S. troops are routinely deployed: Europe, Northeast Asia, and the Persian Gulf. After the losses in Somalia, the United States proved unwilling to intervene outside these zones even to stop genocide.[18]

Defining "Destabilizing" Arms Transfers and "Excessive" Levels of Armaments

Calibrating supply-side arms transfer restraint so to as to promote or maintain stability—defined either as a lack of conflict or as slow rates of change in states' arms inventories—within a regional security complex involves much more difficult calculations about the impact of arms transfers than do considerations of power or profit. It requires judgments about the extent to which arms contribute *to* conflict as opposed to meeting needs that arise *from* conflict. It requires assessments of the interaction between arms and conflict over time and of interactions amongst the members of the security complex. And it requires assessments of the *kinds and amounts* of arms that are potentially "destabilizing" to a particular complex or otherwise "excessive" to recipient defense needs.

That which constitutes a destabilizing arms transfer is hard to pin down. Are transfers destabilizing

- if they introduce qualitatively different military capabilities into a region?
- if they contribute to a spiral of arms acquisitions?
- if they give a state that seeks military preponderance a position of clear superiority?
- if they contribute to the outbreak of conflict?

A paper approved at the December 1998 plenary meeting of the current international mechanism for restraining transfers of arms and related technologies, the Wassenaar Arrangement (about which more later), listed three closely typed pages of such questions. The paper did not indicate which questions might be assigned the greatest weight by governments pondering arms sales, either in a generic sense or in particular circumstances. Nor did the paper suggest means for evaluating the answers to the questions that it posed, for example, how to calculate a regional force balance, or data on ratios of forces historically associated with successful military offense or defense.[19] In failing to do so, however, it was being realistic in two ways: its political context (a voluntary restraint regime operating on consensus) is probably too sensitive to disruption to support a more analytical document, and the issues it addresses have chameleonlike qualities. Consider a few extensions just to the four questions posed above.

Qualitatively Different Capability

What makes a technology "qualitatively different" and what characteristics of "difference" should be considered *a priori* destabilizing so that decision-makers in supplier states know in advance what export requests to disapprove? The answer to the first part of the question would depend in part on the level of detail used in measuring differences. Gross differences would be easy to distinguish, for example, propeller-driven versus jet-powered aircraft, or subsonic versus supersonic aircraft. These are relatively clean distinctions, and verification of compliance with such transfer thresholds would be relatively straightforward. But where such coarse-grained thresholds have already been breached, other, more detailed quality thresholds would need to be set, and the range of choices for defining such thresholds would be broad. Capabilities evolve via generational replacement of whole weapon platforms (aircraft, tanks, ships) or, increasingly, via modernization of key component parts of those platforms. Unless all weapon system modernization is halted, acquisition of new capabilities that eventually add up to a "qualitative difference" will occur as a matter of course. Radar-evading "stealth" capability is an example that applies to platforms. A fifth-generation, 1960s-vintage fighter such as the F-4 Phantom (now a seeming misnomer) has a radar cross section (RCS) of 100 squared meters (m^2), that is, it returns the radar echo of a flat slab 10 meters high and wide. The 1970s-vintage, sixth-generation F-15 Eagle has an RCS just one-quarter as large ($25 \ m^2$). A seventh-generation aircraft like the F-22 Raptor, entering service with U.S. forces after 2000, might be expected to have an RCS no larger than the 1980s-vintage B-2 bomber ($10^{-6} \ m^2$, equivalent to a chip 3 millimeters on a side).[20]

Even if suppliers agreed not to sell new airframes to a regional complex, modernization of existing aircraft would be difficult to stop, as many firms in a number of countries are in the upgrade business and their export controls vary widely in stringency and effectiveness.[21] Extensive hands-on testing of individual equipment might be necessary to verify violations of technology transfer constraints with respect to internal components like radars, communications, fire control systems or, especially, computer software. The cooperation of recipients would be needed for that to happen. In a purely supply side restraint regime, recipients would be the targets, not the creators, of the regime and unlikely to cooperate without external imposition of some other cost of non-compliance (economic sanctions, for example) that would require a further tier of monitoring and enforcement.

If suppliers were to agree to criteria restraining certain technologies, and the necessary verification procedures were established with at least the passive acquiescence of recipients, what might be the security implications of such a restraint regime for the *boundary* members of a security complex? Unless a security complex is wholly surrounded by sea, some members will border on the complex next door. The states in that neighboring complex might or might not be subject to comparable technology restraints. Examples of states on the boundaries of two or more security complexes include Turkey and Pakistan (both discussed in chapter 3).

Can any new technologies or capabilities be identified as *generically* destabilizing and therefore undesirable? One benchmark might be a lack of defensive applications. Virtually every weapon system can play defense as well as offense, even submarines. The one category for which defensive applications are difficult to devise is the ballistic missile, and especially missiles with ranges of several hundred kilometers and up. They can be used for retaliatory terror and therefore serve a sort of deterrent purpose, but terror is best achieved with unconventional payloads. Thus a longer-range missile inventory can betoken not just an offensively oriented strategy but also an interest in WMD.

Military Balance, Military Preponderance

There is no general agreement among analysts as to what maintains regional stability better: an even balance of power or a preponderance of power exercised by a regional hegemon. Because severe quantitative imbalances in conventional arms have given states like Israel and Pakistan incentives to acquire nuclear weapons, creating and maintaining rough balances of conventional military power among adversaries in a region seems, in principle, a desirable thing to do. In practice, however, it is quite difficult to arrive at a mutually agreed balance of power between adversary states. The CFE Treaty was successfully negotiated not when the right balance formula was reached but when the Soviet Union decided to call it quits and agree to equal East-West *numerical ceilings* on tanks and other equipment in the area of limitations, rather than equal combat potential or other fine-tuned measures that are difficult to sell to legislators who do not spend their spare time with quantitative models.

In any case, a rigorous conventional balance would be fiscally impossible in most of Asia, where the physical and economic differentials among states are extreme, unless the predominant power agreed to disarm or to donate resources to its smaller neighbors. Some analysts make a virtue of necessity and argue that a preponderant power—a regional hegemon—is the key to regional stability.[22] Should stability-oriented suppliers therefore promote such hegemony? Presumably they would want to do so only if the would-be hegemon supports the political status quo within its security complex; otherwise their support would only accelerate regional instability. But what if the status quo is a repressive one and the state(s) within the complex that seek to change it are aspiring democracies trying to elude the hegemon's grip, by force if necessary? In that case a supplier focus on stability would run headlong into whatever interests suppliers might have in promoting human security and democratic governance. This is actually not so farfetched a scenario, if one looks down the road at Southeast Asia and its contretemps with China over claims in the South China Sea and makes the assumption that Indonesia will emerge from its current political flux with a working democracy. Similar considerations may apply to Taiwan.

Transfers and Conflict

The quantitative analyses in chapter 2 suggested that the historical relationship between arms transfers and war, while statistically significant at times, is relatively weak overall.

Countries that went to war stocked up on arms beforehand, but not all countries that stock up go to war. Once again, it seems important to weigh the purposes for which war might be waged. War is always costly, but, depending on the political purposes for which it is fought, not always unmitigated evil. Indeed, it may need to be fought to turn back evil. The World War II generation would have a definite view on that point. The 1991 Gulf War fits in this category, as does action by the North Atlantic Treaty Organization (NATO) against Yugoslavia in spring 1999 to halt and reverse ethnic cleansing in the province of Kosovo. In prior decades, Tanzania intervened in Uganda to oust the bloodthirsty Idi Amin (1978), Vietnam invaded Cambodia (1978) to oust the genocidal Khmer Rouge, and the United States invaded Panama (1989) to remove a larcenous military regime that had overturned the results of a democratic election. These are, of course, only the brighter lights in what is a much darker record of warfare and human suffering in the last half-century, most of which occurred in developing countries. The point here is simply that one cannot take a generic position either that imported arms lead to war, or that when they do contribute to such gross instability, the result is to be automatically condemned or even regretted.

Questions of Excess

In addition to judging arms acquisitions on the basis of their impact on regional stability, one can try to judge them in terms of the recipient government's immediate defense needs, which vary widely across states. What a state like Costa Rica requires for defense in its relatively benign security environment will be far less than what an even smaller state like Israel may need to counterbalance a constellation of determined, nearby foes. Such political-military requirements can in theory be measured objectively, but important thresholds of safety are politically, even emotionally, determined. It is extremely difficult—short of running an actual war to test assumptions about equipment, strategy, leadership, and training—to demonstrate conclusively that *this* much capability is necessary for national defense whereas *that* much is excessive. Over time, analysts have devised many interactive models of combat, but what analysts in country A thus determine objectively to be necessary for defense and what the government of country A may agree to buy, leaders in country B may find (or assert) to be threatening, as either prisoners or manipulators of the security dilemma.[23]

There nonetheless would seem to be some *prima facie* cases of excessive arms acquisition in the developing world. Can we know them when we see them? Consider the shah of Iran's amazing arms-buying spree in the early 1970s, related in chapter 3, which probably contributed to his downfall. The shah was embarked upon a program to build Iran into a major power in the Persian Gulf. The United States backed him to the hilt as a regional Horatius who would block Soviet access the region's oil supplies in time of crisis. Viewed through this lens, the shah's goals, while ambitious, were not so much out of place as pursued much too fast, but the shah, who knew he had cancer years before he abdicated, was a man in a hurry.[24]

Iraq's heavy and rapid accumulation of armaments in the 1970s, leading up to its attack on Iran, could be considered a clearer example of armament excessive to national defense needs. Yet upon close examination, several seemingly clear-cut, *a priori* measures of excess capability lose some of their clarity. Saddam Hussein's army had accumulated 2,750 tanks by 1980, half again as many as Iran (despite the shah's best efforts) for a country with a population just one-third as large as Iran's. On the other hand, in 1980 Israel had a population one-fourth the size of Iraq's but a tank fleet nearly half again as large, or six times as much armor per capita, yet Israel was not routinely viewed, outside Arab capitals, as overarmed given the threats that it faced. Did that mean Iraq's tank force was not excessive either? Or must we once again weigh the policies and objectives behind the forces before deciding what constitutes a dangerous excess? In the event, once Iraqi leaders realized that the war they started in 1980 would not be ending soon or ending on their terms, it became clear that the arms they had accumulated prior to the war would not be enough to stave off Iranian forces in a long conflict. Iraq thus placed orders for thousands more pieces of heavy armor and artillery with China, the Soviet Union, and various Soviet allies. By the end of the Iran-Iraq War, it had accumulated twice as much armor, even discounting for war losses, as had been on hand at the start of the war. Yet, prior to Saddam's August 1990 attack on Kuwait, this stockpile did not seem to bother Western observers. Most evidently thought that it still pointed in Iran's direction. Containment of Iran remained a primary goal, certainly of the U.S. government, and Iraq's capabilities before August 1990 were likely not viewed as excessive to that goal. The fact that those forces could also be pointed southward, of course, ultimately became a problem. Given Saddam's history of aggression against neighbors, the international community ought at least to have posted closer watch on Iraqi behavior and signaled in advance that further aggression would entail serious negative consequences. Ironically, however, had they done so, Saddam may have deferred action against Kuwait only to reprise his demands two to five years later, backed by nuclear weapons, perhaps integrated with medium-range ballistic missiles. A ground assault on Kuwait and Saudi Arabia might then have been mounted with the assistance of nuclear blackmail.

Of the two kinds of normative decision criteria for arms transfers just discussed, namely governing norms and concerns about instability or excess armaments, it would seem to be easier to draft a coherent restraint based on the former. Quality of governance is at least measurable, even if contestable by the government being evaluated, and achieving supplier consensus would be another matter. But *a priori* definitions of what constitute "destabilizing" or "excessive" types or levels of arms may be much harder to craft. Definitions generic enough to generate initial supplier consensus yet flexible and detailed enough to apply in specific regional circumstances while maintaining consensus and not, in fact, making things worse in the region in question, simply may not be achievable. The single exception, where generic restraints may remain valid, involves short-warning weapons uniquely suited to offensive purposes and most useful when armed with WMD.

A Short History of Supply-side Restraint Efforts

Actual efforts to restrain the supply of conventional arms to developing states undertaken since the early 1950s have either tried to curtail the distribution of particular weapon systems or dual-use technologies (weapon-based restraint) or tried to curtail transfers to certain states or security complexes (recipient-based restraint). As noted earlier, the latter also incorporate efforts to alter configurations of forces in the interest of regional stability.[25]

Weapon-based Control Efforts

As the Cold War heated up and more developing countries gained their independence and became targets of influence for both East and West, arms transfers became a major tool of influence, as related in chapter 1. By the mid-1970s, the Democrat-controlled U.S. Congress was trying to legislate some weapons- and recipient-based restraints on U.S. arms sales to counter the freewheeling policies of the Republican White House under presidents Richard Nixon and Gerald Ford.[26] The congressionally generated 1976 Arms Export Control Act required that the administration report to Congress on proposed military sales and gave the Congress 30 calendar days to vote against arms deals involving more than $14 million in defense equipment, $50 million in defense articles or services, or $200 million in design and construction services. In practice, however, this legislation resulted in very little congressionally enforced restraint.[27]

Jimmy Carter's CAT

Congress did receive some declaratory support from the Democratic Party's 1976 presidential nominee. During his campaign, Jimmy Carter took a strong stand against the United States' role as the world's leading supplier of sophisticated armaments. In May 1977, now-president Carter issued an arms export policy that declared arms transfers to be an "exceptional foreign policy implement, to be used only in instances where it can be clearly demonstrated that the transfer contributes to our national security interests."[28] However, billions of dollars worth of sales were already in "the pipeline," contracted for but not yet delivered, so the policy could not produce results quickly. It soon ran into barriers of precedent and Cold War politics at every turn.

While the policy was primarily weapon-based, exceptions to it were both weapon- and recipient-based. U.S. allies were exempt from its constraints. The Saudis wanted and got first-line F-15 fighters. The shah of Iran wanted and got approval for advanced early warning radar aircraft but fell from power before they could be delivered.

U.S. officials canvassed allied capitals for support of the Carter arms policy. Officials politely responded to the administration's pitch by saying that they might cooperate in restraining arms supplies if the Soviets did. In late 1977, therefore, the Carter administration initiated discussions on conventional arms transfers with Moscow.

In spring 1978, as the Conventional Arms Transfer (CAT) talks began to take shape, staff at the Arms Control and Disarmament Agency (ACDA) proposed a negotiating approach focused narrowly on types of weapons that might be restrained.[29] The State Department favored a broader political approach that might lead the United States and USSR to "mutual understandings about their rivalry in the Third World." State's approach almost by definition required discussions of regional politics and military balances. It was endorsed by President Carter.[30]

As a political discussion, however, CAT had no chance of succeeding. Either the White House did not think through the implications of a political approach to the CAT talks or the president's national security advisor, Zbigniew Brzezinski, wanted to make sure that CAT would run afoul of other, higher-priority foreign policy objectives. Negotiations with the Soviet Union foundered after four rounds as soon as discussion turned, at Soviet urging, to regions of the globe like the Persian Gulf that the White House considered politically sensitive. The high-placed fear in Washington was that any discussions with the Soviets about the Persian Gulf would be held against the Carter administration should anything happen to the shah or to the rulers of any number of Gulf sheikdoms. As it happened, by late 1978, when discussions in CAT of arms sales to "West and East Asia" were slated to take place, the shah's regime was tottering. Moreover, secret U.S. negotiations to establish diplomatic relations with China were nearing completion. Washington forbade U.S. negotiators to discuss either Asian region with their Soviet counterparts or even to listen to whatever the Soviets might have had in mind to say about these places. As a result, in December 1978, U.S. negotiators were left cooling their heels in windowless meeting rooms in the U.S. embassy in Mexico City, where even the author briefly took up smoking to pass the time (although he did not inhale).

Washington, not Moscow, was responsible for pulling the plug on the talks, but the USSR's interest in them never discernibly rose above giving the United States a chance to get itself in trouble with its major arms clients and oil suppliers. No further evidence of the continuing importance of security assistance to U.S. foreign policy was needed: when restraint efforts ran afoul of *realpolitik,* restraint lost.[31] Without progress in CAT, moreover, Mr. Carter's unilateral arms transfer restraint policy could not be sustained domestically.

The Missile Technology Control Regime

Nearly a decade after ACDA proposed global restraints on selected weapons as an approach to CAT, controlling the transfer of one class of weapons—longer-range ballistic missiles—began to look less fanciful to Washington. This change in attitude was stimulated in part by the so-called war of the cities in which Iran and Iraq rained ballistic missiles upon one another's capitals. It was also stimulated in part by concerns about collaborative ballistic missile programs among states that also were thought to be developing chemical or nuclear weapons. While the missile duels of the Iran-Iraq War did some damage and killed some people, such missiles would have been much more lethally effective had they been equipped with mass destruction warheads. The MTCR was the result: a non-binding, voluntary arrangement that is the best example to date of a *realpolitik* supply-side restraint regime for major conventional weapons. Suppliers support it out of prudent self-interest.

The MTCR was set up in 1987 at U.S. urging and gradually has been expanded to include long-range cruise missiles as well as ballistic missiles. Under the MTCR, the United States and most other missile producers have agreed to suppress the export of ballistic or cruise missiles that can carry a 500 kg payload to a range of 300 kilometers or more. The range constraints encompass missiles like the Soviet-designed Scud-B, used in both the Iran-Iraq War and by Iraq in the subsequent Gulf War. Participants also have agreed to suppress export of the technology used to construct such missiles.[32]

Some analysts view the regime as too limited, particularly with respect to cruise missiles, arguing that more stringent limits would slow proliferation and "provide the United States and other major powers with the 'breathing space' they require to develop effective cruise missile defenses."[33] A 1991 Stanford University study argued that the regime should be extended to long-range strike aircraft, which have a heavier payload, a capability for short-warning attack, and so forth. Strike aircraft remain unconstrained, but in July 1993, regime members did agree to broaden the MTCR's writ to cover shorter-range ballistic missiles like Russia's SS-21 that could be made to fly 300 kilometers with a lightened payload.[34]

By the fall of 1997, 29 countries had formally adhered to the regime. Since the MTCR was launched, five countries have stopped regime-relevant ballistic missile development programs (Argentina, Brazil, Egypt, South Africa, and Taiwan), but others have forged ahead (including Israel, India, Pakistan, North Korea, and Iran). After twice running afoul of temporary U.S. sanctions for shipping parts for its 300-km-range

M-11 missile to Pakistan, in 1994 China agreed (again) to cease and desist. In 1997, on the eve of President Jiang Zemin's state visit to the United States, China also agreed to stop selling Iran its C-802 antiship cruise missiles (which, while powerful, have a range of just 120 km and are not covered by the MTCR).[35]

1990s Proposals for Weapons-based Restraint

In one of the final arms-trade-related acts of the Democrat-led 103rd Congress, the fiscal 1994 National Defense Authorization Act called for a blue ribbon commission to study factors contributing to arms proliferation and what might be done about it. The Presidential Advisory Board on Arms Proliferation Policy was not appointed until early 1995, by which time the Congress had shifted to Republican control and the White House was within a month of announcing its liberalized arms trade policy.

The board completed its work against this difficult political/policy background one year later.[36] Its report hewed to the politically possible, stressing as U.S. policy options a number of weapon-based restraint measures together with controls on dual use technologies. Among its conclusions: sources of advanced conventional weapons will steadily expand; some advanced conventional weapons have destructive properties akin to those of mass destruction weapons; certain transfers can hurt U.S. interests or personnel deployed overseas; and governments' ability to control the spread of technology will decline over time.[37]

To deal with these problems, the report recommended an incremental approach. Such an approach would focus initially on technologies in which the United States had the greatest leverage because there were few substitutes on the market, or where restraint entailed low opportunity costs for U.S. interests abroad. Examples suggested in the report were stealth technologies, advanced missiles, and directed energy weapons. "Weapons of ill repute," a notion dating back to the Carter-era CAT talks, were a second recommended area for restraint, and included "certain incendiary and fragmentation weapons," portable weapons easily appropriable by terrorists, blinding lasers, and antipersonnel mines.[38]

Key component technologies for weapon systems were a third area of concern, but because many have legitimate civilian applications, the report emphasized monitoring of end uses rather than controlling distribution directly. Any credible monitoring system would require "profoundly greater levels of transparency in the international trading system" and an international secretariat to coordinate that monitoring and warn of technology diversion.[39]

Finally, the report stressed that "control arrangements must ultimately focus on the demand side," while noting that regional arrangements to restrain demand are "unlikely to succeed . . . if the major suppliers are engaged in intense competition for arms markets." A supplier cartel is not the answer to competitive pressure, it concluded, because such a cartel (called a "market stabilizing mechanism" by a supporting RAND Corporation study[40]) could put a floor under industry downsizing rather than constrain arms sales.[41]

The advisory board's report recognized the desirability of disconnecting U.S. arms sales policy from economic and trade policy; stressed that demand limitation should be the ultimate goal of U.S. policy; and emphasized that supplier support for regional arms restraint initiatives would be critical to such initiatives' success. But in suggesting the equivalence of certain conventional armaments (large "smart bombs," for example) and nuclear weapons, because the former were now capable of destroying some types of hardened military targets formerly vulnerable only to nuclear explosions, the report probably went too far.[42] The distinction between conventional arms and WMD is what makes antiproliferation efforts in the WMD arena possible. WMD are to be constrained because they are different. Blurring the difference so as to draw advanced conventional weapons into the magic circle of restraint risks having the opposite effect, namely, "conventionalizing" nuclear arms and undermining the case for special, strict constraints on their proliferation and even possession.

Recipient-based Restraint Efforts

Recipient-based restraint regimes are feasible if major suppliers can agree on what constitutes dangerous behavior and can decide which states are to be the subjects of restraint measures. There was little possibility for such agreement during the Cold War, with the exception of sanctions against white minority governments in southern Africa. Elsewhere, one side's rogue was the other side's returning prodigal son.

Since the end of the Cold War, recipient-based regimes have at least been easier to talk about, including big power talks on restraint in the Middle East and multilateral export control arrangements that focus on "states of serious concern." International arms embargoes also have been imposed more frequently since the Cold War's end. Still, these efforts have been only sporadically successful, partly because supplier states have not yet found the arguments for arms sales restraint to be so compelling that they will place other political or economic interests at risk in the pursuit of such restraint.

Arms Restraint and the Middle East

The first post–World War II multilateral effort to manage the flow of conventional arms was recipient-based. In the U.S.-French-British Tripartite Declaration of 1950 on arms in the Middle East, "the three nations agreed to coordinate their sales to Egypt, Iraq, and Israel to maintain a 'rough balance' in the region." Arms transfers into the region from these three powers were regulated by a joint Near East Coordinating Committee. Three years later, however, France was selling first-line fighters to Israel. Two years after that, in 1955, the Soviet Union began a nearly four-decade role as armorer to the Arabs, at first funneling weapons to President Nasser's Egypt through Czechoslovakia.[43] Competing economic and political objectives scuttled this early restraint regime and the Middle East became the destination for a large percentage of arms shipped to developing states over the next 40 years.

In late May 1991, in the aftermath of the Gulf War, President George Bush proposed that the five major suppliers of arms to the Middle East support a number of measures intended to moderate the flow of arms. (Those five—the United States, Soviet Union, France, Britain, and China—are also the permanent members of the United Nations Security Council and make roughly 85 percent of the arms sales to the region.) U.S.-proposed restraint measures included a general code of responsible arms transfers, avoidance of destabilizing transfers, and effective controls on the end-use of items transferred. The five would notify one another in advance of "certain" arms sales, meet regularly to consult on arms transfers, consult ad hoc if one of them believed that supplier guidelines were not being observed, and share annual reports on transfers. Mr. Bush also looked toward an eventual nuclear weapon-free zone in the Middle East, a halt to the production of nuclear materials, and a ban on ballistic missiles.[44]

Over the next year, the "Big Five" held three formal rounds of talks, in Paris, London, and Washington. They agreed to inform one another about transfers of several types of arms, based on categories used in the Conventional Forces in Europe Treaty (namely tanks, armored combat vehicles, artillery, military aircraft, and helicopters), plus naval vessels and "certain missile systems." They also developed criteria for sales to avoid—for example, sales that

- would prolong or aggravate an existing armed conflict, increase tension, or contribute to regional instability;
- introduce destabilizing military capabilities;
- contravene international embargoes;
- be used "for other than the legitimate self-defense and security needs of the recipient states";
- be used to interfere with the internal affairs of a sovereign state; or
- undermine the recipient state's economy.

Note that except for the third point and perhaps the last one, none of these criteria would have been objectively definable, let alone verifiable. In the event, China balked at the concept of prenotification of arms sales, and when the United States agreed to sell 150 F-16 fighters to Taiwan in September 1992, China dropped out of the Big Five process, which then stalled.[45]

Multilateral Restraint: The Wassenaar Arrangement

During the Cold War, the members of the Organization for Economic Cooperation and Development (OECD, consisting primarily of the United States, its European allies, and Japan) used the Coordinating Committee on Multilateral Export Controls (COCOM) to block exports of militarily sensitive technologies to the Soviet Union and its allies. Under COCOM, any member could veto transfers of sensitive technology to the targeted countries by any other member of the regime. COCOM expired in March 1994 before negotiation of a post–Cold War export control regime was complete. Its successor, the Wassenaar Arrangement on Export Controls for Conventional Arms and Dual-Use Goods and Technologies (named for the Dutch town where talks were held), took two years to negotiate. The initial result was accepted by 28 states in December 1995, and included 33 countries as members by late 1998.[46]

The Wassenaar Arrangement is a both a more ambitious and a weaker coordinating mechanism than COCOM. On the ambitious side, it includes many significant members of the group that COCOM formerly targeted (that is, Russia, Ukraine, Poland, Czech Republic, Slovakia, Hungary, and the Baltic states), and it includes an extensive munitions list.

On the other hand, compliance with Wassenaar is voluntary, members cannot veto one another's exports, and there is as yet no strong norm against "undercutting" (delivering goods or services that another member has denied to a targeted state).[47] Members do agree to notify one another at specified intervals regarding deliveries of armaments and dual-use technologies to nonmembers. There is an understanding that Wassenaar's control lists apply particularly to regions of political instability and "states of serious concern" (informally referred to in U.S. government circles as "rogue" or "pariah" states). States must agree to curtail transfers to states of serious concern in order to belong to the regime, but there is no formal agreement as to which states are on the list. Following the regime's December 1996 plenary, the communiqué mentioned only that no Wassenaar member transfers arms or ammunition to Afghanistan "as a matter of national policy." While participants in the December 1997 plenary reportedly reaffirmed that approach, and agreed to exercise restraint in exports to the Great Lakes region of central Africa, no countries or regions were specified in the 1997 communiqué. The United States wanted it understood that states of serious concern included China, Iran, Iraq, Libya, and North Korea. U.S. officials have testified that there is consensus among participants on the latter three, but not on restraining exports to Iran or China.[48] The 1997 plenary also commissioned the study, discussed earlier, to define "criteria for assessing destabilising weapons accumulations," indicating that there were in fact no agreed criteria supporting the arrangement in this area.[49]

The Wassenaar regime includes extensive lists of military and dual-use technologies. The military ("munitions") list covers all the categories of weapon systems restricted by the Conventional Forces in Europe (CFE) Treaty and every other category of military technology conceivable, from ammunition and rocket fuel to exotic kinetic and directed energy weapons. However, France and Russia officially consider the munitions list to be merely "a reference list drawn up to help in the selection of dual-use goods which could contribute to the indigenous development, production, or enhancement of conventional munitions capabilities," suggesting that they do not consider themselves bound under Wassenaar to restrain military goods that do not contribute to "indigenous . . . capabilities."[50] The munitions list is not tiered to distinguish more sensitive from less sensitive technologies.

Wassenaar's dual use technologies list, on the other hand, is a nested series of "basic," "sensitive," and "very sensitive" items. A basic computer system subject to reporting, for example, would be one that is capable (among other things) of operating in extreme heat or cold, one that is radiation hardened, or one that is capable of performing more than so many hundred million theoretical operations per second (Mtops). A sensitive computer would be one with, say, three times faster processing capacity, and a very sensitive computer might be one that is five times more capable than that.[51] The sensitivity varies with the military applications for which an item is theoretically useful. A "very sensitive" computer might be capable, for example, of simulated testing of nuclear weapons.

At the 1998 plenary, the United States obtained other participants' agreement to waive export controls on certain data encryption technologies while gaining their agreement to impose export controls for the first time on encryption techniques contained in mass-market software. Controls were lifted on 56-bit encryption (the higher the number, the stronger the encryption), while controls were "authorized" (but not mandated) on exports of mass-market software that uses more than 64-bit encryption. The American Electronics Association noted that mass market software had previously been exempt from encryption limits because it "was generally acknowledged" not to be susceptible to export controls, and that the Internet made such software "even more uncontrollable." The ease with which software can be transmitted over the Internet and the global reach of any workstation plugged into the Net leave large gaps in enforcement.[52]

U.S. industry considered itself unilaterally disadvantaged once again by rules that were unlikely to achieve their stated national security purpose and pressed for further relaxation of U.S. export rules. And just 13 months after urging the members of

Wassenaar to institute controls on mass-market encryption, the U.S. government relented, permitting U.S. companies "to export any encryption product around the world to commercial firms, individuals and other nongovernmental end-users" without a license. A new category of "retail" encryption products—those widely available in the market—were to be exportable to any end-user including foreign governments (except for the seven governments regarded by the United States as supporters of terrorism, namely Cuba, Iran, Iraq, Libya, North Korea, Sudan, and Syria). The Commerce Department's announcement treated the 1998 Wassenaar encryption accord as a sort of lesser-included element of the new, liberalized U.S. regulations, but it is hard to view that accord as anything more than a futile gesture quickly overtaken by the march of technology and the pressure of the marketplace.[53]

Under Wassenaar, members are supposed to share information on arms and technology transfers. The exchanges are confidential to the member governments and thus are designed more to build cooperative relations among the supplier-members than among countries in the regions to which members send listed technologies. The semi-annual exchanges on military transfers to nonparticipating states are evidently much less detailed than the associated munitions list would suggest, being confined to seven general categories (tanks, armored combat vehicles, artillery, combat aircraft, attack helicopters, warships, and missiles and missile launchers).[54] Thirty members of the regime participated in the first exchange, in September 1996; Russia, Ukraine, Slovakia, and Bulgaria had not passed national enabling legislation in time to participate in that round.[55] A "small group" comprising the largest arms suppliers (France, Germany, Italy, Russia, the UK, and the United States) was supposed to create more detailed guidelines for arms-related data exchange, but progress was slow.[56] At the 1997 plenary, France and Russia blocked a proposal by the United States and Britain to expand the munitions list to include categories for transport helicopters and aerial refueling vehicles.[57]

The communiqué of the December 1998 plenary noted that participants were exchanging increasing amounts of information confidentially and by promoting such transparency were advancing the arrangement's aim of preventing (undefined) "destabilising accumulations." Participating states reaffirmed that they would exercise "maximum restraint" in making transfers "in particular to regions in conflict."[58] Hours after the 1998 plenary closed, however, came public notice of a $150 million Russian sale of combat equipment to Ethiopia, then still observing a cease-fire in its border war with Eritrea. Bulgaria sold 140 tanks to Ethiopia in 1998, and 80 more to Ethiopia and Uganda, which was suspected of acting as a pass-through for equipment destined for rebel forces fighting in Sudan and the Democratic Republic of Congo. The former was mired in a long-running civil war, while the latter erupted in conflict in August 1998, with fighting that involved the forces of a half-dozen African states, arrayed both for and against the government. In February 1999, active fighting between Ethiopia and Eritrea resumed and both sides claimed to have caused (while denying that they suffered) thousands of casualties.[59] Wassenaar's voluntary provisions, in short, seem weaker than some participants' economic need.

In addition to information on arms deliveries, the semi-annual data exchanges are to include license denials for all tiers of dual use items. License *approvals* for sensitive and very sensitive items are also to be notified semi-annually, but *denials* of such items are to be notified "preferably within 30 days but no later than 60 days of the date of denial."[60] Reporting on denials is designed to alert other members of the regime (a) to requests for sensitive technology from targeted states; and (b) to actions taken to limit such access. Ideally, what one member denies, no other member will provide (otherwise, making and posting a denial would merely give other members a hot sales tip). Reporting on dual use transfers has been less forthcoming than on conventional armaments, perhaps because states are already under an implied obligation to make highly aggregated reports on their arms transfers to the UN (see chapter 5). In the event, only half of the regime's members reported dual use transfers in the first data exchange.[61]

The key to Wassenaar's functionality is consensus among its members as to who the target states and/or regions are. Among Washington's designated pariahs, Iran is the one that most strains the consensus. U.S. policy and legislation seek a much tighter trade embargo on that country, for emotional as well as rational political reasons, than other states are willing to support. Because the arrangement is voluntary, its members have to see some benefit to compliance. Where benefits do not outweigh the costs of trade foregone, compliance will be difficult to sustain, as the Ethiopia-Eritrea war demonstrates.

International Arms Embargoes

A distinctly recipient-based control regime, an arms embargo tries to terminate flows of armaments to a specific country or countries. It may be the only sanction imposed, or it may be one of a package of broader economic and trade sanctions, as in the case of United Nations sanctions against the white-minority regime of Southern Rhodesia starting in 1965, or against the apartheid regime of South Africa, starting in 1977. UN arms embargoes have also been instigated in the 1990s against Angola, Liberia, Rwanda, Somalia, and the former Yugoslavia in response to those countries' civil wars. Trade sanctions incorporating arms embargoes have been imposed by the UN on Libya (for harboring terrorists) and Iraq (for its aggression against Kuwait in 1990–1991, and to encourage continuing compliance with UN-mandated arms control measures).[62]

Arms embargoes require the close cooperation not just of arms supplier countries but also of the target's near neighbors. Southern Rhodesia continued to receive arms from South Africa, which in turn managed to build up an indigenous arms production capability while under international embargo. The more recent sanctions regimes applied to sub-Saharan African countries were instituted only after conflict had raged for some time and already taken most of its victims. In most instances, these countries' factions were already armed to the teeth owing either to substantial Cold War shipments of arms or to porous borders and a lively illicit trade in easily portable light weapons.

The Bosnian case highlights the bluntness of embargoes as instruments of policy. The UN's embargo, imposed on all of the former Yugoslavia, allowed neither military nor moral distinctions to be made among the factions fighting within its former province of Bosnia-Herzegovina. Aggressors and victims alike were to be denied arms. In such cases, the tighter the embargo, the greater the advantage to the aggressor. Conflict may in fact end faster for being more lopsided, but at high moral cost.[63]

As a result, the Bosnian embargo was selectively adhered to by countries sympathetic to one or another of the local belligerent factions. Indeed, in November 1994 President Clinton, acting under direction from the U.S. Congress, declared that the United States would no longer enforce the embargo with respect to Bosnia, whose beleaguered Muslim faction was viewed as a victim of aggression primarily by neighboring Serbia and Serbs resident in Bosnia.[64]

The 1991 UN embargo on arms to Iraq may be the best-supported internationally, but by the end of the decade it, too, showed serious strains. Both France and Russia were pressing to ease sanctions, the former to get back into the oil business with Baghdad and the latter to recoup the equivalent of several billion dollars owed Moscow for arms supplied to Iraq during its war with Iran. In early 1998, a crisis brewed over Iraqi refusal to allow United Nations weapon inspectors charged with dismantling Iraq's WMD programs to inspect its "presidential sites." France ruled out cooperating with planned U.S. and British air strikes. A last-chance diplomatic mission led by the UN's Secretary-General Kofi Annan produced a new accord that satisfied Washington temporarily and allowed UN arms inspections to resume, but Iraq continued to obstruct the inspectors' work, essentially blocking new inspections from August onward. In December, Iraq shut down inspections, the inspectors withdrew, and U.S. and British fighter-bombers struck a wide variety of Iraqi military facilities.[65] Iraq then moved to the international back burner, along with Africa's wars, as first the U.S. impeachment debates and then the crisis in Kosovo absorbed U.S. and other NATO members' attention and resources.

Wherever one or more major powers have had clear political stakes in the outcome of a conflict (if only to reinforce a stalemate), rapid arms transfers rather than arms embargoes have been the preferred policy option. Thus, both the United States and the Soviet Union supplied a high volume of arms to the belligerents in the October 1973 Middle East War. The United States orchestrated a selective quarantine of Western arms to Iran in the 1980s while it provided non-weapon support for Iraqi war efforts, and other suppliers showered Baghdad with arms.

In Angola, a UN embargo had little bite because the targeted Union for the Total Independence of Angola (UNITA) had access to diamond mines, aircraft, airstrips, and willing buyers in other countries, plus willing sellers of black market arms. As government forces were bearing down on UNITA in late 1994, a new peace accord pressed by the international community saved it from defeat; in 1998, a recalcitrant UNITA remained the principal obstacle to peace in Angola, the UN withdrew the last of its peacekeepers, and the country returned to war.

In short, arms embargoes may be of some use when applied to interstate conflict situations but will be respected only in those rare instances where the conflict is trouble enough to attract international attention but not so as to engage the major powers on competing sides of the war. Embargoes may be of more use, as in the case of Iraq or Libya in the 1990s, as international punishment for past transgressions. But as the impatience of French and Russian officials indicates, the consensus favoring a postconflict embargo may be difficult to sustain as economic and other pressures mount for "parole."

In internal conflict, international arms embargoes have rarely been either militarily effective or morally sustainable. In almost no case have they been applied sufficiently early to deny the means of war to a potential aggressor faction. The dilemma of preventive action applies: without evidence of aggression, it is difficult to generate support for an embargo; with such evidence, it is difficult to design one that does any immediate good.

Seeking a Code of Conduct for Arms Transfers

In a quiet parallel to the international campaign to ban land mines that has been spearheaded by nonprofit, non-governmental organizations (NGOs), a number of such groups in the United States and Europe recently have been pressing governments on the notion of a "code of conduct" for arms recipients. Such a code would add explicit considerations of political morality and human security to the more usual economic and power-political criteria that now guide arms sales. Specifically, as proposed by Nobel Peace Prize winner and former president of Costa Rica, Oscar Arias, and as championed by a large number of non-governmental organizations, such a code of conduct would prohibit transfers of "weapons, munitions, sub-components, and delivery systems," "sensitive military and dual-use technologies," or "military and security training" unless a recipient country meets all of the following criteria:

- compliance with international human rights standards (including those banning genocide, torture, or extra-legal/arbitrary execution);
- compliance with international humanitarian law (including the Geneva Conventions of 1949 and the Protocols of 1977; granting access to non-governmental humanitarian organizations in time of emergency; and cooperation with international tribunals);
- respect for democratic rights (signified by free and fairly contested elections with secret ballots, freedom of expression, and civilian control of the military);
- respect for international arms embargoes and military sanctions decreed by the UN Security Council or regional organizations to which a country belongs;
- participation in the UN register of conventional arms;
- commitment to promote regional peace, security, and stability (no arms to countries at war, unless recognized as in compliance with the self-defense principles of UN Charter Article 51, or functioning in a UN-mandated operation; no arms beyond those needed for legitimate self-defense; no new technologies; recognition of other states' borders; no advocacy of national, racial, or religious hatred);
- opposition to terrorism (signified by ratification of the relevant international conventions and cooperation in apprehending/extraditing terrorist suspects); and
- promotion of human development (signified by governmental health and education expenditures greater than military expenditures, unless justified by "exceptional needs").[66]

Political support for such a code has seemed to grow over time. As of mid-1995, bills embodying language very similar to the international code had attracted a number of co-sponsors in the U.S. Congress but failed to reach the floor of either chamber. In 1997, the code passed the House of Representatives without a dissenting vote as an amendment to the fiscal year (FY) 1998 State Department budget authorization bill. Legislation containing the code of conduct also was introduced in the Senate.[67]

The draft code of conduct required that the president annually certify those countries meeting the code's stringent political and behavioral requirements. Only certified countries would be eligible to receive U.S. arms or military aid. Like most such legislation, it contained a provision for presidential exemptions and for emergency situations. The Congress could pass a law nullifying a presidential exemption, much as it can currently deny a proposed sale under the Arms Export Control Act. Historically, the Congress has rarely exercised that power.[68]

Nine months after passing the House, the code of conduct language was deleted from the FY 1998 State Department bill by a House-Senate conference committee. In the 106th Congress, somewhat modified language still acceptable to code of conduct supporters was introduced into the FY 2000 State Department Authorization by the House International Relations Committee and passed the Congress in November 1999. The bill as passed established criteria by which states' democratic credentials and human rights performance must be evaluated, including promotion of democracy, respect for human rights, non-aggression, and participation in the UN conventional arms register; required that the president certify annually which states do and do not meet those criteria as part of the annual foreign assistance budget request; prohibited arms sales to states that do not meet them, unless the president grants an exemption on national security grounds; and required that the United States open international negotiations on an arms transfer restraint regime.[69] The exemption provision would help to avoid embarrassing moments with major U.S. arms recipients that remain allergic to democracy, such as Saudi Arabia, or whose human rights record is guarded at best, such as Turkey.

Across the Atlantic, the new Labour government in Britain, chairing the European Union (EU), tabled a European Code of Conduct with the support of the government of France. The EU code, as adopted in May 1998, is more diffident with respect to adherents' obligations than is the Arias code, above, and its preamble acknowledges "the wish of EU Member States to maintain a defence industry as part of their industrial base."[70] Export licenses are to be refused, however, if they would violate,

- international law (international arms embargoes, treaties);
- current voluntary restraint regimes like the MTCR; or
- EU member states' "commitment not to export any form of anti-personnel landmine."

Licenses are also to be refused where "there is a clear risk that the proposed export might be used for internal repression," including major violations of human rights, or where exports "would provoke or prolong armed conflicts or aggravate existing tensions or conflicts in the country of final destination." Since we do not have ready means beyond subjective judgment for measuring the impact of transfers on conflict, this last criterion might or might not restrict arms transfers to areas of conflict. Arms transfers dispatched to help the recipient win more quickly might even be encouraged.

In making arms transfer decisions, EU members are also admonished to "take into account" such things as territorial claims that a would-be recipient has pursued in the past by force; the buyer's record in support of terrorism or international organized crime; its commitment to non-proliferation of WMD; and whether the proposed export would "seriously hamper the sustainable development" of the recipient. Annual national reviews of code implementation are to be shared confidentially among ministers, but not with parliaments or public, limiting accountability.[71]

The campaigns to implement a code of conduct for arms sales in both the United States and the EU have had to accept political compromises that reflect their respective political realities. In the United States, were the Arias code to become law or policy unilaterally, it would face the same sort of buffeting that President Carter's unilateral restraint policy faced in the late 1970s. Cold War pressures are gone, to be sure, but competitive pressures and national interests in access to petroleum remain strong. A number of prominent U.S. arms recipients probably would continue to qualify for arms under a stringently interpreted code. Indeed, Israel, South Korea, Argentina, and Chile are all rated as "free" countries by the widely respected Freedom House surveys.

But just 12 other developing countries, of the 106 whose data form the basis for the analytical work in this book, are rated "free." Taiwan, Thailand, Malaysia, Singapore, and Kuwait are considered "partly free," while Egypt, Saudi Arabia, and the United Arab Emirates are "not free."[72] Thus the president either would have to make sufficient exemptions to the code to undercut it severely, or the United States would have to end arms transfers to perhaps two-thirds of its current major clients.

Such terminations are what many code advocates would prefer. Yet, while denying all arms to states with mixed records on democracy and human rights might be normatively satisfying, it might also leave those states open to intimidation by better-armed neighbors, especially ones that build their own military equipment, as in the case of Southeast Asian states and China. If restraining sales to "partly free" Southeast Asian states were to encourage China to pursue its territorial claims in the South China Sea, for example, supplier restraint itself would have been destabilizing to that security complex.

Even if the United States and Europe both were to adopt a strict code of conduct, other suppliers such as China and Russia are unlikely to do so. Current Chinese leaders care little for Western notions of human rights and even view them as a direct threat to continued Communist Party dominance, if not to China's territorial integrity. Acknowledging the rights of Tibetans, for example, would imply some acknowledgment of their claimed rights to political autonomy or independence.

Russia, on the other hand, is hungry for foreign exchange. Russian arms sale decisions are unlikely to give much weight to human rights principles unless there are financially tangible quid pro quos, such as Western bank loans, investments, or debt forgiveness, involved. Western states and international financial institutions are unlikely to pay Moscow to forego arms sales, however.

Over time, both China and Russia may come to see the wisdom of a restraint regime that sells arms only to countries that treat their people well and do not threaten neighbors. Until then, however, sellers that scrupulously support a code of conduct would be leaving large segments of the global market open to suppliers untrammeled by such considerations. Buyers excluded from deals with some suppliers by the code would lose access to some cutting edge weapons, but they might partially bridge the gap with smuggled high-tech subcomponents with which to upgrade their acquisitions. The history of smuggled technology in support of Iraq's WMD program, among others, suggests that technology export controls would have to be both stringent and pervasive to prevent substantial black market end runs around official, code-filtered technology transfer channels.[73]

The code of conduct for arms transfers is one example of a broader development strategy known as "conditionality," the delivery of assistance or services providing certain conditions are met by the recipient. Specifically, the code is an example of "allocative conditionality," where donors "announce general criteria that recipients must meet to be eligible for aid at all, or to qualify for high levels of aid: for instance, minimum standards of human rights protection, or progress toward better governance."[74] The record of performance for such conditionality in economic aid packages is mixed.[75]

Compelling countries to change their policies or behavior, as many applications of conditionality try to do, is much more difficult than trying to deter unwanted behavior. Withholding arms sales to punish bad governments would be a form of compellence. Such termination of arms sales to bad actors might be sufficient achievement for some code of conduct advocates. It would be better, however, if the would-be recipient changed its policies or behavior in an effort to qualify for the desired arms—that is, if it became more democratic and did a better job on human rights.

This points to two implications of the code of conduct. Sales termination alone, without recipient state reform, would not do anything directly for the local people whose interests the code is supposed to serve. Moreover, if arms roughly equivalent to those denied by code adherents remain available from other suppliers, then adherence to the code of conduct may just divert market share to non-adherents and conceivably leave the recipient government's citizens *worse* off than before the code was applied. If, on the other hand, a government did undertake reforms in an effort to qualify for arms, code supporters would find themselves in the awkward position of offering guns, in effect, as rewards for humane behavior and progressive governance.

Despite such shortcomings, however, it is hard to make the reverse argument that, because there are shortcomings in the code of conduct logic, the United States and other industrial democracies *should* sell arms to countries with abysmal human rights records or *should* sell arms to countries that bully their neighbors or violate UN arms embargoes. Still, the political morality embodied in the code comes up hard against the political realities that U.S. foreign policy must deal with on a daily basis while furthering the country's material interests. These cross-pressures may help account for the oscillating support given the code in Congress: widespread support in the House in 1997, followed by a quiet demise in a closed-door committee in 1998, and then partial resurrection and codification in 1999.

Common Threads:
Non-reciprocity and Non-participation

Supply-side efforts to constrain the proliferation of conventional weapons and related technologies must contend not only with the forces of supplier and recipient national interest and the forces of market economics but also, as noted at the start of this chapter, with the fact that all supply-side control efforts to date involve a basic lack of reciprocity between buyers and sellers. Unlike arms control accords between the United States and Russia, or regional agreements like CFE, or global treaties like those banning biological or chemical weapons, which imposed equal obligations on all signatories, where conventional arms transfers are concerned, "developed countries seek restrictions on developing countries that they do not honor themselves."[76]

No consideration was ever given to reciprocity in the 1978 CAT talks or to participation of recipient states in those talks, even though CAT involved the strategic weapons of the developing world. States of the Middle East were not party to the Perm Five talks to restrain transfers to their region. The MTCR does not require that supplier-participants destroy or otherwise constrain their own inventories of ballistic or cruise missiles, nor does it offer any "incentives for compliance to those it is directed against."[77] Whether motivated by *realpolitik* (like MTCR) or by normative concerns (like the code of conduct), supplier-originated, supply-side arms transfer controls typically establish two or more tiers of status. To the developing countries who are usually the targets of such restraint measures, such policies may smack of paternalism at its worst.

The history of international relations outside the arms control arena is replete with examples in which states or groups of states have exerted conforming pressure on other states, usually when the latter have failed to follow accepted international law or practice—committing aggression, sponsoring terrorism, or violating trade agreements, for example. That pressure can take the form of diplomacy, tariffs, sanctions, military brinkmanship, or war. But by and large those wielding the pressure are bound by the same set of rules as those being pressured and are making an effort to bring the outlier(s) back into conformity with prevailing rules and norms. Here again, supply-side arms transfer restraint measures differ.

Brad Roberts at the Institute for Defense Analysis has summed up the critique of supply-side "strategies of denial," noting that export controls offer diminishing returns:

> By seeking to enshrine a permanent division between the seekers and the possessors of advanced technologies, export controls undermine the political foundations of regimes that many states in the developing world have joined not because of their security benefits but because of the increased access they would bring to specific technologies with economic benefits. . . . The inherently discriminatory character of nonproliferation regimes can actually be a stimulus to proliferation, because it suggests a desire among the "haves" to keep the "have nots" from getting something to which they have just as much right as any state.[78]

Observations and Conclusions

The arms market is a competitive one, the more so as it has shrunk from its peak values in the 1980s. The United States has an advantage in that market, having a reputation for both the most advanced military technology and a good record of after-sale support. With its substantial market share, the United States might try to set an example of restraint in arms transfers similar to that attempted by the Carter administration in the 1970s, when circumstances were less propitious, but in all likelihood arms buyers would just turn to other sellers for substitute goods and services. The U.S. government recognizes this fact and thus pursues what restraint efforts it does in multilateral fora, which are voluntary until and unless formalized by treaty. Neither current forum, the MTCR or Wassenaar, has been so formalized. Both have members or lurkers that violate these regimes while promising (informally) to abide by them. Unless regime members are willing to impose some sort of cost for such violations, or provide some sort of direct benefit for compliance, the violations will likely continue.

While supplier defection is clearly a problem with supply-side restraint regimes, recipient support is almost by definition to be counted out. Unique among arms control regimes, supply-side arms transfer regimes attempt to apply restraints not only to states that have agreed to them but also to would-be arms recipients that have *not* agreed. Arms suppliers would strenuously object were a different group to impose such an arrangement on them. (In fact, it did and they did. The group was the Organization of Arab Petroleum Exporting Countries and the regime was the Arab oil embargo. The industrialized states objected, but not enough to stop doing business with Arab oil producers when the embargo ended. Energy addiction is a powerful motivator.) The lack of reciprocity in supply-side restraint means that those who wish to implement restraint can count on no cooperation from those to whom restraints are applied and can anticipate concerted efforts to circumvent the regime in very creative ways. In short, any state, group of states, or international organization that wishes to impose supply-side restraints must be prepared to enforce them vigorously for the duration of the regime.

On the other hand, arms transfer limitations can be valuable adjuncts to political reconciliations and a useful tool for expressing the displeasure of the international community with a dangerous member, provided once again that the members of the community are willing to work at enforcing the sanctions imposed. Like all sanction regimes, arms embargoes and similar tools entail costs to more than just the target(s) of sanctions. There is no standing international mechanism to compensate third parties for trade losses suffered while sanctions are in force, and lost arms sales might well be last on many people's list of goods and services for which compensation might be considered. Lacking such compensation, however, states that are losing arms sales revenues will be tempted to circumvent sanctions. The longer the regime stays in place, the greater the temptation, as the Iraqi case suggests.

Realpolitik reasons for arms transfer restraint are most germane to countries that are likely to undertake international military interventions at some distance from home. That is, they are of most interest to the United States. Eventually they may be of interest to European states, but not for some time, if at all. Russia is unlikely to be interested or able to send substantial forces beyond its immediate region for long while, and neither is China. So the world's only superpower, the king of the hill, the number-one salesman of lethal goods and services also has the most to gain from arms transfer restraint. One suspects, however, that such potential gains, which are prospective and still abstract, do not stack up particularly well in policy circles when compared to the tangible near-term benefits in revenues and jobs gained from sales of weapons abroad.

As to normative criteria for restraining arms sales, a coherent restraint regime based on a concept like the code of conduct would be more straightforward to craft than one based on more slippery, difficult to define, terms such as "destabilizing" or "excessive" levels of armaments. Such terms are much more useful as post-facto rationalizations for denying arms to countries one does not like than as criteria for making objective decisions about what is reasonable for other governments. The tenets of the code of conduct are explicitly political, as are the tenets of *realpolitik*. Unlike *realpolitik*, the message of the code is not, "I refuse to sell you weapons now because I might want to flatten you later," but there is a built-in implication that only democratic regimes have a right to defend themselves and thus, in practical terms, a right to remain sovereign. If one believes that sovereignty really does arise from the consent of the governed, then declining to sustain, with lethal armaments, governments that are based otherwise is at least politically consistent. The code does implicitly offer weapons in exchange for good governance, and that implied reciprocity may be unintentional, but compared to the other supply-side measures it seems a long step in the right direction.

CHAPTER 5

Regional Measures to Reduce Demand for Armaments

I n the international arms market, supply meets demand, which arises both from institutional interests and from the welter of security concerns, internal and external, real and imagined, reviewed in chapter 3. Trying to curb this market from the supply-side alone is not likely to be effective, yet altering demand for arms is not easy, nor can it be expected to happen overnight. Where there is at least a modicum of cooperation in two or more states' relations, various tools and techniques known collectively as "confidence- and security-building measures" (CSBMs) may ease the demand for new armaments, reduce the likelihood of conflict, and complement arms control agreements.

CSBMs are likely to share the credit for improvements in regional security relations with other policy dynamics that help to transform hostility into peace. High-level, systemic changes like the end of the Cold War can have positive ripple effects within regional security complexes. Third-party mediation or direct negotiation may help to settle outstanding territorial disputes. Changing levels of military or, in some states, religious influence, or the death or displacement of aggression-minded (or peace-minded) leaders may alter the domestic politics and foreign policies of a state. Greater economic openness and growth may reduce internal tensions that might otherwise find or be directed toward external targets. Open, legitimate government—a combination of institutional democracy and respect for the rights of individuals and minority groups—may also promote national and regional stability. Where such variables appear to be important components of mitigating the demand for arms and the prospect of conflict in a regional security complex, their roles will also be highlighted here.

In general, however, in this chapter we will be emphasizing the roles that CSBMs have played and can play in regional security. The first part of the chapter lays out competing definitions of CSBMs and what may be preconditions for their success. It is followed by a segment that assesses the United Nations (UN) Register of Conventional Arms, a kind of global CSBM for arms transfers inaugurated in the wake of the 1991 Gulf War. The chapter then evaluates measures specific to individual regional security complexes, starting with Europe, where the CSBM concept originated in the 1970s.

Confidence-building measures can make a useful contribution to conflict prevention where interstate relations are uneasy but fall short of open hostility, and can play a useful post-conflict role—in the form of peacekeeping and related measures that help to rebuild order and trust, not only between states but also between groups within states.

Official CSBMs can be usefully supplemented by the work of non-governmental (civil society) organizations. Official and unofficial measures alike work better where hostility is confined to the level of governments—to ruling elites and issues of relative power—rather than diffused into civil society and tied to fundamental issues of belief or identity. The effects of CSBM programs on levels of demand for conventional armaments or on levels, likelihood, or frequency of regional conflict is not readily measurable in aggregate terms because the combinations of variables that influence governmental decision-making are different in each security complex and even within the various states of each complex. This variability and the size of the conceptual tent that holds CSBMs make theories about these measures and their effects on arms imports and conflict difficult to formulate. In general terms, political change is a usually necessary precursor to major movement in either CSBMs or conventional arms control. Yet the process of building regional peace and security is sufficiently iterative that modest CSBMs can be instrumental in nudging the process forward at relatively low initial risk to the participants. In some circumstances they can keep a touchy relationship from decaying unnecessarily due to tactical actions or errors that may otherwise snowball.

Setting the Scene for CSBMs

Among analysts who write about confidence-building measures, there are several functional definitions for CSBMs. Some authors limit their definitions to measures with military security functions, while others include measures with basically political or cultural impact. Objectives common to all, however, are avoidance of misunderstanding, avoidance of conflict, voluntarism, and reciprocity.

Definitions of Terms

Ted Greenwood at the Sloan Foundation focuses on military security measures. Included in his definition of CSBMs are *mutual security pledges* (such as recognition of de facto international borders or agreement that avoiding war should take precedence over efforts to change the political status quo); *transparency measures* (such as military-to-military contacts or direct inspection of military capabilities); and *measures to cope with crises or accidents* (such as hotlines or regional risk reduction centers).[1] American University's Jack Child suggests that effective CSBMs should promote *military transparency and openness,* minimizing the possibility of deceit with respect to military capabilities or intentions. They should also promote *predictability and reliability;* aggression should be quickly evident to other participants and warning sufficient to permit effective defense. They should entail *reciprocity and balance,* with matching commitments on all sides. Finally, they should promote *adequate communications,* not just technical channels for messages but readiness to receive them.[2]

Other authors mix military and political undertakings. The Stimson Center's Michael Krepon describes an incremental, three-stage process that starts with *conflict avoidance* (using high-level communications links or measures to provide early indications and warning of hostile intent) until core grievances can be addressed by diplomacy. In stage two, the parties involved move on to *confidence building* (buffer zones along borders, foreign observers at military exercises) once core grievances have been resolved or at least finessed. In stage three, they turn to what he dubs *peace strengthen-*

ing measures (using such things as constraints on the size or location of military exercises, or intrusive military inspections).[3]

Marie-France Desjardins at London's International Institute of Strategic Studies divides the universe of CSBMs into those intended to enhance military security and those intended to enhance political cooperation, noting that any one set of measures is unlikely to achieve both ends. The security emphasis, she argues, reflects the original European notion of CSBMs and includes such things as *exchanges of military observers* for exercises, *military rules of the road* (for example, various "avoidance of incidents at sea" agreements), *on-site inspections* of military capabilities, and *restraints on military operations or readiness* to make surprise attack difficult. To be effective, such measures require strict implementation and verification of compliance. If implementation is lax or compliance cannot be verified, she cautions, such measures can even prove politically counterproductive. Moreover, they can be the focus of "exaggerated hopes" and wishful thinking on the part of their advocates, and the frequent assumption that CSBMs are easier and less costly to negotiate than arms control may undermine prospects for the latter sort of accord while not necessarily replacing it with a better product.[4]

Desjardins notes that CSBMs designed to *enhance political cooperation* arose outside Europe, and emphasizes confidence-building as process, usually a personalized process of high-level political communication and interaction, rather than the creation of products like a hotline or inspection scheme. Regular leadership interaction is intended to make regional political behavior more predictable, and the hope is that the mood of cooperation thus created will spill over into regional security policies and practices.[5] But in emphasizing personalities over institutions, this approach requires stable, long-term political leadership and, indeed, a kind of clubby, paternalistic model of governance in which there are no deep grievances between members of the club to get in the way of the process.

I include traditional UN peacekeeping operations, a kind of *post-conflict reassurance* measure, within the definition of CSBMs. Previously, I have noted that the presence of neutral peacekeepers

> provides a mechanism for limiting the consequences of disputes that inevitably arise when implementing a cease-fire or political settlement between groups who remain basically hostile to, or mistrustful of, one another. International peacekeepers can verify, for example, that one side's willingness to lay down its arms is reciprocated by the other. They can monitor the repatriation of dispossessed populations and property, and provide the public security and administrative coherence necessary to restore the basic amenities of life. They can verify that election processes are free from bias and voter intimidation, and that election outcomes are valid reflections of votes cast. . . . Peacekeeping is [therefore] a confidence-building measure, providing a means for nations or factions who are tired of war, but wary of one another, to live in relative peace and eventual comity.[6]

Peacekeepers can also be deployed in advance of conflict, to monitor borders where "tensions arise from mutual mistrust." Like post-conflict peacekeepers, they provide "military confidence-building, but without the usual intervening war." However, where such preventive deployments are designed to deter military action—that is, in situations in which potential threats flow only one way—they are reassuring only to the worried party, much like any other military force might be and, not being symmetric in their effect or desired by both parties, fall outside the realm of CSBMs. That is also the case for peacekeeping's more robust cousins, humanitarian intervention and peace enforcement, which apply military force to coerce desired behaviors from local parties, albeit in the interests of creating peace.[7]

Preconditions for CSBMs: Evidence from Europe

Cathleen Fisher, at the Henry Stimson Center, looked at political-cultural preconditions for CSBMs based on Europe's first-out-of-the-gate experience. Among the factors facilitating their development she noted centuries of *shared history, cultural affinities, and religious ties* that together bridged, to some extent, the ideological divide between the region's Cold War blocs. She also noted the existence of *strong states* with stable governmental systems; *joint experience with multilateral security institutions* such as NATO and the Warsaw Pact, which developed separately but in rough parallel to one another; *strong public and elite support* for CSBMs despite deteriorating top-level political relations; and a *desire to minimize the risk of nuclear war* through miscalculation or accident. European CSBMs were also *"embedded in a broader strategy* of East-West economic cooperation and political dialogue."[8] To these elements might be added the fact that European CSBMs involved a relatively large number of middle-sized powers, plus two superpowers who led the respective blocs and tended roughly to balance one another out.

Moreover, the European mix included some neutral and non-aligned states that would be caught in any continental nuclear holocaust but that had no direct leverage over the combatants' force size or doctrines. These states had incentives to devise some form of tension-reduction that would not undercut either competitor's sense of military security. In contrast to NATO, where deterrence concepts dominated, the neutral and nonaligned states of Europe had incentives to search for measures of reassurance—things that could be done to reinforce the joint perception that Armageddon would not occur today or tomorrow.

These preconditions are difficult to find in regions outside Europe. Elsewhere, religion may be a fault line, as it once was in central Europe, rather than a common referent. States may be weak and unstable and civilian control of the military ineffectual. Experience with regional institutions may be limited. There may be only one dominant regional power. Nuclear weapons, if present, may inflame national passions rather than induce caution. These differences have not prevented CSBM regimes or peacekeeping operations from being implemented in a number of other regional security complexes, but do suggest that such regimes will need particular care and feeding and may be the victims of unreasonable expectations. In the case of peacekeeping, and especially those operations tasked with implementing a peace accord in the aftermath of civil war, the key preconditions for success are the willingness of local faction leaders to accept compromise political objectives in the interest of peace, and the active support of neighboring states for the peace implementation process.[9]

Finally, efforts to enhance communications and interactions among states in a region sometimes benefit from being started in non-governmental settings that operate in parallel with or set the stage for official activities. Controversial issues can be discussed and controversial participants engaged in such discussions without committing governments to anything. These so-called track II activities have been especially prominent in east Asian explorations of CSBMs.

Global Transparency:
The UN Register of Conventional Arms

If part of the problem in maintaining national security is the sheer difficulty of knowing what the neighbors are doing and planning and how strong their forces are, then presumably better mutual knowledge would at least lead to better policy by weakening the grip of the security dilemma. Whether it leads to better relations depends on what the neighbors are really doing. If all states in a security complex favor the status quo, mutual transparency in defense matters should allow each to feel more secure at lower levels of armaments and expenditures. This is the logic behind recent efforts to increase the transparency of the trade in major weapon systems by means of a voluntary central data register maintained at the United Nations. This section reviews the register's origins, objectives, and first few years in operation. It then turns to an assessment of the register's relative utility, and problems that arise when one or more states in a security complex are dissatisfied, and do not, in fact, support the status quo.

Origins and Five-Year Development

The concept of a UN arms register dates back to the 1960s, when arms transfers were bound up in the competitive secrecy of the Cold War. Near its end, in December 1988, the General Assembly asked the secretary-general to study ways of promoting transparency in armaments, and the resulting Expert Study Group submitted its report a few months after the end of the Gulf War, when interest in arms trade restraint was at its height. On December 9, 1991, with 150 votes in favor, the General Assembly passed resolution 46/36L, establishing a voluntary armaments register. Member states were invited to submit annual reports on their imports and exports of arms in seven categories (battle tanks, armored combat vehicles, large-caliber artillery systems, combat aircraft, attack helicopters, warships, and "missiles and missile launchers"; see table 5.1). Background reports on current military holdings and national arms production were also invited. Expert panels continued to meet to discuss ways of improving the register in future years (more categories, greater detail, etc.).[10]

Initial replies from states for calendar year 1992 were due to the UN's Office of Disarmament Affairs by April 30, 1993. Of 185 UN member states, 83 submitted replies by the deadline, another six by mid-1993, and two more not until the end of 1994. Thirty-three of the original states reporting also provided background information on military holdings or national arms production.[11] For only about one-quarter of the total transfers reported did exporter and importer data agree on types and numbers of weapons changing hands; in two-thirds of the cases, only the exporter or the importer reported a particular transaction. Some of these mismatches occurred because only one side of the transaction submitted a report to the UN. In other instances, states' interpretations about what constituted an arms delivery, what date to give it, and what arms to include in the register accounted for the discrepancies. The "warships" category was particularly troublesome (just 11 percent supplier-recipient agreement) partly because of confusion about what sorts of ship transfers were to be reported. The "missiles or missile systems" category also proved problematic because it lumps together a weapon launcher and its ammunition, which is not done in any other category, and because it sets a reporting threshold range of 25 kilometers, which effectively excludes reporting transfers of antitank missiles and shorter-range air-to-air

Table 5.1 UN Register of Conventional Arms: Reporting Categories

Weapon Category	Description
Battle Tank	Tracked or wheeled self-propelled armored fighting vehicle with high cross-country mobility and a high level of self-protection, weighing at least 16.5 MT unladen weight, with a high muzzle velocity direct fire main gun of at least 75mm caliber.
Armored Combat Vehicle	Tracked or wheeled self-propelled vehicle, with armored protection and cross-country capability, either: (a) designed and equipped to transport a squad of four or more infantrymen, or (b) armed with an integral or organic weapon of at least 20mm caliber, or an antitank missile launcher.
Large Caliber Artillery System	A gun, howitzer, artillery piece combining the characteristics of a gun and a howitzer, mortar or multiple-launch rocket system, capable of engaging surface targets by delivering primarily indirect fire, with a caliber of 100mm and above.
Combat Aircraft	Fixed-wing or variable-geometry wing aircraft armed and equipped to engage targets by employing guided missiles, unguided rockets, bombs, guns, cannons, or other weapons of destruction.
Attack Helicoptors	Rotary-wing aircraft equipped to employ anti-armor, air-to-ground, or air-to-air guided weapons and equipped with an integrated fire control and aiming system for these weapons.
Warships	Vessel or submarine with a standard displacement of 850 MT or above, armed or equipped for military use.
Missiles and Missile Launchers	Guided rocket, ballistic or cruise missile capable of delivering a payload to a range of at least 25km, or a vehicle, apparatus or device designed or modified for launching such munitions.

Source: UN General Assembly, *Transparency in Armaments,* Resolution 46/36L, December 9, 1991, Annex.

and ground-to-air missile systems. The first-year level of agreement between exporters and importers in the missile-related category was just 13 percent.[12]

Virtually all states in Europe submitted data on imports and exports, whereas few states in sub-Saharan Africa or the Middle East did so. Major importers failing to file included Saudi Arabia, Iraq, Syria, Bangladesh, Kuwait, and Taiwan (which is not a UN member). Major exporters were better-represented. China, France, Germany, Italy, Russia, the United Kingdom, and the United States all filed reports. In their review of the register's first year, Edward Laurance and Herbert Wulf blamed regional instability for the failure of most Middle Eastern states to file reports with the UN.[13]

In the second year of the register, covering transfers made in 1993, just 34 states submitted data to the UN by the initial reporting deadline of April 30, 1994, but 54 additional states submitted data over the succeeding year. Of these 88 participants, 19 were new to the register in 1993, roughly matching 22 states that dropped. Only five of the top ten arms importers among developing states submitted reports to the register, but states accounting for 95 percent of arms exports did participate. Most transfers in 1993 were intra-European, dominated by the weapon "cascades" allowed under the Conventional Forces in Europe Treaty (see below). No Arab state submitted a report and only one in four states in sub-Saharan Africa did so.[14]

Gross discrepancies between exporters' and importers' numbers continued to plague the enterprise. Close students of the process tended to attribute these variances to idiosyncratic national interpretations of the register's requirements rather than to deliberate efforts to deceive.[15] If they are right, variances should diminish over time as countries get used to the reporting requirements. To date, the UN Secretariat has no mandate to query or correct data discrepancies before they are published.

For the third year of the register, covering transfers in 1994, 46 countries submitted replies by the initial deadline of May 31, 1995, and 85 had done so by the end of the year.[16] Of these, 48 were "nil" reports (no reportable imports or exports). Of the 201 reported arms transactions, 30 percent involved one non-participating country (usually the importer); 40 percent were acknowledged by just one side (usually the exporter) even though buyer and seller both submitted reports to the register; in 5 percent of the cases both sides reported the transaction but disagreed on quantity; and in 25 percent there was complete agreement on content and quantity.

The rate of complete agreement varied by exporting country. Nine in ten German transactions (mostly with other participating countries in Europe) were confirmed by the importers, for example, while rates of agreement for Poland and Moldova were zero, as they reported sales only to non-participating states.[17] Altogether, 20 countries named in exporters' reports did not file with the register. These included the entire Persian Gulf security complex and most Middle East countries, and a scattering in South Asia and West Africa.

The UN convened a 23-nation Group of Experts in 1994 to review both the operation of the register and several proposals for strengthening or expanding it. On all main issues, the only consensus among these nationally appointed experts was to leave the status quo unchanged. Ideas rejected included expanding the register to include national holdings and production of weapons, increasing the level of detail (so that readers might see not just numbers of tanks sold, for example, but types or models as well), giving the UN Secretariat a greater role in questioning member states about data discrepancies, and revising the weapons categories, including the roundly disliked missiles category.[18]

In August 1996, the UN Secretariat issued its report on register submissions for transfers made in 1995. By that time, 93 countries had submitted data, and three more submitted reports by July 1997. Of these, 44 were nil. The Arab League once again failed to participate. Two dozen states provided extra information on their equipment holdings and several added data on arms production as well. Most NATO members and states hoping to join NATO did this, as did other developed industrial democracies (Switzerland, New Zealand, Japan, and Austria). Among developing states, only Argentina, Brazil, Mexico, Armenia, and Azerbaijan submitted holdings data.[19]

As of mid-October 1997, 90 countries had sent reports to the register for transfers made or received in 1996, of which 40 were nil.[20] Countries that had reported military holdings previously did so once again and were joined by Luxembourg, Slovakia, Sweden, and the Former Yugoslav Republic of Macedonia. No new developing countries opted to report on holdings or production, and regional participation in the register continued to vary widely. Europe could claim nearly complete participation, but African participation continued to drift downward, with just eight states sending reports for 1996. Of Middle Eastern states, Israel submitted reports to the register, whereas the Arab League issued a "consolidated reply," without data, that criticized the register as neither "balanced nor comprehensive" because it failed to include "data on . . . high technology with military applications," holdings of weapons of mass destruction, or any criticism of Israel for "occupation of Arab territories." Israel noted its continuous participation in the register since 1993, the failure of most of its neighbors ever to participate, and its objections to any changes in register coverage until such participation increased.[21] In other words, the time-tested Middle East politics of stalemate were exported to the register.

Overall participation changed very little for transfers made in 1997. Ninety-two states submitted reports, with 12 that had not participated in 1996 replacing another 12 that dropped out or failed to report. Qatar broke ranks with the rest of the Arab League and sent in imports data for the first time, but otherwise the Arab boycott of the register continued. Iran, the only Persian Gulf state to submit reports to the register consistently, had not done so for this reporting year as of June 1999, but its previous reports were received by the UN as much as 18 months after the nominal due date. Patterns of agreement between exporters and importers were not much changed from earlier years: one-third of the transactions involved non-participating buyers; a little less than a quarter of the transactions were not reported by participating importers; 14 percent involved differences in reported quantities; and 29 percent showed complete agreement (see table 5.2 for details by supplier). Twenty-seven buyers named in exporters' reporting did not submit reports of their own.

The UN's Group of Governmental Experts on the UN Register convened once again in 1997 but could not reach consensus on formally expanding it to include military holdings or model-type data, or on something as seemingly simple and necessary as modifying the definitions of some weapons categories, even though the combined missiles-and-launchers category is roundly recognized to be essentially useless in its present form. The group did agree to recommend that a more prominent icon for the register appear on the UN's home page on the World Wide Web; two years later, however, that icon is nowhere to be found and researchers must know where to look for register-related data that the Secretariat posts on the Web.

Assessing the UN Register

After six years of reporting, what can be said about the value of the UN register and the politics of global transparency? On the plus side, the register may be doing that "UN thing" the organization has done in other controversial issue areas, such as human rights, establishing hortatory international standards of behavior well in advance of its member states' willingness actually to abide by those standards. The transparency standards are on the books, and the reporting requirements may exercise some subtle impact on state policies, provide a yardstick by which to judge states' behavior over time, and even contribute somewhat to reducing the sway of the security dilemma. The existence of the register may also encourage the adoption of its reporting practices and objectives regionally, as the Organization of American States did in 1997 (more below).

Table 5.2 UN Conventional Register: Table of Exporter/Importer Agreement

Supplier	Case Counts by Supplier				Percentage Distributions			
	NP	No Rpt	Differ	Agree	NP	No Rpt	Differ	Agree
Australia	1	4			20	80		
Belarus	4	1	1		67	17	17	
Belgium				1				100
Canada	2				100			
Czech Republic	1	1			50	50		
Finland				1				100
France	9	4	1	2	56	25	6	13
Germany	1			7	13			88
Greece				2				100
Israel	1	2	1		25	50	25	
Italy	1	3		1	20	60		20
Netherlands	3	1		?	50	17		33
Poland	1				100			
Romania	1				100			
Russia	5		1	6	42		8	50
Singapore				1				100
Slovakia	1	1			50	50		
South Africa	3	1	1		60	20	20	
Spain				2				100
Sweden		2		1		67		33
Switzerland				1				100
Turkey		1				100		
Ukraine	3	3		2	38	38		25
UK	5	1	2	7	33	7	13	47
USA	11	11	15	9	24	24	33	20
Case totals	53	36	22	45				
Percent of cases	34	23	14	29				

Notes: NP = Importer is non-participant in register; No Rpt = Importer did not report transfer; Differ = Differing importer and exporter reports; Agree = Importer and exporter reports agree. Source: "The 1998 UN Register of Conventional Arms," *Basic Reports* No. 67, November 27, 1998, pp. 4–11.

Countries importing substantial amounts of arms can easily refuse to submit reports to the register but at the risk of being exposed by their suppliers, who thus far have been much more complete in reporting their deliveries than recipients have been in reporting their acquisitions. This supplier willingness to provide reports may be due to the register's coverage of deliveries—that is, completed transactions rather than agreements, which exporters might be quite reluctant to report so as to protect their competitive position.

The UN does not require data on models of weapons delivered, which can indicate just how much firepower a country may be acquiring. Major suppliers other than Russia and Ukraine do tend to volunteer basic model data, however. The UN also encourages voluntary reports on national military equipment holdings, which may turn out to be a more important innovation than the reports on arms transfers themselves, but relatively few countries outside the European groups and Central Asia's newly independent states have as yet reported on holdings.

The stalemate in the register's experts group accurately reflected the wide political gulfs that separate Arabs and Israelis in particular and northern and southern governments generally. North and South tend to have antithetical policies regarding the register at any given time, although the two camps have essentially reversed positions on some issues since the register was first developed. The North was initially against expanding it to encompass national holdings and production, for example, and the South was for it, arguing that reports covering transfers alone were "discriminatory:" Such reports would not reflect true trends in armaments for such countries as Israel, apartheid South Africa, or the developed industrial countries, which made many of their own weapons. Over time, however, northern countries led by the rapidly disarming European Union began to press for more expansive transparency measures. southern countries felt a sudden draft in their sovereign prerogatives and cloaked themselves in the register's status quo. Given the seeming impossibility of negotiating changes in the reporting scheme that are satisfactory to all factions among UN members, it seems remarkable that this enterprise was launched at all. The momentum to that end generated by the Gulf War was brief, but it was real.

Military transparency is a desirable objective when one seeks to promote reductions in tension between basically pro-status-quo powers, and the register may contribute to an improved sense of security in some places simply by opening previously secretive transactions to semi-public semi-scrutiny. However, the delays of 12 to 18 months that are common between weapon delivery and telling the UN about that delivery may vitiate the register's potential contribution to regional stability.[22] It cannot offer much in the way of warning of dangerous regional arms buildups unless the builder has a five- to ten-year timeline for aggression in mind, plus a strategy of relying on imported weapons (rather than imported technology and local production), plus transparency minded suppliers. Moreover, the interests of a potential aggressor's neighbors may not be well served by military openness: the object of military intelligence is, after all, to divine the capabilities of potential adversaries. Transparency may even be dangerous when there are bad guys around who are willing to use violence to change the status quo. The Gulf Arabs, in other words, have reason beyond their confrontation with Israel not to be unduly open about their imports and inventories, so long as Saddam Hussein runs Iraq and zealots still have the upper hand in security matters in Iran.

Finally, although the official rhetoric that launched the register offered it up as a means of avoiding excessive and destabilizing regional arms buildups, there are no working UN benchmarks for either phenomenon, and such benchmarks are, as discussed in chapter 4, exceedingly difficult to define under the best of circumstances. Because of the UN's complex push me-pull you politics, the register's minders and its periodically convened experts are unlikely to devise acceptable benchmarks, so this objective of the register is unlikely ever to be realized. Thus, flawed and limited though it may be, the register is likely to retain its current form unless another Gulf War–like crisis shakes policy loose again.

Regional Confidence Building and Arms Control

While conflict-*preventive* CSBMs are thought of as less demanding than arms control measures, they are, as indicated earlier, not so simple, straightforward, and politically neutral as to be equally applicable anywhere, anytime. *Post*-conflict CSBMs like peacekeeping operations are not themselves conflict resolution measures, but by helping prevent the recrudescence of violence, they can facilitate such measures. Either form of CSBM might help to increase a conflict's "ripeness for resolution" by promoting a sense that political compromise is possible, survivable, and possibly beneficial.[23]

CSBMs and Arms Control in Europe

Conventional forces visibly symbolized the Cold War's East-West standoff in Europe and stood in complex relation to the nuclear forces stockpiled in quantity by both sides. Sometimes conventional forces were seen as a firewall against nuclear use, sometimes as a tripwire to ignite it, and sometimes as the mop-up units that would enter the wasteland once the fireballs and crushing pressure waves died away. The enormous size of Soviet conventional forces gave NATO every incentive to look for ways to limit those forces. The Soviet Union, in turn, wanted Western recognition of Europe's post–World War II borders and especially those of Poland, which had lost a substantial chunk of territory to the Soviets. Moscow in turn "gave" the Poles a chunk of Soviet-occupied Germany, which set the Oder and Neisse rivers as the new Polish-German border.

In the latter 1960s, Chancellor Willy Brandt of West Germany launched his policy of *ostpolitik* intended to normalize relations with Eastern Europe. The policy bore fruit in 1970, when West Germany signed treaties with the USSR and Poland recognizing "Oder-Neisse" and followed up with treaties that regularized relations with East Germany and improved access to Berlin, whose walled-off western half was a Western outpost surrounded by powerful Soviet field armies.[24] The rest of NATO also began to look harder at the possibility of detente, or an easing of tensions, with the Soviet-led Warsaw Pact.

By the early 1970s, U.S.-Soviet relations had begun to thaw, their bilateral nuclear arms control talks were making progress, and the time seemed right to open talks on European security issues. Moscow still wanted political talks and NATO still wanted to focus on force reductions, so two separate negotiations were launched in 1973. Although NATO viewed the force reduction talks as meatier and more relevant, in fact the political talks produced results sooner and ultimately undermined Soviet control in Eastern Europe and, indeed, the Soviet Union itself.

Conference on Security and Cooperation in Europe (CSCE)

Convened in Helsinki, Finland, in July 1973 with 33 European countries plus Canada and the United States participating, CSCE addressed three unofficial "baskets" of issues: security in Europe; economics, science, and the environment; and humanitarian cooperation. The Helsinki Final Act, signed in July 1975, provided for continuing discussion of these issues at regular intervals.

The security basket held ten principles of interstate relations, from the usual respect for sovereignty and inviolability of borders (the major Soviet goal), to respect for human rights and fundamental freedoms *within* states, which the USSR accepted only reluctantly. The security basket also stipulated a series of confidence building measures designed to reduce the risk of war via miscalculation. Each signatory pledged to provide 21 days' advance notice of military maneuvers involving more than 25,000 troops; volunteered to provide notice of other maneuvers and major military movements; and volunteered to invite the other side to send observers to its maneuvers.[25]

The humanitarian basket encouraged "freer movement of people, ideas, [and] . . . access to published and broadcast information, journalism, and various forms of cultural and educational cooperation."[26] Over the next 14 years, these principles helped to flood Eastern Europe with blue jeans, rock music, and a yearning for self-governance that accelerated the end of communist rule in the region. Far from cementing the status quo, as Soviet leaders and many U.S. critics of CSCE believed in 1975, the Helsinki principles proved to be subversive stimuli for change.[27]

Periodic conferences also strengthened the military CSBMs first devised at Helsinki. In 1986, the mandatory notification period for major military exercises was doubled to 42 days and the threshold number of troops was halved, to 13,000; foreign observers became mandatory for large exercises; members of CSCE had to provide an advance annual calendar of planned exercises requiring prenotification; and all military CSBM provisions could be verified by requests for on-site inspection that could not legally be refused. This was the first Soviet agreement to such inspections on its own territory. At the time, Mikhail Gorbachev had been general secretary of the Soviet Communist Party for about one year.

Four years later, Gorbachev had pledged to pull Soviet troops from Eastern Europe and CSCE gave itself a modest set of permanent institutions: a Council (of member state foreign ministers), a Secretariat (in Prague), a Conflict Prevention Center (in Vienna), and an Office for Free Elections (in Warsaw).[28] CSBMs were also strengthened to increase transparency of military forces, equipment, and budgets through annual exchanges of information that could be verified through "evaluation visits." Other new measures agreed to in 1990 and refined in 1992 increased military-to-military contacts, set up a CSCE communications network, and provided for annual implementation assessments while reducing the mandatory notification thresholds still further (to 9,000 troops). By the time these new measures went into effect, however, the political and military contexts in which they functioned were changing rapidly: the old central European military stalemate had ended, Germany was reunified, the Soviet Union had broken up, and tough, hardware-based arms control measures had been agreed to in the Conventional Forces in Europe (CFE) Treaty.

CSCE changed with the times, even changing name, from Conference to *Organization,* the better to reflect its status as a fixture in European security affairs. It has begun to take on some operational tasks, although its first planned field operation, a peacekeeping force to be interposed between Armenian and Azeri forces fighting over Nagorno-Karabakh, was stillborn. OSCE's first actual operation found it responsible, under the Dayton Accords, for the conduct of national (1996, 1998) and municipal (1997) elections in Bosnia-Herzegovina, a very difficult task made more so by its lack of full control over the electoral process.[29]

Negotiations on Conventional Force Reductions

At about the same time that the first full session of the CSCE opened in Helsinki, talks between NATO and the Warsaw Pact on force reductions in Central Europe opened in Vienna. Unlike CSCE, these "mutual and balanced force reduction" (MBFR) talks would produce no agreement; rather, they became an exercise in head-banging futility in which the two sides could not agree even on exchanges of data on how many troops they had in the "region of concern" (which covered the Benelux countries, the Germanies, Poland, and Czechoslovakia). Because U.S. administrations used the prospect of mutual force reductions to forestall congressional action to reduce U.S.

troops in Europe unilaterally, and because Soviet leaders prior to Gorbachev had little incentive to reduce their military stranglehold on the region, neither side had a compelling interest in reaching an actual MBFR accord.[30]

As early as 1980, however, France proposed extending the geographic area covered by the talks "from the Atlantic to the Urals," so that arms limitations would reach into the Soviet Union itself. Such talks would address NATO concerns about the large forces stationed on Soviet soil that could move west rapidly in any NATO–Warsaw Pact conflict. The idea went nowhere, however, until Soviet policy and outlook changed.

Soviet general secretary Gorbachev endorsed the concept of Atlantic to the Urals (ATTU) negotiations in April 1986. Ten months later, preliminary agenda-setting talks got underway. They made slow progress until December 1988, when Mr. Gorbachev announced at the United Nations that the USSR intended to withdraw its military forces from Europe. Four months later, on March 6, 1989, the formal negotiations opened, and on November 19, 1990, members of NATO and the Warsaw Pact signed the CFE Treaty, marking the formal end of the Cold War confrontation in Europe.

Meanwhile, East and West Germany had been reunified. The Warsaw Pact soon dissolved and the Soviet Union thereafter broke into a number of independent states. CFE's Joint Consultative Group (JCG), a body designed to iron out wrinkles in treaty implementation, negotiated the reallocation of German and Soviet treaty limited equipment (TLE) quotas. The post-Soviet quotas were parceled out among eight new states located within the CFE's area of application (Russia, Ukraine, Belarus, Kazakhstan, Moldova, Georgia, Armenia, and Azerbaijan).[31] The new arrangements were then formally agreed to by an "Extraordinary Conference" of CFE's members in Oslo, in June 1992. Eight months later, Czechoslovakia's quotas and obligations had to be divided between its two successor states, the Czech Republic and Slovakia, and ratified by another Extraordinary Conference convened at Vienna.

As an alternative to destroying TLE, the parties were allowed to convert it to civilian use, transfer it to other regions or, in the case of NATO, arrange for it to "cascade" from the central region of Europe to NATO members on the "flanks," namely, Norway, Spain, Portugal, Greece, and Turkey. By the end of 1993, the United States, Germany, the Netherlands, and Italy had transferred 2,450 tanks, 1,274 other armored combat vehicles, and 482 pieces of artillery to the flank states.[32] Germany sent excess RF-4E Phantom reconnaissance aircraft to Greece (27) and Turkey (46), and sent 50 Alpha Jet combat trainers to Portugal.[33] CFE rules required those recipients to use cascaded equipment to replace, rather than augment, their national armed forces. Nonetheless, recipients like Greece and Turkey ended up with substantially upgraded armories. In retrospect, the irony is that CFE moved weapons out of areas where disputes had been resolved to areas where they had not.

Although CFE left large numbers of heavy weapons in the ATTU (78,800 treaty-limited items on each "side"), it also led to the removal or destruction of substantial amounts. NATO members removed more than 27,300 weapons from the ATTU (not more than 16 percent of which "cascaded" elsewhere in NATO). Half of the weapons removed or destroyed belonged to Germany (which, after reunification, was "crushing for two"). The renamed "eastern" group of states eliminated about 37,500 weapons, about half of which were Russian.[34]

While technically an arms reduction measure rather than a CSBM, the CFE Treaty nonetheless incorporated extensive measures to enhance regional military transparency and to verify initial numbers of TLE, their destruction, and states' continued compliance with treaty limits, and thus build confidence in the integrity of the treaty regime. CFE's on-site inspection provisions allow treaty members to stage "challenge" inspections within the ATTU that the target state cannot legally block. The treaty parties also exchange data annually on force size and structure, and notifications are required of any changes in force size or "character." Continuing data exchanges and on-site inspections give European military security matters a great deal of predictability.[35]

CFE's success was propelled by fundamental policy changes in Moscow under Mr. Gorbachev's leadership. Absent the USSR's decision to abandon its military hold on Eastern Europe, CFE would not have gotten past the initial agenda-setting stage and would have accomplished little more than its plodding predecessor. In short, it was a result rather than a cause of major changes in East-West relations. The sources of those changes lay deep inside the governing philosophy of Marxism-Leninism and the failure of central economic planning to keep pace with a world of exploding open markets and increasingly open information. Yet while altered political intentions were necessary to undo Europe's long military standoff, they were not sufficient. CFE was a crucially important next step that eliminated the capability for rapid, offensive, conventional military action, and gave tangible evidence that the long military standoff was not just less dangerous but actually gone.

By the late 1990s, as NATO moved to accept new members that were former members of the Warsaw Pact (Poland, Hungary, and the Czech Republic) it was clear that CFE needed fundamental adaptation to the new European security scene. Armed with a mandate from the December 1996 OSCE summit, the treaty parties began to negotiate an "adapted treaty" in Vienna. Just over two years later, they had agreed on a basic framework to replace the CFE's collective force ceilings (western and eastern) with national ceilings on ground TLE for each of the 30 signatory states. To accommodate the stationing of foreign forces within the treaty's central region, "territorial" limits were set for each signatory that are about one-third higher than the limits on national forces. To legally host larger amounts of foreign TLE, a signatory would have to reduce its own forces to keep the joint total under the territorial ceiling. The new framework allows for limited deployments in excess of those ceilings for purposes of military exercises or peacekeeping operations and still higher "exceptional temporary deployments" in crises. Various compromises were worked out to assuage Russian concerns about having NATO members nearer its territory, and to reduce the concerns of CFE "flank" states, such as Norway, Georgia, and Turkey, regarding Russian deployment options under the new accord. Symbolizing the new agreement's compatibility with existing regional security structures, it was slated to be opened for signature at the November 1999 OSCE summit in Istanbul.[36]

Open Skies Treaty

In May 1989, President George Bush picked up an idea from the Eisenhower administration and called for an "Open Skies" regime in which countries would open their air space to one another in ways sufficient to permit selective gathering of intelligence on military capabilities and activities. NATO's North Atlantic Council endorsed the concept in late 1989, laying out the basic elements of an Open Skies regime. On March 24, 1992, in Helsinki, Finland, the Treaty on Open Skies was signed by 27 states, including all members of NATO and former members of the Warsaw Pact, plus Soviet successor states Georgia, Belarus, and Ukraine.[37]

The treaty allows signatories to overfly one another's territory in specially designated aircraft carrying optical, infrared, and synthetic aperture radar sensors. (Signals intelligence equipment, which picks up radio-frequency emissions and communications, is forbidden). Each signatory has an annual "active quota," or total number of flights that it has the right to conduct over other parties' territory each year, and a "passive quota," or total number of overflights it must accept from other parties. Larger states have higher passive quotas: when the treaty is fully operational, for example, the United States will accept up to 42 overflights per year. A state may require that its own designated Open Skies aircraft be used to conduct observation missions over its own territory (called a "taxi" flight because the observed state is providing transport for the observers).

Each signatory submits its overflight requests for the coming year to the Open Skies Consultative Commission (OSCC), the body with representatives from all signatory states that oversees implementation of the agreement. In this instance, it tallies the total requests against each member's passive quota. Requests in excess of quotas have to be reduced, the cuts being worked out among the requesting states for endorsement by the OSCC. Overflight missions must submit and follow a flight plan, remain within certain altitude limits, and so on. Pictures and other sensor data acquired during the flight are available to any signatory for the cost of image reproduction.[38]

The treaty will enter into force as soon as 20 signatories have ratified it. Among the 20 must be those states subject to at least eight overflights per year. They include the United States, Canada, France, Germany, Italy, Turkey, and the United Kingdom, all of which ratified by November 1994, and Belarus, Russia, and Ukraine, none of which had done so as of summer 1999. Pending entry into force, the OSCC has been meeting several times a year to work out implementation details, and trial overflights have been conducted. Ukraine, for example, has conducted two trials over U.S. territory (one "taxi" flight and one using its own aircraft, in April 1997). Russia has conducted one flight over the United States, in August 1997, and the United States conducted a "taxi" flight over Russia two months later. The United States conducted a total of 10 joint trial flights in 1998 and planned 15 more for 1999.[39]

The holdup on Open Skies has been the Russian Duma, whose communist majority was as averse to ratifying arms cuts and transparency measures as the U.S. Congress is to funding the United Nations. The attitudes in both legislatures reflect a sort of domestic backlash against the wave of immediate post–Cold War internationalism and a growing, nationalist parochialism whose instinct is to circle the wagons in both economic and military affairs. A lesson for other regions and especially for newly democratizing countries, where the fresh voices in national politics may be discordant ones, is that domestic political dynamics have the potential to stall and even unhinge multilateral agreements reached by chief executives in brief periods of diplomatic euphoria.

Open Skies is a novel experiment in military transparency that runs counter to the military instinct to deceive the enemy and maintain the element of surprise. Yet because the United States and (funds permitting) Russia have the capability largely to replicate from space anything that an Open Skies aircraft can do, reciprocal overflights are not likely to be particularly revealing bilaterally. However, they will open up new sources of data to all the other signatories of the accord. Anything that American cameras pick up over Russia will be accessible to Germany or Georgia; anything that German airborne radars pick up over Belarus will be accessible to Russia and Poland. In the bad old days, the Soviets were willing to share data with the United States in the course of implementing strategic arms control accords on condition that such data not be publicly released. Such conditionality does not apply to Open Skies and that leveling effect will take some getting use to. Moscow objected to the first Open Skies proposal in 1955 partly because it feared exposing severe military weakness. Russian opponents of the current agreement may fear much the same thing.

Conflict and Confidence-Building in Sub-Saharan Africa

Known in the public mind as a region of failed or failing states and predatory civil wars, sub-Saharan Africa does have many problems. Its borders were decreed by outsiders rather than fought over by local groups, so they reflect colonial deals rather than ethnic domains. But despite some conflicts that remain unresolved, southern Africa, in particular, is better off at the end of the 1990s than at the beginning. Mozambique seemed to be making a steady recovery from decades of vicious internal strife, Botswana and Namibia remained stable, and the new South Africa made both a relatively bloodless transition from apartheid to majority rule and a peaceful transition from one elected president (Nelson Mandela, 1994–1999) to another (Thabo Mbeki, 1999–). South Africa's rejoining the rest of the region breathed new life into the Southern African Development Community (SADC), sometimes described as the "backbone of peacebuilding" in this security complex.[40]

SADC grew out of the long struggle against South African apartheid. At the Lusaka Summit of the nine Frontline States in April 1980, just after majority rule had been achieved in Zimbabwe, a loosely structured Southern African Development Coordination Conference was inaugurated. Although the region's political-military struggle was a subtext for this coordination effort, security was not formally on the institution's agenda and it was not a legally binding arrangement for its members. Namibia joined the group after it became independent in 1990. In 1992, it was transformed by treaty to emphasize the promotion of common values, good governance, democracy, and respect for human rights within the subregion, and it changed from a "coordination committee" to a "community." The 1992 treaty endows the organization's annual summit meetings of heads of state with the power to make legally binding commitments for its 12 members, and to impose sanctions for non-compliance or non-payment of financial contributions. Sixteen "sector coordinating offices" address issues as diverse as agriculture, finance, and tourism.[41] As of mid-1999, 14 states were members of SADC: Angola, Botswana, Democratic Republic of Congo (DRC, joined in 1997), Lesotho, Malawi, Mauritius (1995), Mozambique, Namibia, Seychelles (1997), South Africa (1994), Swaziland, Tanzania, Zambia, and Zimbabwe.

A year prior to South Africa's joining, SADC established an Interstate Defense and Security Commission. In 1996, the commission drafted and a summit meeting approved what is known as the Organ on Politics, Defense, and Security. It has a sweeping mandate to "safeguard . . . the subregion against instability arising from the breakdown of law and order, interstate conflict, and external aggression; promote political cooperation . . . ; develop a common foreign policy . . . ; cooperate fully in . . . conflict prevention, management, and resolution . . . ; mediate interstate and intrastate disputes . . . ; promote and enhance the development of democratic institutions . . . ; [and] develop a collective security capacity . . . and a subregional peacekeeping capacity."[42] Seen as the "successor to the Frontline States," this institution has functioned separately from the rest of SADC, responsible only to its own summit meetings, and chaired by President Robert Mugabe of Zimbabwe.

In 1997, Mandela chaired the regular SADC summit and expressed his strong support for eliminating this parallel arrangement, considered by some analysts to be a vestige of the old days when South Africa itself was the target of regional security concerns and Zimbabwe was a leading Frontline state with greater relative regional influence.[43] With the advent of majority rule in South Africa, the dynamics of southern Africa's security complex have changed. Its old struggles, driven by the existence of apartheid and opposing efforts to preserve and destroy it, have given way to complex interethnic feuding that spills across borders and to the need to fight rampant crime, small arms trafficking, and the very "breakdown in law and order" noted in the organ's founding document. Of course, economic growth, political accountability, and societal stability are linked. To head off future instability and conflict, local economies must be built up, opened to trade, and made attractive to outside investment while governments must grow beyond the postcolonial pattern of personalized, one-man rule. In both respects, old pariah South Africa is looked to as a model by reformers and resented as a threat by remaining old-style strongmen like Mugabe.[44]

The war that refuses to die in Angola and the slow-motion collapse of the DRC into war-torn pieces have especially strained relations among SADC's members and brought military and security issues to the fore. The Angolan war, like one of those trick birthday candles that reignites when blown out, has outlasted the Cold War (having started while Angola was a Portuguese colony) and three peace agreements (1975, 1991, and 1994). In each case renewed fighting has cost hundreds of thousands of lives. The Angolan government's principal opponent, the Union for the Total Independence of Angola (UNITA), has bankrolled itself with the sale of alluvial diamonds, relatively abundant in the areas it has controlled. Although subject to UN embargo from July 1998, sales apparently continued. Condemned by both SADC and the Organization of African Unity for its foot-dragging implementation of Angola's peace agreements, UNITA took advantage of the government's military involvement (and overextension) in the internal conflicts of two neighboring states, Congo-Brazzaville and the DRC, to take the offensive in late 1998. The government's military and economic positions grew increasingly precarious, and it charged at a SADC summit that neighboring Zambia had given material support to UNITA, including heavy arms.[45]

Next door to Angola, after 32 years of corrupt rule by Mobutu Sese Seko (25 of them subsidized by the West) the DRC barely functioned as a state. Because its collapse carried grave implications for SADC and the stability of south-central Africa, a little background is in order.

For years, central authority hardly reached the vast hinterlands of the DRC, an area larger than Western Europe. Basic infrastructure was allowed to crumble as the country's political and military "leaders" pirated public resources. In the eastern part of the country, deep ethnopolitical fissures complemented the terrain of the African Rift Valley and the Great Lakes that fill its deeper chasms. On the far side of the Rift, the small states of Rwanda and Burundi, with their lopsided populations of Hutu (heavy majority) and Tutsi (dominant minority), had erupted periodically into bloody communal violence ever since independence. A particularly violent, government-orchestrated outbreak tore Rwanda apart in 1994, even as a UN-supervised peace accord was being implemented, ostensibly ending a three-year Hutu-Tutsi civil war. This time, the perpetrators aimed to destroy Rwanda's Tutsi minority and any Hutu that stood in their way. By the time the Tutsi-led forces of the Rwandan Patriotic Front (RPF) swept into the country from the Ugandan border, where they were based under the peace agreement, as many as 800,000 people had been butchered. As the RPF moved forward, much of the Hutu population in northern Rwanda fled into the DRC (then Zaire), assuming wrongly that the RPF would engage in tit for tat reprisals. Those responsible for the genocide—the Forces Armées de Rwanda (FAR) and their militia allies, the *interhamwe* ("those who fight together")—fled as well.[46]

The ex-FAR went about reestablishing itself as a fighting force based in the Zairean camps, its members fed and cared for as refugees by the international relief community. Canadian scholar Howard Adelman has traced how it financed and rearmed itself and set about planning to finish the genocide with the 18,000 to 48,000 troops at its disposal, plus an equivalent number of *interhamwe*. The Zairean government failed to prevent hostile cross-border actions against Rwanda, and the *interhamwe*, earning cash by running drugs, paid Zairean soldiers to attack local Tutsi, so-called Banyamulenge (not a tribal but a place referent: "people from Mulenge"), from late 1995 onward. In May 1996, Mobutu made the harassment policy official, and thereafter sought to force the Banyamulenge out of Zaire. Although long resident in North and South Kivu provinces, the Banyamulenge were regarded as outsiders and fair game for ethnic cleansing. They turned to the RPF, now the recognized government of Rwanda, and to its Rwandan Peoples Army (RPA) for support. When the provincial government in South Kivu finally issued an expulsion order, the Banyamulenge and the RPA attacked the camps used as bases of operations by the ex-FAR and *interhamwe*, and pursued the fighters and their dependents farther into Zaire. The bulk of the refugees (some 640,000) went back to Rwanda, but the RPA teamed up with a long-time, if desultory, Mobutu opponent named Laurent Kabila, and built a rebel force that within six months had ousted the Mobutu government.[47]

In the first year of his rule, Kabila's armed forces were Rwandan-led and roughly half Banyamulenge in composition. Growing restive with his image as a proxy for foreigners, Kabila ordered all Rwandan and Ugandan troops out of the country in July 1998. Within a week, the second "rebellion" in as many years had broken out, starting as the first one had in the eastern provinces. The RPA and Banyamulenge banded together again, with assistance from the Ugandan Army (Uganda also suffered from the depredations of guerrilla raiders based in the DRC). Substantial numbers of Congolese army forces and former members of Mobutu's military elite joined them. In September, official Congolese radio stations began to exhort listeners to kill Tutsi, in language reminiscent of broadcasts in Rwanda at the time of the genocide there.[48]

Kabila's remaining military capabilities rapidly collapsed and he invited neighboring states to send troops. Within weeks of the outbreak of the new rebellion, Zimbabwe, Angola, and Namibia had dispatched troop contingents to the DRC. Zimbabwe committed up to 8,000 troops, Angola 6,000 (until reduced by the resumption of large-scale fighting in Angola itself), Namibia 2,000, Chad 1,000, and Sudan perhaps 1,000. Kabila also made common cause with the ex-FAR/*interhamwe*, which the RPA's 1997 campaign had failed to eliminate.[49] SADC, in effect, found itself fighting side by side with the Rwandan *genocidaires*.

Not satisfied with SADC's response either to the new war in the DRC or to the old one in Angola, Zimbabwe's Mugabe formed a new defense alliance in early April 1999 with Angola, Namibia, and the DRC. Some commentators saw this move as deepening the split within SADC between members with troops on the ground in the Congo and those like South Africa who sought a negotiated settlement. Others saw it, in combination with repressive domestic policies in Zimbabwe, as indicative of the dangers posed by an arbitrary ruler to SADC's evolution as a community of free and democratic states.[50]

As conflict engulfed Angola and the DRC, SADC took steps to strengthen its ability to field peacekeeping forces. The first multinational peacekeeping exercise in southern Africa, Operation Blue Hungwe (April 1997) involved roughly 1,000 troops and took place in Zimbabwe, in conjunction with a training workshop for 70 military officers from SADC members held at the Zimbabwe Staff College to familiarize them with United Nations peacekeeping rules and procedures. A much larger follow-on exercise in April 1999 involved 4,000 soldiers and police from 12 SADC countries, plus civilians from international and non-governmental organizations (NGOs). It was hosted by South Africa at its National Defence Force Battle School at Lohatla, in the Northern Cape. The 16-day Operation Blue Crane, planned for two years, simulated the interposition of forces between fighting factions in a mythical civil war. About half the cost of the exercise was subsidized by 11 European countries, Canada, the United States, and India. Air transport to and from the training ground was provided by the United States, Belgium, France, Germany, Italy, and India. France and India sent naval vessels to participate in the leg of the operation that simulated a maritime refugee exodus, near the South African port of Durban. The exercise demonstrated that the SADC's disparate militaries could plan and work together and cooperate with civilian components of a peacekeeping mission, and even deal with the press, played within the simulation by a group of working journalists. While the exercise did not give SADC the "real and usable capacity" for peacekeeping that spokespersons had hoped for, it laid a better foundation for such an enterprise, provided the necessary logistical capacity could be developed to support it. An "informal SADC objective" is to have a functioning "peace brigade" by 2002.[51] For now, SADC members would depend upon outside support to move and sustain, if not help to equip, such a force.

The need for better planning, doctrine, and training for peacekeeping was brought home in part by SADC's intervention in landlocked Lesotho following contested election results (79 of 80 seats in parliament went to the ruling party in the May 1998 vote), political unrest, and a military mutiny on September 15 that led the government of Lesotho to ask for aid. It came a week later in the form of 1,000 troops from South Africa whose arrival sparked unexpected military resistance. It also touched off riots, arson, and looting in the capital, Maseru, in many instances of South African-owned businesses, which the intervention force was not primed to control. The operation grew eventually to include 3,500 troops—nearly twice the strength of the Lesotho Defence Force (LDF)—from South Africa and Botswana and did not fully withdraw until the following May. SADC mediators used the intervening period to help construct an Interim Political Authority (IPA) for the country with members drawn from all 12 of Lesotho's political parties. New elections were to be held by mid-2000, featuring proportional representation to replace the winner-take-all system that contributed to the original unrest. A smaller force of 200 to 300 military trainers from Botswana and South Africa, plus a small number of Zimbabwean instructors, replaced the peacekeepers to undertake a one-year program to professionalize the LDF.[52]

The law and order problems presented by Lesotho, while difficult, are at least of manageable proportions: Lesotho is tiny, surrounded by South Africa, and its troubles are largely indigenous. The DRC, by contrast, is enormous. Movement is difficult in many areas except by water or air, as roads have fallen into disrepair. It is an ethno-linguistic stew of 50 million people that has not been effectively governed since independence in 1960. Its neighbors are mostly unstable polities with their own coups, wars, and ethnic strife to worry about. The DRC's role as a staging area for chaos led Rwanda and Uganda to foment and support two rebellions in the DRC the better to secure their own borders against armed insurgents. Their efforts created a regional war with the DRC as the playing field (or, more accurately, forest). By spring 1999, the war had divided the DRC into several equally ungoverned and ungovernable parts, as the rebellion itself broke into factions and their supporters began cutting separate deals with Kabila through intermediaries like Libya. Mediation efforts sponsored by SADC, with Zambia's President Chiluba in the lead, threatened to founder over Kabila's refusal to meet with and thus acknowledge rebel leaders (much preferring to treat his difficulties as foreign invasion rather than insurrection), and over Angola's charge of Zambian support for UNITA.[53]

All sides recognized that any eventual peace in the DRC would probably have to be supervised or policed by somebody, but volunteers—indeed, the basic capacity—to undertake the task were scarce. A December 1998 proposal by Thabo Mbeki, then South Africa's deputy president, that the fighting forces already in the country police themselves in the event of peace was rejected at the time by all parties.[54] Such a force, even if backed by a UN mandate with a South African-managed command structure deriving leverage from control of the supply spigot, would be difficult to implement and has no precedent in peacekeeping. The opportunities for such an operation running off the tracks would be legion, from disputes over assigned patrol areas to illicit exploitation of those areas (especially ones noted for diamonds or gold). Moreover, the presence of large, irregular forces in this conflict—not just the main opposition *Rassemblement Congolais pour la Democratie* (RCD) but the rival *Mouvement de Liberation Congolais* (MLC), plus the ex-FAR, the *interhamwe,* Mayi Mayi militias, and Sudanese-supported guerrilla armies preying on Uganda—meant that peace implementation would require either that the national armies involved in this conflict turn on and arrest or fight their erstwhile irregular allies, or that the latter would continue to go about their destabilizing business. Without at minimum a border force to stop (or, more realistically, reduce) incursions into Uganda, Rwanda, and Burundi from DRC territory, the cycle of violence and refugees in the Great Lakes region would continue.

Southern Africa is a study in contrasts: democracy and autocracy, order and chaos, cooperation and conflict. More than any other region outside of Europe, southern Africa has been moving in the direction of organized cooperative security measures, despite the chaos reigning in the DRC and Angola. It benefits from the example offered by its largest power, South Africa, whose policies, politics, and regional leadership since the mid-1990s have stood in stark contrast to those of Nigeria and the shadow it cast over western Africa for much of the decade. The good news, at the end of the decade, is that one can indeed see democracy, order, and cooperation that did not exist at its start. Progress will consist in maintaining those gains and avoiding wholesale slippage.

Tools to reinforce and spread the gains include NGOs that are actively engaged in humanitarian relief efforts, conflict resolution training, and efforts to build up the civil societies of a number of states. NGOs provide practical training for a new generation of democratic activists and potential leaders, and work to promote women's rights and economic self-sufficiency through myriad microenterprise projects. Southern African academics have formed the Southern African Human Rights Network to lobby governments on rights issues. Those autocratic leaders most threatened by development of strong and autonomous civil society have begun to take note of the "threat" it can pose, and to blame "foreign" interests and money for its growth. Such objections may be the truest indicator of real progress.[55]

Aging CSBMs in the Middle East[56]

Long known for the intransigent nature of its religious, ethnic, nationalist, and territorial disputes, the Middle East is nonetheless the locus of some of the earliest and, on their own terms, most effective of operational confidence-building measures. On the Golan Heights between Israel and Syria, the peacekeepers of the United Nations Disengagement Observer Force oversee a military thin-out zone. An international peacekeeping entity independent of the UN manages the Multinational Force and Observers, which oversees a similar zone in the Sinai. These thin-out zones cover the main axes of attack used by Arab forces in the October 1973 Middle East War. They have held up despite serious Israeli-Syrian fighting in Lebanon in 1982 and ongoing Israeli exchanges with Iranian-backed Hezbollah guerrillas operating out of the Syrian-controlled Bekaa Valley in Lebanon.[57]

Supplementing the ground thin-out zones are regular MFO helicopter inspections of the Sinai, and U.S. aerial reconnaissance flights known as "Olive Harvest," data from which is shared with the local governments. These measures collectively reduce the threat of surprise ground force attack, partly by increasing early warning of such attacks. As such, they reduce the need for regional, and especially Israeli, armed forces to maintain costly and ultimately dangerous high levels of alert. By seeing to it that forces remain separated at borders, these operations also reduce the probability of inadvertent conflict.

Since the end of the Gulf War and the onset of the Madrid Peace Process, regional governments (less Libya, Lebanon, Syria, Iran, and Iraq) have participated in arms control and regional security (ACRS) discussions. Initially the talks involved four working groups on exchange of military information and advance notice of "certain military activities"; on search and rescue and minimizing incidents at sea; on a regional communications center; and on various "conceptual issues" related to regional security. In late 1993 the first three "operational" groups were merged into one forum on communications, information exchange, and maritime issues. In 1994, the conceptual group drafted a declaration of principles for interstate relations in the region loosely modeled on the Helsinki Final Act of 1975. After much further wrangling, however, the declaration was downgraded to a "statement" on arms control and security.

ACRS participants disagreed over the geographic scope of the "Middle East." Israel includes Iran, for example, which has not been part of the ACRS process, and refused to discuss nuclear issues "until Iran was integrated into a regional regime." The talks eventually broke down in 1995 over Egypt's insistence that participants first sign "existing nuclear, chemical, and biological weapon treaties, and allow international inspections," and Israel's insistence (in addition to inclusion of Iran) on more modest CSBMs as prerequisites to any moves toward denuclearization.

Despite the political maneuvering that ultimately brought the talks to a halt, there was interim progress on a number of fronts: a regional communications network, conflict prevention center, maritime accords, and prenotification of military exercises involving more than 4,000 troops or 110 tanks (along with exchange of unclassified data on military structure, publications, organization, and personnel). This work remains on the shelf, as it were, to be taken up again should political circumstances permit.

Meanwhile, non-governmental track II discussions continue. Institutions involved in track II discussions include the Stockholm International Peace Research Institute, the UN Institute for Disarmament Research, the University of California's Institute on Global Conflict and Cooperation, Columbia University-based Gulf/2000, Sandia National Laboratory's Cooperative Monitoring Center, and Search for Common Ground. These efforts constitute "a CSBM in their own right," argues Middle East analyst Jill Junnolla, "as they provide a rare opportunity for informal contact among Arabs and Israelis." Track II workshops have dealt with the nuclear issue, Islamic extremism, arms control verification requirements, and non-military issues of common interest like pollution control.[58]

Shaky CSBMs in South Asia

The South Asian security situation suffers from all of the conflict-promoting elements that dog other regions, while having rather fewer conflict-mitigating elements than most. On the "promoting" side, there are territorial and jurisdictional disputes, religious and ethnic enmities, overly influential militaries, and gross asymmetries in military power. On the mitigating side, there are trappings of democracy in several states. But there are no external balancers regularly engaged in the region's political-military affairs. (Pakistan uses China as a political-military balancer only insofar as Beijing is willing to supply weapons.) Moreover, principal antagonists Pakistan and India are weak states with unstable governments and much social unrest, and each charges the other with fanning that unrest. In the principal jurisdictional dispute, over Kashmir, three powerful motivators—state sovereignty, religious piety, and political self-determination—are in the driver's seat and fighting for the wheel. Moreover, both India and Pakistan possess nuclear weapons, held under command and control procedures whose quality and reliability are largely unknown outside those states.

While major interstate war has not recurred within the region since 1971, measures to reduce the likelihood of such recurrence, while on the books, are either moribund or honored mostly in the breach. Each regularly directs blame for this situation toward the other.

Indo-Pakistani Conflict Avoidance Measures

The Simla Accord of 1972, which followed by six months the Indo-Pak war that split up the original, geographically bifurcated Pakistani state, contained all the right words for building peaceful bilateral relations. The two sides pledged, among other things, to settle their differences by peaceful means, to refrain from unilaterally altering situations currently in dispute (including in Kashmir), to refrain from the threat or use of force, to respect one another's territorial integrity, and to prevent "hostile propaganda" directed against the other.[59] But not until 1990 did the two states embark upon joint efforts to implement recognizable CSBMs at the instigation of the United States.[60] In

July 1990, the Indian and Pakistani foreign secretaries sat down for the first of a series of meetings that would yield agreements on:

- advance notice of military exercises, maneuvers, and troop movements;
- use of the 1971 Direct Communications Link (DCL) between the Directors-General of Military Operations (DGMOs);
- measures to prevent air space violations and to permit overflights and landings by military aircraft;
- not attacking each other's nuclear installations and facilities in the event of war (agreed to in 1988 but not ratified until 1991); and a
- joint declaration on the prohibition of chemical weapons.[61]

The advance notice measure applied to division- and corps-level exercises taking place within a certain distance of the line of control in Kashmir and of the international border elsewhere. It and the other measures make up an interesting list of CSBMs. Unfortunately, compliance has been spotty and instances of bad faith abound, undermining the utility of these measures.

The Direct Communications Link, previously used infrequently because neither DGMO wanted to be the first to "give in" and call, was to be exercised weekly under the 1990 accords. When it was used prior to 1990, however, "disinformation was often relayed," contravening the basic purpose of such a link and predisposing a listener not to believe information being sent from the other side.[62]

The air space accord of April 1991 prohibited fixed-wing combat aircraft from flying closer than ten kilometers to the international border. Unarmed helicopters, transport, and observation aircraft may approach no closer than one kilometer. However, air space violations, which are useful for testing the other side's air defense dispositions and responses as well as for gathering current border intelligence, still occur, albeit at a much-reduced rate since the accord went into effect (from more than eight per month to fewer than two per month). In exchanging lists of nuclear facilities, both sides reportedly withheld information on at least one nuclear installation.[63] Finally, when the global Chemical Weapons Convention entered into force in 1997, India admitted that it had maintained a stockpile of chemical weapons (contrary to the letter and spirit of the 1990 declaratory CSBM).

Despite the current track record, some analysts looking at the South Asian political morass and the potential risks that nuclear weapons and missile proliferation may pose in any future military crisis still advocate step-by-step CSBMs as the principal way to cultivate a culture of cooperation between Indian and Pakistani political leaders. Such experience, they argue, may chip away at the edifice of hostility while reducing day-to-day risks of conflict.[64]

Research by Sheen Rajmaira confirms the need to keep chipping. Rajmaira examined conflictual and cooperative behaviors exchanged by India and Pakistan between 1980 and 1990 and found a "trend of increasing conflict" over the period, at the end of which Indo-Pak relations were in fact generally acknowledged to be at their lowest point in decades. From her analysis, she concluded that "India and Pakistan treat conflict and cooperation as relatively distinct types of foreign policy behavior" rather than as different points on a single behavioral continuum. Both tend to view conflict as the long-term norm, and to view cooperation as a short-term phenomenon which has little long-term, evolutionary impact. This suggests a potentially dangerous dynamic . . . [in which] extreme sensitivity to conflict coupled with the relative inattentiveness to cooperation might fuel a rapid and unpredictable escalation of conflict between the two rivals.[65]

Her conclusions seemed to be confirmed by an otherwise hard to fathom sequence of events in fall 1998 and spring 1999. In October, India and Pakistan held their first talks in 35 years on the status of Kashmir, more or less in parallel with talks on nuclear issues. The following February, India's Prime Minister Vajpayee inaugurated direct bus service to Pakistan, riding the first bus himself to an elaborate diplomatic reception on the Pakistani side of the border. Seven weeks later, however, both states tested long-range ballistic missiles, and in late May, Indian troops found Pakistan-supported Islamist guerrillas—in what had to have been an elaborately-planned operation—occupying mountain positions in India's part of Kashmir.[66] Intense, if localized, fighting erupted (see chapter 3).

Further experience with CSBMs, Rajmaira argued, might help "develop a long-term memory for cooperative behavior" as one of the few practical pathways toward sustainable peace. However, the efficacy of such measures also hinged on greater governmental stability and relief from the challenges posed by "ethnic and separatist movements within both countries [which] augment the paranoia of both sides"[67]— just the sort of challenge posed by the fighting in Kashmir.

SAARC and Regional Asymmetries

South Asia's gross power asymmetries and India's reluctance to deal as an equal in consensus-based multilateral institutions show very clearly in the region's relatively meager multinational endeavors. One way to induce a culture of cooperation in such circumstances may be to circumvent the security field entirely and focus on economic, social, environmental, and cultural matters. In a number of regions, this has been accomplished partly through official multilateral fora and partly through non-governmental mechanisms that create opportunities for exchange at the level of civil society. The South Asian Association for Regional Cooperation (SAARC), a concept originally put forward by Bangladesh in the late 1970s, was set up in December 1985 as the venue for annual meetings of heads of state or government to discuss non-security matters. SAARC's charter explicitly precludes the raising of "bilateral or contentious issues," largely at the insistence of India, which prefers bilateral dealings that use "its size and power to greatest advantage."[68] But the other states in the association are either so small, so poor, so hostile, or so beset by internal conflict that multilateral cooperation on an equal footing is difficult in any case.[69]

In SAARC's first six years, the heads of state/government meetings took place annually, lapsing for a while to every other year. The ninth summit in May 1997 enabled the prime ministers of India and Pakistan (Nawaz Sharif, then just reelected, and Inder Gujral, head of a new coalition government) to meet conveniently in a third country and agree to restart the foreign secretaries' meetings that had produced the existing crop of Indo-Pakistani CSBMs. The foreign secretaries met in June and agreed to a potentially historic formula for discussion of Kashmir together with other issues that divide the two countries, in the "integrated manner" long sought by Pakistan.[70] Unfortunately, by the end of the year, both prime ministers were embroiled in political crises and the Gujral government fell, to be replaced by a coalition led by the Hindu nationalist Bharatiya Janata Party. The two countries went on to test long-range missiles and nuclear weapons in 1998, and further progress on confidence-building stalled for several months.

Political CSBMs in Southeast Asia

In 1967, at the height of the Vietnam War, the governments of Thailand, Singapore, Malaysia, Indonesia, and the Philippines formed a consultative mechanism with a subtext of resistance to communism in the region but whose agenda never explicitly focused on the subject or on military security at all. By including Indonesia, the Association of Southeast Asian Nations (ASEAN) also aimed at regional reconciliation following Indonesia's abandonment of its policy of "confrontation" with Malaysia (the policy went over the side with the Sukarno regime in 1965). The new Suharto government saw in ASEAN a way to "exchange [Indonesia's] erstwhile role of regional troublemaker for that of constructive partner." ASEAN's small secretariat has been located in Jakarta for the past two decades.[71]

The Objectives of ASEAN

ASEAN is unique among multilateral political arrangements in the region for having been undertaken at regional initiative and for involving no states outside the immediate subregion. Several members of ASEAN maintained bilateral security ties (with the United States, for example), and there remained political-military tensions between them but within ASEAN's meetings confrontation was assiduously and by and large successfully avoided. The theme, instead, as the old song lyric goes, was "getting to know you, getting to feel free and easy." Indeed, the phrase in Malay is *musyawarah-mufakat* ("consultation and consensus"). The objective of ASEAN's ongoing, informal, but high-level interaction is to raise the comfort level in regional intergovernmental relations, on the assumption that leaders' personal comfort will translate into higher national comfort levels.[72]

In 1976, ASEAN's members codified their relationship to a degree with the Treaty of Amity and Cooperation for Southeast Asia. Signatories pledged to respect one another's sovereignty, to avoid interference in one another's internal affairs, and to renounce the use of force in settling disputes.[73] Having ruled out all the fun stuff of international relations, what was left? What was left was the slow and steady building of familiarity among leaders of relatively stable governments and the notion of confidence building as process rather than product.[74] That each of the original members of ASEAN was a distinctly status-quo power meant that they shared interests in regional stability—avoidance of boat-rocking—and in the tempestuous setting of the Cold War, this was a useful objective that supported steady economic development.

Nonetheless, the Cold War framed ASEAN's original political environment. Once communism ebbed as a force for radical change and U.S. political-military commitments to East Asia seemed poised to ebb along with it, the members of ASEAN faced a minor crisis of relevance and security: relevance because all the states of the Asia-Pacific area were becoming more or less "non-aligned," and security because the United States had been the major political balancer in the region. If the United States drew back, what would Japan and China do? Long memories of Japanese behavior in Southeast Asia during the Pacific War (and Japanese history books' tendency to brush by it) raised worries about future Japanese security policy. The rapid growth of Chinese economic power, several unresolved jurisdictional disputes (including Taiwan and claims in the South China Sea), and regional perceptions of a China out to right the wrongs inflicted upon it for 150 years raised more tangible and shorter-term concerns.[75] From mid-1997 onward, the Asian economic crisis and ASEAN's clubby, see-no-evil, speak-no-evil approach to its members' domestic policies—some of which helped pull the others' economies down the drain—suggested to some observers that the institution had outlived its usefulness.[76]

The biggest challenge to ASEAN may lie in the very diversity of its expanded membership. As of April 1999, when Cambodia was finally admitted to full membership, its long-term goal of including all ten Southeast Asian nations was fulfilled. Realizing that goal, however, has meant extending membership to a military dictatorship (Burma), which complicates ASEAN's relations with the European Union; to a declining communist state (Vietnam); and to a quasi-democracy with sort-of-ex-communist leadership (Cambodia under Hun Sen). None of them joined ASEAN to have its domestic systems scrutinized or challenged by its neighbors. Nor do Singapore or Malaysia cotton to outside criticism. At the same time, however, the increasingly democratic and populist regimes in the Philippines and Thailand see a need to move beyond the institution's original conventions.[77] In July 1998, Thailand's Foreign Minister Surin Pitsuwan proposed what he called "flexible engagement," whereby members could offer constructive criticism of policies that "affected another country or offended its principles."[78] At the December 1998 ASEAN summit in Hanoi, Philippines president Joseph Estrada lobbied on behalf of democracy and greater respect for human rights.[79] If Indonesia, the historical heavyweight in ASEAN, emerges from its current period of turmoil with a similar democratic orientation, the cracks in ASEAN's carefully plastered facade may prove irreparable, but the cause of openness in Southeast Asian politics and economics would be advanced.

The ASEAN Regional Forum (ARF)

ASEAN foreign ministers meet annually, and the custom following those meetings has been to invite representatives from states of the wider Asia-Pacific basin to postministerial conferences to promote dialogue on common economic interests. Foreign Secretary Raul Manglaupus of the Philippines first raised the notion of a multilateral security dialogue at the 1989 postministerial. The notion was pursued initially through NGO channels, where ideas considered too politically sensitive or avant-garde for official channels could be discussed with relative freedom (and the discussions' significance could be denied by governments). Through NGO seminars in 1990–1991, the notion of a regional dialogue on security-related questions was discussed within ASEAN itself, and in January 1992, ASEAN prime ministers endorsed "external dialogues in political and security matters" at the annual post-ministerial.[80] The Council for Security Cooperation in the Asia Pacific (CSCAP), a network of "regional security-oriented research institutes," was founded in July 1992 in Kuala Lumpur to pursue the NGO dialogue in an organized fashion.

The new official dialogue was slow to start. At a May 1993 postministerial meeting, Singapore advocated bringing China, Russia, Vietnam, Laos, and Papua New Guinea into what had been, up to that point, a "Western-aligned" forum. Two months later, the foreign ministers of ASEAN and its dialogue partners agreed to set up a new ASEAN Regional Forum (ARF), an expanded version of the familiar post-ministerial meeting structure, "to develop a 'predictable and constructive pattern of relationships in the Asia-Pacific.'"[81] In 1994, the first ARF convened in Bangkok, chaired by Thailand, the ASEAN member that had served as chair of the institution's own annual meeting. Eighteen foreign ministers met for dinner and an "unfocused" discussion.[82] The custom of ASEAN-member chairmanship of ARF meetings continues.

In between the 1994 and 1995 governmental sessions, an informal "intersessional" network of seminars mixing government officials, academics, and researchers discussed confidence building, peacekeeping, and preventive diplomacy. Meanwhile, out in the world, Chinese troops were discovered building military structures on a Philippines-claimed reef in the Spratly Islands, South China Sea. The aptly named Mischief Reef caused a brief flare of tension between Beijing and Manila. Chinese sensitivities over the status of Taiwan also were highlighted a month before the second ARF meeting, when Taiwan's President Lee Teng-hui was granted a U.S. visa for an "unofficial" visit to the States.

The second ARF session convened in Brunei in August 1995 with Cambodia participating for the first time. Singapore tabled a concept paper urging implementation of concrete regional CSBMs on two tracks: those easier to implement officially in the near term (track I), and more difficult issues to be discussed in the NGO forums (track II). A follow-on track I Intersessional Support Group discussed confidence building and transparency in defense policy, and track I Intersessional Meetings discussed peacekeeping operations and coordination of search and rescue missions.[83]

In between the second and third sessions of ARF, Indonesia and Australia confirmed that ARF would not be a substitute for bilateral security ties. In December 1995, the two countries surprised their neighbors by signing a bilateral security treaty that called for regular high-level consultations on "common security" issues and mutual consultation "in the case of adverse challenges to either party," with the possibility of joint measures in response.[84] Also, in March 1996, before Taiwan's presidential elections, China chose to target long-range ballistic missile tests very near Taiwan's major shipping ports. The United States responded by dispatching a pair of aircraft carrier battle groups. The most forceful response in the region by the American military since the August 1976 "tree cutting incident" in the Korean demilitarized zone, it may have helped to reassure Asian governments that the United States remained engaged in important Asian political-military affairs.[85]

China's assertive, even bellicose, actions regarding Taiwan formed the political backdrop for the third ARF session in Jakarta in July 1996, although the issue of Taiwan was too sensitive to address openly. India and Burma participated in the 1996 session for the first time, bringing the total to 21 countries.[86] India's admission helped prompt discussion of criteria for adding new participants. The criteria decided upon effectively ruled out Central Asian and Latin American participation by requiring that new prospects have a "demonstrated . . . impact on the peace and security" of Northeast or Southeast Asia or Oceania. The meeting endorsed the intersessional group's report on confidence building, which called for high-level security dialogues, annual publication of defense policy statements, support for the UN Register of Conventional Arms, information exchanges on disaster relief, voluntary observation of military exercises, and support for global arms control regimes. Participants also endorsed recommendations on peacekeeping training issues, with a weak collateral endorsement of UN standby force arrangements (through which countries indicate forces that may be available for voluntary call up to future peacekeeping operations). The Chairman's Statement (the only written proceedings of an ARF session) concluded that in 1996 the "participants . . . displayed a high degree of comfort in their interactions," indicating progress (as ASEAN measures it) toward ARF's basic objective.[87]

In 1996–1997, track I intersessional meetings covered disaster relief, confidence building, search and rescue, and peacekeeping. Track II meetings addressed preventive diplomacy and non-proliferation. China hosted the CSBMs meeting; its report endorsed voluntary sharing of information on defense policies and conversion, on UN register participation and the like, but broke no new ground. Discussion of maritime CSBMs, potentially useful in the Spratly Island dispute, was deferred to the next intersessional cycle.[88] Overall, the CSBMs agenda did not advance much.

The July 1997 meeting in Malaysia brought defense officials into this conclave of foreign ministries for the first time. The resulting Chairman's Statement stressed that the Asia-Pacific security situation "continues to improve," and that the process of "dialogue and cooperation is gaining momentum," implicitly attributing the improvements to ARF. But discussion was dominated by events in Cambodia, where the political faction led by Hun Sen had mounted a coup three weeks earlier. Conditions in Burma also figured prominently, discussed for the first time in the presence of Burmese representatives. Participants continued the work of the "core" intersessional group on confidence-building measures in November 1997 and March 1998 meetings, which produced summary tables of CSBMs implemented to date: "exchanges of perceptions" on regional security; meetings of heads of national defense colleges; a meeting on the military roles in disaster relief; and exchange of information on voluntary notice of military exercises and invitations to foreign observers. Other CSBMs included high-level military to military contacts and seminars on how to produce defense white papers, efforts that continued during the 1998–1999 intersessional, along with workshops and training sessions on peacekeeping, military medicine, and disaster relief. The first track II meeting to be held in Russia under the ARF umbrella addressed security and cooperation in the Asia-Pacific, at Vladivostok, in April 1999. The chairman's report from the sixth ARF, held in Singapore in July 1999, stressed that the "evolutionary approach" taken by the forum would "continue to move at a pace comfortable to all ARF participants."[89] That is, consensus and inoffensiveness would remain the forum's watchwords.

ARF: Is a Non-barking Dog So Bad?

While tangible progress in operational or institutionalized security cooperation may be hard to attribute to the ASEAN Regional Forum, the process of regular dialogue— both in the various multilateral meetings and in the bilateral exchanges that the meetings encourage—is itself a confidence-building measure of some value. To Sherlock Holmes, a non-barking watchdog was an important clue that helped solve a murder mystery. To countries of the Asia-Pacific, a relative lack of serious interstate security incidents in such a large and populous region is equally worth noting. The ongoing dialogue and intersessional skull sessions that ARF stimulates are at least neutral to that stability and may even contribute positively to it.

The low-key, face-saving, confrontation-avoiding ASEAN model used by ARF has clear limitations. Whatever progress it makes toward conflict resolution can be easily undermined by actions like the occupation of Mischief Reef, the firing of missiles near Taiwan, the coup in Cambodia, another deadly riot in Rangoon or political instability in Indonesia. The United States and its forward deployed military forces act as regional power balancers. Absent the United States, states proximate to China would be left the basic options of appeasement or turning to Japan as balancer.

Still, having ARF is better than not having it, and certain understandings and habits of cooperation may be nurtured as a result. The forum is "a modest contribution to a viable balance or distribution of power within the Asia-Pacific by other than traditional means" and represents as much of a security cooperation regime as the regional traffic will bear.[90] ASEAN's insistence on maintaining control of the forum keeps it from appearing to be one or another great power's pawn, yet limits the extent to which the major powers view ARF as a relevant and effective tool of diplomacy and conflict management. ASEAN's dilemma is how to make ARF a more effective tool for engaging a rising China without having it slip out of ASEAN control. With its work promoting military to military contacts and modest CSBMs, and its spreading network of groups, meetings, and seminars, the 22-member ARF may come to be of practical relevance independent of its smaller "parent." Yet, not having to rotate chairmanship of the institution to any of the major participating powers may be too politically convenient to give up, even if its parent institution ASEAN ultimately fails to rebound from the economic crisis.

Dialogue in Northeast Asia

The dominant issues in the Northeast Asian security complex, as related in chapter 3, date back to the founding of the People's Republic of China (PRC) and the Korean War. The status of Taiwan, which Beijing considers a province of China and hence an internal Chinese matter, has been a sensitive issue for the PRC since 1949. To date, therefore, despite the periodically volatile nature of relations between the two entities, China-Taiwan CSBMs have not been the subject of discussion in official international channels, nor have track II initiatives paid much attention to the question. Kenneth Allen, at the Stimson Center, has worked to energize discussion of what he calls "cross Strait CBMs," pointing out that some modest informal measures already exist, and that a series of small and useful steps might be taken before any breakthrough on the political level. The two air forces have already agreed tacitly to avoid military flights across the center line of the Taiwan Strait, which the other side would consider provocative. In conjunction with a direct shipping arrangement using specified cargo ships, there is a low-level hotline to use in the event of maritime emergency. Further "communications measures" might be added, including more frequent exchanges of delegations of retired military officers and security experts, with active duty officers eventually added to the program. Further direct communication links might be used voluntarily to notify the other side of upcoming military exercises; such exercises might be limited in size or location, and joint or third-party observation measures might verify compliance. As Allen concludes, "Beijing and Taipei have a choice of renewed confrontation . . . or gradual accommodation and resolution of their differences," and if they choose the latter course, "CBMs will be an essential means for reconciliation."[91]

Since 1950, Northeast Asia's other intramural struggle, on the Korean peninsula, has been characterized by hostility and war preparation—efforts to build confidence in combat victory rather than confidence in peaceful coexistence—and espionage and assassination have been the modal forms of communication. The Statute of the Worker's Party of the Democratic Peoples Republic of Korea (DPRK) commits the party and the armed forces it directs to the "liberation" of the entire country and construction of a self-dependent communist society. The party's tight control of society and its doctrine of self-dependence (or *juche*) contribute to North Korea's almost uniquely insular international outlook, fortress mentality, and universal indoctrination of the populace to serve the state as one.[92]

The withdrawal of Soviet assistance in the late 1980s, however, ushered in a period of economic decline that has yet to reach its nadir. Best estimates are that North Korea's gross domestic product has shrunk by about 4 percent per annum since 1988, meaning that its economy in 1997 was only about two-thirds what it was in 1988. A combination of bad economic practice and bad weather conspired to push North Korea into famine conditions from 1996, and reports emerging from NGO relief missions inside the country, as noted in chapter 3, painted a picture of widespread malnutrition and increasing starvation.[93] The length and depth of the crisis finally seemed to convince North Korean leaders of the need to open up slightly to the outside world, including a dialogue with the World Bank for assistance in implementing economic market reforms along the lines adopted by China.[94]

South Korea, meanwhile, has been moving toward increasingly open democracy and markets, especially in the wake of the economic crisis. New president Kim Dae Jung declared an amnesty for most political prisoners, pardoning two former presidents convicted of bribery and of participation in a 1980 massacre respectively. He also upheld an agreement with the International Monetary Fund for restructuring Korea's economy in exchange for substantial assistance with the country's international debts.[95]

The Agreed Framework on Nuclear Issues

Militarily and politically stalemated for decades, with U.S. troops helping to hold the line, the Korean confrontation took a potentially nasty turn in 1993 when U.S. intelligence suggested that despite a 1992 agreement with the International Atomic Energy Agency to monitor its nuclear facilities, North Korea was siphoning off plutonium in the spent fuel from its nuclear research reactors in sufficient quantities to make a nuclear fission weapon. Moreover, the North was believed to have a ballistic missile under development capable of hitting Japan.[96]

Talks on the nuclear issue between Washington and Pyongyang, and a key intervention by former U.S. president Jimmy Carter in June 1994, led to an October 1994 Agreed Framework between the two countries. North Korea agreed to freeze its nuclear "research" program, to accept secure storage of existing spent reactor fuel, and to dismantle its current reactors in exchange for new, safer, and less proliferation-prone power reactors to be financed internationally. To carry out the agreement, the Korean Peninsula Energy Development Organization (KEDO) was established by the United States, Japan, and the Republic of Korea in March 1995. KEDO's main task is to contract and pay for the construction of the new power reactors, and in the meantime to oversee supply of up to 500,000 metric tons of heavy fuel oil per year to six North Korean thermal power generation plants.[97] The agreement was reached without major input from the South Korean government, which is expected to fund 70 percent of the total $4.6 billion cost. Japan is to fund another 20 percent. South's quasi-governmental Korean Electric Power Corporation (KEPCO) is the prime contractor for the light water reactors. Infrastructure development at the reactor site began in August 1997.[98] Fulfilling South Korea's part in the financing of the reactor agreement—which is a combined arms control and confidence-building measure, broadly defined—will depend critically on the country's rebounding from its late 1997 financial crisis.

In the shadows of the Agreed Framework, and in the aftermath of North Korea's long-range missile test in August 1998, representatives from Washington and Pyongyang met to discuss the missile program and missile technology proliferation. By March 1999, four rounds of talks had produced no breakthroughs, and by mid-year North Korea appeared to be preparing to test an even longer-range missile, an event postponed by energetic diplomacy and agreement by North Korea not to test its long-range missiles while negotiations with the United States continued and some U.S. economic sanctions were eased.[99]

Four-Party and Two-Party Talks

In April 1996, the United States and South Korea proposed joint talks with North Korea and China to craft a permanent peace treaty to replace the peninsula's 1953-vintage Military Armistice Agreement—which the South never signed and the North repudiated in 1992. About a year later, North Korea responded favorably, and preliminary discussions were held on the campus of Columbia University in New York City in August and November 1997. Formal negotiations began in Geneva, Switzerland, in December 1997, with a second round in late March 1998 that the DPRK quickly broke off when the United States refused to put U.S. troop withdrawal from Korea on the table.[100] Thereafter, the North remained in the room without having U.S. withdrawal on the agenda, but continued to insist that such withdrawal was its first priority. By the fifth round, the participants were able to talk about CSBMs "in detail" but not able to agree to any.[101]

In April 1998, while awaiting a date for resumed four-party talks, North and South Korean officials met in Beijing to discuss humanitarian issues (principally famine relief and family reunification), resuming a dialogue disrupted by the death of DPRK founding leader Kim Il-Sung in 1994. They met for a second time in late June 1999, also in Beijing; the North demanded and got 22,000 tons of fertilizer from the South as a quid pro quo for the second round of two-way talks. The meeting went ahead despite a naval gun battle one week earlier in which one North Korean gunboat was sunk with all hands after intruding into South Korean-controlled waters.[102]

Finally, in late May 1999, former U.S. secretary of defense William Perry led a U.S. delegation to Pyongyang for 15 hours of talks with North Korean leaders. Perry presented a package of proposals that offered normalization of diplomatic relations, economic assistance, and an end to sanctions, in exchange for North Korea agreeing not to make or possess nuclear weapons, agreeing to inspections to verify compliance, to halt development and sales of ballistic missiles, and to refrain from military provocations against South Korea. A full week of U.S.-North Korean talks in Berlin, Germany, in January 2000 led to North Korean acceptance of further discussions in the United States in the spring.[103]

Track II Diplomacy

Paralleling and in some sense preparing the way for official talks have been unofficial, track II meetings and electronic networks sponsored by American NGOs. Since 1993, the University of California's Institute on Global Conflict and Cooperation (IGCC) has sponsored several informal meetings of the Northeast Asia Cooperation Dialogue (NEACD), featuring government officials from China, Japan, South Korea, Russia, and the United States acting in their private capacities. (North Korea, while invited, did not send representatives to any of the eight working sessions held through November 1998). Each national delegation presents its perspective on general security issues, emphasizing "what has changed in the most recent eight months," and a second session presents military perspectives on regional security issues. A third session addresses a non-security issue of common interest (for example, environment, food supply, energy). According to IGCC, in their discussions of regional security, participants have come to prefer the term "mutual reassurance measures" over the more common but "conceptually too narrow" confidence building measures. At the seventh session, after two years of discussion, participants agreed to a set of "principles of cooperation in Northeast Asia."[104]

The nonprofit sector being the competitive intellectual market that it is, other venues are also evolving for informal discussions. In October 1997, the Georgia Institute of Technology's Center for International Strategy, Technology, and Policy (CISTP) held its first Northeast Asia Military-to-Military Security Dialogue, in cooperation with the Russian Institute for Strategic Studies. Retired senior military officers and "security experts" from Russia, China, Japan, South Korea, and the United States discussed "cooperative regional security, military transparency, information accessibility, [and] confidence building measures." North Korean representatives were invited but did not attend. CISTP's objective is to develop personal relationships that have some chance eventually to influence the establishment of a Cooperative Regional Security Regime.[105]

The North Pacific Working Group of CSCAP held four meetings through November 1998, the last one in Beijing, with North and South Korean representatives among the 40 individuals participating, as always for an NGO-sponsored event, in their private capacities. The group discussed security implications of the economic crisis (no consensus) and the Korean security situation (expressing concern about its "overall intractibility").[106]

The Nautilus Institute, based in California, sponsors the Northeast Asia Peace and Security Network (NAPSNET), which distributes regional news, commissions and distributes (on the Internet) short regional security studies from North American and Eurasian authors, and hosts online discussions of their work.[107]

More ambitious, technologically, is a project sponsored by IGCC that it calls "Virtual Dialogue—Northeast Asia." Supporting electronic dialogue among NEACD participants is to be a network of hardware and software that IGCC calls "Wired for Peace." Taking as given that China and North Korea remain allergic to free and unfettered flows of electronic information within their societies, it aims to provide secure electronic data channels between and among NEACD participants, so that the expensive face-to-face meetings may be supplemented by rapid and frequent electronic interchange. Envisioned as more than an e-mail net, "Wired for Peace" is intended to tap sophisticated workgroup technology developed at University of California at Davis and Lawrence Livermore National Laboratory. The ambitious concept contemplates real-time interactive data sharing, "breakout rooms" to facilitate the equivalent of corridor conversations in between plenary discussion sessions, document translation services, multiple indexing of documents, and a "recorder and rapporteur" function. IGCC hopes to cross-link the system with the data collections of the Stockholm International Peace Research Institute and CSCAP.[108]

South Korean scholar Moon Chung-In has concluded that just such track II initiatives might help Koreans overcome the substantial obstacles that remain in the way of peace, but only if they first let go of their single-minded focus on reunification on terms favorable to one side. Political/ideological indoctrination on both sides of the thirty-eighth parallel stresses unification as "the only avenue to national survival and prosperity," which means that CSBMs, arms control, and peace building initiatives tend to be seen as elements of the other side's (malevolent) unification strategy. Recognition of each government by the other, Moon argues, and acceptance of the de facto situation of "two separate states" are the necessary preconditions for peace. After mutual recognition, arms control and peace-building measures then could be considered on their own merits.[109]

These sentiments seem to be echoed in Kim Dae Jung's new policy of "constructive engagement." While remaining determined to resist aggression by the North, it "extends an olive branch to Pyongyang by renouncing a policy aimed at absorbing the North or provoking its collapse" and seeking instead eventual reunification via "cooperation and confidence building." The new policy also seeks to separate economic relations from politics, so that trade and investment might evolve even if political relations do not.[110] North Korean perceptions that the United States has not yet made the same transition, that is, still hopes for the ultimate, near-term collapse of the communist regime, may reduce the North's willingness to cooperate in the various two- and four-party talks, or with international aid agencies seeking to alleviate famine conditions.[111]

Moon argues that forming the necessary community of technical expertise in arms control and CSBMs should also be a high priority for the two Koreas. Electronic information networks and online libraries (like those being formed by IGCC) could help to build the necessary knowledge base, while a community of shared expertise might be built via unofficial conflict resolution workshops:

> Given that the Korean conflict is not concerned with interests arising from control over scarce resources, but over values, identity, recognition, and security needs which are all in unlimited supply, track two diplomacy workshops aimed at controlled communication could be very useful in making breakthroughs in the current inter-Korean arms control stalemates.[112]

CSBMs, Arms Control, and Democracy in Latin America

In contrast to most of the world's developing regions, Latin America was an early escapee from European colonialism, its major metropoles, Portugal and Spain, being early victims of imperial overstretch. The mainland states, mostly independent by 1821, struggled for more than a century to reconcile the ideals of republican governance with the authoritarian traditions of their Catholic inheritance that favored landed elites and military hierarchies. Largely immune from the late wave of European imperialism that carved up Asia and Africa in the mid- to late nineteenth century, the states of Central and South America had a long period in which to work out most major border issues, sometimes by war (as in Mexico's loss of Texas and, ten years later, the rest of what is now California and the American Southwest to the United States). In the first decades of the twentieth century, the United States took it upon itself to promote good governance and greater respect for the prerogatives of American companies both in the Caribbean and among the smaller states of Central America. By 1948, republican tendencies were sufficiently ascendant that the 21 signatories of the Charter of the Organization of American States (OAS) could declare in its preamble that "representative democracy is an indispensable condition for the stability, peace and development of the region." At the same time, they took pains to ensure that the OAS charter forbade outside intervention in the internal affairs of its member states.[113] For the next four decades, Cold War priorities meant that promoting democracy took second place to shutting out communism, and that even democracy could be sacrificed in the name of that task. Both outside interventions and internal military takeovers were justified by the need to keep communism at bay in the hemisphere. Moreover, argues Tom Farer, in post-Castro Latin America the tenor of military takeovers changed from emergency interruption of constitutional government to indefinite suspension

> of both elected regimes and the frequent companions of such regimes, freedom of speech and freedom of association. Uninhibited by the dysfunctional constraints of electoral and constitutional government, they would proceed methodically to heal a feverish national society by cauterizing its leftist infections and enclosing it in a new political economy.[114]

Yet, by the mid-1980s, much of South America enjoyed at least some of the institutional elements of democracy, and by the mid-1990s the same was true of Central America as well.[115] Preserving and building democratic rule, consistent with the aspirations of the OAS charter, became a greater collective interest of governments in the region.

In the particular political and historical context of Latin America, the confidence in "confidence-building" has come to mean, (a) assurance that the U.S. military, in particular, would not be used to intervene unilaterally in another country's domestic affairs, and (b) assurance that legitimate civilian rule would not be take away by a coup d'etat. Perhaps the best way to achieve (a) in the post–Cold War era is to promote (b). Moreover, if the "democratic peace" thesis is correct (that democracies tend not to fight one another[116]), then the best way to promote arms restraint in the region is to promote sustainable democracy, which should lay the groundwork for restraint regimes.

There have been several initiatives to constrain conventional armaments in Latin America, including an abortive measure to limit armaments in the mid-1970s and efforts in the 1990s linked to support for democratic governance. In 1995, the OAS began to consider CSBMs as such for the first time.

Declaration of Ayacucho (1974)

In 1974, in an effort to head off an arms race with the right-wing military regime of Augusto Pinochet in Chile, the head of Peru's left-wing military government proposed that the countries of the Andes commit themselves to "create conditions which permit effective limitations of armaments and put an end to their acquisition for offensive warlike purposes in order to dedicate all possible resources to economic and social development."[117] Eight Andean countries signed the subsequent Declaration of Ayacucho. Follow-on talks addressed measures to "freeze existing ratios of weapons-to-manpower and levels of military expenditures in relation to gross national product." They also addressed monitoring measures, demilitarized zones, and thin-outs of military forces along common borders. The talks were inconclusive, however, partly because Brazil, the region's largest country and neighbor of all but two Andean states, was not a part of the process. This undercut its potential effectiveness.[118]

In 1978, the president of Venezuela attempted to reenergize the stalled Ayacucho process during the United Nations Special Session on Disarmament, and Mexico's foreign minister proposed an entity comprised of all states of Latin America and the Caribbean to set limits on the acquisition and transfer of arms in the region. Representatives of 20 Latin and Caribbean states (including Cuba) met in Mexico City in August 1978 and "agreed to exchange information on weapon purchases and work toward a regime of restraints."[119] No operational regime resulted, but Latin America as a whole has seen relatively modest levels of arms imports in the past two decades.

Central American Peace Process (1983–1991)

Much of Central America was torn by conflict in the 1980s. Beginning in 1983, some Latin American states worked to promote an end to conflict. The efforts of the Contadora Group (Colombia, Mexico, Panama, and Venezuela) emphasized interstate CSBMs aimed at controlling arms trafficking, use of one state's territory to destabilize another, and other "interference in the internal affairs" of states in the region. Although the region's conflicts were primarily internal, they were heavily affected by the Cold War, and ideological issues loomed large (or were made to do so as a way of deflecting attention from what were mostly fights about the distribution of land, wealth, and access to government). "Interference" applied equally to Soviet or Cuban support for a leftist regime like Nicaragua's and to the U.S. creation and training of the counterrevolutionary guerrillas sent to overthrow it.

The Contadora principles were largely folded into the successor "Esquipulas" peace process (both were named for their first meeting sites), which furthered a regional peace plan devised by Costa Rican president Oscar Arias in 1987. The plan included a five-state Central American Security Commission intended to address "security, verification, control, and limitation of weapons." The Esquipulas process, abetted by United Nations mediation, helped to end the civil wars in Nicaragua and El Salvador in 1989 and 1991. Jack Child argues that one of the most important CSBMs that Esquipulas provided was an increased rate of high-level communications among Central American leaders.[120] The endgame of the peace process in both states was assisted by UN peacekeeping operations, and in Nicaragua by an OAS-sponsored mission to demobilize the U.S.-supported "contras."

Santiago Commitment to Democracy (1991)

As early as 1959, OAS ministers of foreign affairs agreed that "harmony among the American Republics can be effective only insofar as human rights and fundamental freedoms and the exercise of representative democracy are a reality within each one of them," and that the "existence of anti-democratic regimes constitutes a violation of the principles on which the Organization of American States is founded."[121] Three years later, they directed that sentiment at Fidel Castro's Cuba, the only avowedly communist country in the hemisphere, when ousting it from the OAS. As Acevedo and Grossman note, members of the OAS "considered that collective action against illegitimate governments did not constitute a violation of the principle of non-intervention," at least in the context of the Cold War.[122]

The next step toward collective defense of democracy came three decades later, when the Cold War had ended. In June 1991, the OAS General Assembly approved Resolution 1080, which called for a meeting of foreign ministers, or convocation of a special session of the assembly, within ten days of "any occurrences giving rise to the sudden or irregular interruption of the democratic political institutional process or of the legitimate exercise of power by the democratically elected government in any of the Organization's member states. . . ."[123] The resolution did not commit members to collective action in defense of democracy, but established a mechanism whereby they were at least obliged to discuss the ramifications of what would previously have been considered the internal affair of a member state.

The Santiago Commitment was tested three times in the next two years: by the coup in Haiti (July 1991), and by the "self-coups" (*autogolpes*) in Peru (April 1992) and Guatemala (May 1993). Haiti had conducted its very first democratic presidential election in December 1990, with the populist Jean-Bertrand Aristide elected overwhelmingly by the country's extremely poor public. Shortly after his ouster by the Haitian military, OAS ministers met and condemned the coup, demanded Aristide's restoration, and invited OAS members to suspend economic, commercial, and financial ties with Haiti. These voluntary sanctions were honored mostly in the breach, and UN sanctions fared little better. Ultimately, Aristide was restored to office with the help of 20,000 U.S. troops, assisted by contingents from other states and supported by UN Security Council Resolution 940.[124]

When Peruvian president Alberto Fujimori suspended democratic government in order, he argued, the better to fight both political corruption and terrorist guerrillas, the OAS called for the reestablishment of democracy, but did little else. When Guatemalan president Jorge Serrano suspended the country's constitution, OAS condemnation energized local opposition, international economic sanctions alarmed the business community, and the military intervened to reverse his action. Since economic sanctions were also applied to Peru, analysts suggest that the differences in international impact in the two situations might be attributed to the fact that Fujimori's action was locally popular while Serrano's was not.[125]

As a corollary to the Santiago Commitment, in 1997 the OAS charter was amended by the Washington Protocol, which allows the suspension from the organization of any member whose democratically elected government has been overthrown by force.[126]

Central American Democratic Security Treaty (1995)

At a July 1991 meeting of Central American presidents, the government of Honduras unveiled a comprehensive proposal for regional disarmament and confidence-building that would set ceilings on military inventories and troop levels; require pledges from regional governments on the non-use of force in interstate disputes; and seek a pledge from the United States not to repeat its support for irregular fighting forces in Central America.[127] At a December 1995 summit in San Pedro Sula, Honduras, the presidents of Costa Rica, Honduras, El Salvador, Guatemala, Nicaragua, and Panama signed a Treaty on Democratic Security in Central America, which drew upon the earlier Honduran concept. The signatories pledged to achieve a "balance of forces" through arms limitation, to exchange annual reports on the presence of foreign military advisers on their territories, to cooperate in the fight against illegal drug and arms trafficking, and to reintegrate refugees and displaced persons in the region.[128] The agreement also emphasized "consolidation of democracy, the rule of law, and development," while encouraging a "regional arms control arrangement that promotes transparency."[129] Finally, it provided for a pooling of local resources to support a Central American Police Academy, based in El Salvador, for which the United States pledged technical and training assistance.[130] Since rising crime rates are a near-universal menace in postconflict Central American societies, this new joint effort at training effective civilian police is a welcome step.

Human rights groups like the Arias Foundation criticized the 1995 agreement, however, for its failure to advocate steeper cuts in military spending or steps toward regional demilitarization. Two countries in the region, Costa Rica and Panama, currently have no military forces, but in the others the military remains a powerful political institution wary of further measures to downsize it.[131]

OAS Declarations on Confidence and
Security-Building Measures (1995, 1997)

In 1992, the government of Chile proposed a hemispheric meeting on CSBMs. In March 1994, the OAS held its first meeting of governmental experts on the subject, and the first meeting of the hemisphere's defense ministers in July 1995 also addressed CSBMs. The following November, Chile got to host the conference that it had proposed. The resulting "Declaration of Santiago" committed OAS member states to move toward advance notice of military exercises, to participate fully in the UN Register, to exchange of information on defense policy and doctrine, and to invite foreign observers to military exercises.[132]

Two years later, in June 1997, the OAS General Assembly nudged the organization toward a regime of "advance notification of major arms acquisitions," defined as those covered by the UN Register. The same meeting reauthorized or expanded OAS peace-building missions in Guatemala, Nicaragua, and Haiti. In November, OAS members signed the Inter-American Convention against Illicit Production of and Trafficking in Firearms, Ammunition, Explosives, and Other Related Materials.[133]

Overall, the OAS approach toward CSBMs remains tentative, yet any movement toward transparency in military doctrine or spending, advance notice of arms transfers, or advance notice and foreign observation of military maneuvers is a step ahead for a region whose states historically have been very leery of any measures that chip away at sovereignty or give the slightest justification for outside intervention in their affairs. The long history of U.S. military intervention, from the Spanish-American War through contemporary drug eradication and monitoring efforts, reinforces resistance to transparency measures and arms control. An even longer tradition in many Central and South American states of military careerism, involvement in politics, and periodic seizures of the reins of government makes efforts to open up the region's security sector difficult and slow-going. But governance in the region has been largely civilianized and democratized over the past decade, and as these forms are consolidated, greater transparency in military affairs seems likely to follow, as does greater cooperation in dealing with the transnational security issues (guns, drugs, organized crime, economic migrants, and refugees) that have supplanted interstate threats as the principal concerns of contemporary governments.

Observations and Conclusions

What, in the end, are the objectives of all these confidence-building initiatives? They are the avoidance of war, the reduction of tensions and misunderstandings that could lead to war—partly by inducing policy changes among countries participating in CSBMs—and the freeing of resources for purposes other than buying guns and paying

Table 5.3 Overlapping Ranges of Conflict and Reassurance

Spectrum of External Conflict/Cooperation	Spectrum of CSBM/Reassurance Measures
Total war (WMD).	No cooperative interaction.
Medium-scale conventional war (Iran-Iraq, Ethiopia-Eritrea).	No cooperative interaction.
Low-scale conventional war for limited objectives (Falklands).	No cooperative interaction.
Large cross-border operation (Israel-Lebanon)	No cooperative interaction.
Minor border clashes (N-S Korea at sea).	Declaratory CBMs; policy white papers.
Subversion, outside sources (India-Pakistan, East-West standoff in Cold War).	Unofficial issue discussions; nonmilitary CBMs: trade, cultural, social, sporting. Official discussions on issues of common interest (not necessarily security).
Competitive arms acquisitions (India-Pakistan).	Easy military communications CBMs: officer exchange visits; port visits; UN Register. Steps of direct political-military consequence: hotlines, advanced notice of weapon tests and exercises. Steps that alter military readiness or increase attack warning: de-alerting, border thin-out zone, no-fly zone; host a peacekeeping operation.
Cooperative balance-of-power (ASEAN/ARF).	Steps that require field cooperation: joint search and rescue, contribute troops to peacekeeping, piracy suppression. Steps that alter military capabilities: voluntary arms import limitations (quantity or quality).
Pluralistic security community (NATO, less Greece and Turkey; possibly South America).	Steps that verifiably alter military capabilities: mandatory arms import/force size limitations, with mutual compliance inspections. Arms reductions, with ongoing verification. Joint operational training, planning.
Amalgamated security community (USA, goal of European Union).	Common defense doctrine, policy, forces.

troops to stand ready to resist others' guns and troops. Do they facilitate these objectives or are they themselves facilitated by political changes whose origins lie elsewhere? The details of the answer vary by security complex. Properly used communications links, for example, might keep a mini-crisis from escalating to major proportion. Greater personal interaction between military leaderships may add a human dimension to decision making that undercuts the maintenance of routine enemy images and thus also promotes more effective crisis management. Operational CSBMs can reduce concerns that military exercises actually cover mobilization for war. Yet each of these effects requires an initial political/perceptual orientation that is able to empathize, at least a little, with one's counterparts in the confidence/reassurance enterprise, and that may not be able to withstand levels of interparty conflict that exceed some critical threshold. Possible thresholds and the sorts of CSBMs they allow are suggested in table 5.3, which compares an extended version of the external conflict scale used in chapter 2 with confidence-building, reassurance, and arms control measures discussed in this chapter, roughly ranked by depth of impact on military capabilities and operations.

Neither confidence-building nor arms control measures are likely to be pursued by two or more states actively engaged in warfare. Certain kinds of CSBMs become politically possible if expressions of mutual hostility do not exceed, say, minor border clashes. "Subversion" comes in many varieties, from support of guerrilla fighters and terrorists to broadcasts of balanced international and local news that closed societies would prefer that their people not hear. Even non-military measures such as cultural, social, and sporting exchanges, viewed as forms of CSBMs by open societies, can be corrosive of certain polities. Competitive arms acquisitions might well continue in the face of CSBM regimes—from communications measures to border thin-out zones—that are directed at keeping military forces from swinging into action inadvertently. That does not make such CSBMs the enemy of arms control: countries that are not ready to negotiate the size or shape of their militaries may more willingly agree to reversible operational constraints, that is, they may opt for what John Ruggie calls a "cooperative balance of power."[134] Countries that do reach accords to limit and reduce armaments may find that corollary CSBMs reinforce those agreements. Europe's CFE and OSCE exemplify such a symbiosis. Security complexes that are capable of negotiating and implementing such regimes in all likelihood have evolved into "pluralistic security communities"—that is, they have reached a level of political accommodation that precludes armed conflict to settle differences, although the states involved retain their own independent foreign and security policies and their own military forces.[135] (An "amalgamated" security community, by contrast, is one with unified policy and forces, such as the United States, which amalgamated in 1789 with ratification of the Constitution and has remained so except for the period of the Civil War [1861–1865].)

Where underlying relations are hostile and driven by issues of identity that affect the relations of the populations at large, as in the case of India and Pakistan, CSBMs can be negotiated but may not be implemented, or they may be implemented in bad faith. Pushing such measures where implementation is likely to be thus distorted can be counterproductive.

Indeed, extremism is, almost by definition, a show-stopper for many CSBMs, as it is for most forms of compromise. Interpreted as unwillingness to tolerate or to share power, space, or legitimacy with alternative belief/value systems or their adherents, it takes many forms. They include racism (as in Rhodesia and apartheid South Africa), religious fanaticism (as with Iranian hardliners, the Taliban in Afghanistan and Pakistan, Hindu extremists in India, or the Irish Republican Army), hypernationalism (as in Serbia), cloistered ideologues (North Korea), fascist dictatorships (Burma, Iraq), and plutocratic greed (Haiti).

Are there, however, ways in which CSBMs can be implemented so as to reduce tensions or the risk of conflict between or among states whose motivations may be less extreme, yet whose relations are difficult? One approach is to involve official third parties, either state- or international organization-based. Third parties have run the effective peacekeeping and monitoring missions on Israel's borders for a very long time, for example, and the data they gather is shared equally among the states concerned. This approach won't work where one or more states' views of the dispute do not admit of "outside interference," and it will be problematic where one side in a dispute has been named the aggressor by the international community (for example, Iraq since 1991).

It is in the great middle ground of the security dilemma, between security community and low-level conflict, that CSBMs function best. They are being used by regional groupings like ARF and SADC, and perhaps now the OAS, to increase the strength, frequency, and variety of interstate communications and to magnify common values and interests. In the case of ARF, multilateral meetings and bilateral exchanges may help to minimize risks of war by accident or escalation from some otherwise minor incident. In southern Africa, SADC is struggling to build a capacity for peacekeeping while coping with the internal hemorrhaging of two of its larger members. In the Western Hemisphere, the OAS may be able to build on democratization and recent settlement of a number of remaining border disputes to create regimes of CSBMs and arms control that reinforce Central and South American progress toward pluralistic security communities.[136]

With regard to arms acquisitions, effective CSBMs may help to render them less dangerous to regional stability because channels of communication on doctrine and policy and discussions between political and/or military leaderships can help to clarify intent.

The utility of CSBMs also might be assessed from a counterfactual direction: what might relations be like in the regions where CSBMs have been implemented if no such efforts had occurred? What states or regions would be better off without regular high-level meetings of leaders, mid-level information sharing, notice of military exercises, or military exchange visits? Would security and stability improve? Would arms flows decrease? In the middle ground, bringing CSBMs to a halt would likely serve neither regional stability nor individual states' security interests.

Unofficial track II meetings can facilitate discussion of sensitive matters without committing government face, status, or clout to the effort. Like a butterfly bandage on a serious cut, they may not be the primary source of healing for serious regional rifts but may still help to avoid needless scarring.

On the other hand, CSBMs that escape official channels can be harmful to the health of some ruling orders. Information transparency, cultural exchanges, or tourism, for example, can reinforce the stability of a legitimate status quo but can undermine the stability of one that is not. In Europe, greater openness and emphasis on human rights eventually helped cause Eastern bloc governments to fall and states to fissure. So North Korean leaders are not irrationally worried about letting outside aid agencies have the run of the country, or about letting tourists speak freely with the local population.[137] At this writing, there are thousands of South Korean tourists and businessmen in the North seeing sights and cutting deals. It remains to be seen whether official diplomacy, military deterrence, or the steady flow of cash and conversation that these exchanges bring will have the greater impact on the peninsula in the end.

CHAPTER 6

Conclusions and Implications
for U.S. Policy

The arms trade is a particularly vexing problem for those who would curtail it, as it both reflects and exploits the fundamental political organization of the globe. Politics is about power, status, and security, and armaments contribute power, status, and security to a state and its military institutions. Since the state historically has been the ultimate user of organized violence as well as society's ultimate defender against violence from other states, conventional arms are to it what a good set of pipe wrenches are to a plumber, namely the tools of its trade. The distribution of those tools at a discount in the first quarter-century of Cold War that followed World War II served the political interests of one or the other camp in that competition. Cash sales thereafter rode the boom in oil prices, and the arms trade became a lucrative economic enterprise as well. Because that trade has served both the political and economic interests of supplier states, it has proven particularly resistant to supply-side controls. Because the arms trade has catered to recipient states' security wants and needs as well, demand-side restraint has been slow to arise. Since the late 1980s and the end of most subsidized arms sales, the most effective constraint on demand has been economic recession, not political action or reconciliation.

Global conventional arms control is rendered even more difficult by the fact that the governments of nation-states are not well adapted to an environment of globalized networks of relations. Founded on an exclusionary premise, tasked to protect a fixed amount of physical territory from outside incursion, and organized in bureaucratic hierarchies, governments are much less flexible in how they conduct their affairs than are market-based commercial firms and highly networked civil society organizations—two classes of global actors with growing clout.[1] The zero-sum notions of national autonomy and security that underlie nation-state *realpolitik* contrast with the positive-sum thinking that propels commercial economic market growth and private, nonprofit groups' work in, say, democratization, human rights, or public health. Authoritarian governments whose base of support has long been narrow, weak, or divided, fear that dissatisfied constituents will find new strength in connectedness to like-minded outsiders, including wealthy expatriate communities. Such fear may be well founded, since exposure to Western goods and ideas helped to shatter the communist governments of Eastern Europe and the Soviet Union.

At the beginning of the twenty-first century, recipient states enjoy a global arms buyers' market that is freer than at any time in the last 50 years. Sales competition is such that the major arms supplying states, which are also the permanent members of the United Nations Security Council, function more consistently as market competitors than they do as joint stewards of international peace and security. Of the major suppliers, only the United States has sustained military operations outside its home region, so the others can afford to behave without regard for the *realpolitik* reasons for arms transfer restraint, especially the risk of facing one's own weapons in a future regional intervention. Normative reasons for restraint have, to date, fared little better.

Although conventional arms suppliers sell a dangerous commodity, it is one whose impact is highly context-dependent. Arms to a would-be aggressor may increase the likelihood of war by altering its estimates of success, while arms to a would-be victim of aggression may reduce that likelihood proportionally, but not always. Conventional arms transfers are thus a difficult subject to manage, even more difficult than alcohol, to which we made analogy in the introduction to this book: we know how much liquor will make the average person drunk and we set law enforcement standards accordingly. We cannot as readily measure, in advance of trouble, what makes a regional arms balance unstable or accumulations too great. Such case-dependence makes management regimes difficult to craft, implement, and sustain on any sort of quasi-scientific basis, even where there is political will to do so.

This chapter is about steps that might be taken in such an imperfect, interconnected world to enlist the support of the major arms supplying states for regional regimes of stability and security and to encourage states in the security complexes of the developing world to generate such regimes. If anything is clear from the preceding chapters, it should be a sense that armaments alone—either supplied or withheld—neither create nor sustain the kind of stability that developing regions need to join and benefit from economic globalization, the open and transparent markets that it necessitates, and the open polities that it implies. On the other hand, to ignore completely the role of armaments in fostering regional instability would clearly be a mistake: Iraqis did not roll into Kuwait on bicycles. Angolans do not fight each other with hedge funds. In a world that aspires to be orderly, lethal commodities require serious and orderly management, within a decision-making framework that suggests what to manage and when.

The remainder of this chapter discusses several such frameworks of increasing political and technical difficulty but potentially increasing efficacy as well, if efficacy is defined in terms of promoting regional peace and stability. Remaining within the realm of status quo politics and institutions, only limited changes in the arms trading system and its consequences are possible. However, the costs of such modest achievements would themselves be modest and the potential benefits should therefore not be dismissed lightly. Greater changes could be had by strengthening or extrapolating from current restraint measures like the Missile Technology Control Regime (MTCR) and Wassenaar Arrangement to increase their reciprocity and to promote more proactive partnership among the world's democracies. Such partnerships could aim to limit the access of bad actors to the tools of war or to sanction warlike behavior while supporting the growth of the global community of democracies, a longer-term solution to problems posed by proliferation of conventional and other armaments.

For a synopsis of these possibilities, see table 6.1. Owing to the interplay between conventional armaments and weapons of mass destruction (WMD), table 6.1 includes measures that may be equally applicable to controlling both sorts of weapons.

The chapter concludes with recommendations for U.S. policy.

Table 6.1 A Matrix of Restraint/Stability Measures

Objective/Focus		Measures in Place or Readily Actionable (coverage/range/sponsor)	Technically or Politically More Difficult Measures
Types of weapons	Complete systems	UN Register of Conventional Arms MTCR (limiting trade in longer-range missiles) CFE Treaty (heavy weapons; Europe, FSU) Ottawa Convention (antipersonnel landmines) Biological/Toxic & Chemical Weapons Conventions Treaty of Tlatelolco (nuclear weapons-free zone) Conv. on Conventional Weapons ('inhumane weapons) Non-Proliferation Treaty (new nuclear weapons states)	Ballistic missile ban (for missiles of 100km+ range not already restrained by international agreement).
	Basic technologies	Wassenaar Arrangement (hi tech/dual use) London Suppliers Group ('nuclear tech) Australia Group (chem/bio tech)	Make Wassenaar enforceable. Continuous post-sale end-use monitoring of sensitive technologies.
Specific actors	Outlaws and active state sponsors of terrorism	Iraq sanctions (by UN, US) Iran sanctions (by US) Libya sanctions (by UN '92–'99; US '86+) North Korea sanctions (by US) Agreed Framework (by N. Korea & US)	Unilateral counter-proliferation (rollback of outlaws' advanced conventional or unconventional weapon capabilities).
	Authoritarians	Burma sanctions (by US, EU) Cuba sanctions (by US)	Code of Conduct for arms transfers.

(continues)

Table 6.1 *(continued)*

Objective/Focus		Measures in Place or Readily Actionable (coverage/range/sponsor)	Technically or Politically More Difficult Measures
Situations	Former fracture zones	OSCE (Europe, FSU) Open Skies Treaty (US, Europe, FSU) Santiago Commitment to Democracy Central American Democratic Security Treaty *Supplier support for Latin American democratization, military professionalization, and arms import restraint*	Democratic Security Partnership (Gompert). Pluralistic security communities in Central and South America, southern Africa, elsewhere.
	Current fracture zones	Yugoslavia arms embargo (by UN) Peace implementation in Bosnia, Kosovo (by NATO) Iraq no-fly zones (by US, UK) Cyprus, Middle East peacekeeping (by UN, MFO) *Support for Israel-Lebanon, Israel-Syria peace.*	Joint Western defense of Persian Gulf energy sources. Concerted international effort to settle the Kashmir dispute and conflicting South China Sea territorial claims.
	Areas of chronic instability	Liberia, Somalia, UNITA arms embargoes (by UN) *Supplier support for restraint measures in central Africa*	Peace enforcement by major powers in collaboration with regional powers.

Modest Initiatives I:
Encouraging Regional Efforts to Promote Peace

Andrew Pierre makes the argument that the search for supply-side measures to restrain the arms trade came in reaction to the failure of recipient states to create restraint regimes of their own.[2] The assessment of confidence and security building measures (CSBMs) in chapter 5 suggests, however, that several regional security complexes are now or may soon be in a position to implement restraint measures to promote peace and stability. Two levels of action are needed to facilitate such initiatives. On the first level are actions to create the political conditions that may lessen demand for conventional armaments. On the second level are actions that build on those conditions to craft regimes of restraint in arms acquisition, military spending, military end-strength, or weapons inventory. Major arms suppliers should be on record as prepared to honor such initiatives.

First-level Actions

Based on the positive experiences of Europe and South America and the cautionary experiences of the Middle East and South Asia, necessary first-level actions to promote regional peace include member states' satisfaction with territorial borders and a measure of mutual political accommodation. Put another way, revanchist claims, deep ethno-religious enmities, and absolutist agendas generally make unbuildable foundations for regional regimes of arms restraint. Certain modest conflict-avoidance and confidence-building measures can help to improve relationships in which security dilemmas, misperceptions, or miscommunications play a role in igniting or sustaining strife, but usually something that renders a relationship at least somewhat less conflictual needs to precede even such limited steps. In Europe, it was a two-step process of agreement on postwar borders (1971), followed 15 years later by Mikhail Gorbachev's internally driven Soviet foreign policy changes.

Where border problems are manageable and there is at least some flex in states' foreign policy objectives, stable peace is most readily built among competent, responsive, and accountable governments. Elements of such governance include honest provision of public services, with effective anticorruption measures that separate public service from private gain; non-discriminatory provision of public services and protections, to prevent or mitigate unrest among a state's constituent groups; and a level, stable, and law based economic playing field to encourage investment and promote economic growth. International support for good governance and the economic growth to sustain it may take the form of debt relief and/or development assistance linked to "democracy, the rule of law, good governance and respect for human rights and for core labor standards." Contributors may expect such funds to be invested in social infrastructure (schools, public health).[3]

The value of political openness and accountability can be seen in the wake of East Asia's financial crises of 1997–1998. Where crony capitalism had relied too much on fickle international finance and its finicky definitions of propriety and fiscal accountability, such global exposure eventually burst some economic bubbles. The countries that recovered most quickly were those like Thailand and South Korea that cleaned house politically and economically, instituting both more responsive, democratic government and more transparent financial dealings.[4] Recalling terms used in chapter 2, these countries upgraded their "security software," especially the subroutines for governmental legitimacy. With broader agreement on the government's right to rule, they could also take necessary, if difficult, steps to improve its capacity to govern effectively, especially in the economic sphere. Both countries were fortunate in already having fairly well integrated societies, but their experiences suggest that those who advocate democracy as the necessary complement to open market economics as a recipe for regional stability are on the right track.

The path to prosperity for a developing country was once also thought to include an indigenous defense industry. Armaments once incorporated unique, cutting-edge technology that young states hoped to spin off into economic bounty. Today, however, cutting-edge defense industries frequently adapt rapidly changing civilian technology to boost military systems' performance. Civilian information technologies and other electronics applications (for example, in sensors, guidance, and the decision-making adjuncts known as "expert systems") are evolving so rapidly that the global market has become far superior to government channels as the source of leading edge technology, industrial innovation, and profit. Acquisition of various kinds of civilian technology, whether through foreign direct investment or local entrepreneurship, may now offer the sort of high-tech boost to local industry that used to be sought first from defense production.

Exceptions to the trend favoring non-military investment might be found among those countries that the general commercial market, in its profit-making wisdom, may consider too risky, unstable, or otherwise inopportune for investment—states that are mired in local or regional feuds or corrupt practices—in favor of calmer and less kleptocratic climes. In such cases, governments can try to exploit the buyer's market in armaments to wring technology transfers, sales offsets, or countertrade deals from sellers. In both the short and longer term, however, sneakers or circuit boards have greater third-party market potential than co-produced fighter-aircraft widgets.

In addition to competent, responsive governance and productive economies, highly professional militaries are key to a program of regional peace and stability. The University of North Carolina's Stephen Biddle argues that in states in which the military rules or is feared by governing elites as a threat to their rule, "the threat of political violence by the military creates powerful disincentives for military professionalism. [N]o regime can safely permit officers of unknown loyalty to attain high rank. To do so is to risk violent overthrow either directly . . . or through unopposed rebellion of disaffected groups within the society."[5] Techniques to control the risk of military intervention in politics include "frequent rotation of commanders, suppression of horizontal lines of communications within the military hierarchy, divided lines of command, isolation from foreign sources of expertise or training, and exploitation of ethnic divisions in officer selection or combat unit organization." Moreover, without an ethic of "disinterested dedication to duty," a military organization may have difficulty accomplishing even "mundane tasks such as equipment maintenance and . . . efficient staff procedures." When it is called upon to go into battle under modern, high-technology, high-tempo conditions in which events are not likely to go according to plan, such an organization is likely to suffer badly at the hands of an equivalently modernized military with a professionalized officer corps. The latter sort of forces and their states, Biddle argues, will "prosper disproportionately in the military competition of the future."[6] In short, professionalization pays off in military terms, and political involvement is corrosive of professionalism. An apolitical military can do a better job securing its country's interests.

It is easier for the military to stay out of politics if civilian officials are competent and effective at their tasks and act in the public interest, especially in matters of defense and national security. South African defense analyst Laurie Nathan argues that in emerging democracies, civil and military leaders need their own versions of internal CSBMs to build mutual trust and respect. For example, the government itself must operate within the law, refrain from using the armed forces for partisan purposes, respect their professional input and "corporate interest in defence policy making," and provide the wherewithal they need to do their jobs effectively.[7] Joao Honwana, senior researcher at the Centre for Conflict Resolution, Cape Town, argues further that, "just as democratic civil-military relations are essential for democracy to succeed, so the democratization of the state and society are critical to the establishment of a democratic civil-military interaction."[8] That is, reliably civilian governance must be democratic governance; otherwise one authoritarian institution is as good—or as eligible—as any other to take the helm of state or to act as its unquestionable internal "guardian."

Second-level Actions

Where basic political foundations for restraint have been laid, countries within a security complex may find it in their interests to codify such restraint. South America, for example, seems ripe to pick up the Ayacucho arms restraint initiative where it was dropped in the late 1970s, perhaps with a nudge from track II groups like the Carter Center's Council of Freely Elected Heads of Government, which argues that states in the region ought to forego a costly round of largely cosmetic arms modernization and focus their energies instead on the real threats that face their societies. Narcotics cartels, for example, remain a blot on the region. Having undermined Colombia, they are doing the same in Mexico. Drugs and the organized crime issues they entail perhaps typify the sorts of new security issues that otherwise stable regions face in decades to come, issues that require extensive regional cooperation, including major efforts to stem the corrupting influence of drug money.[9] Although, as chapter 5 indicated, there is some movement to face the new agenda, there were also rumblings of interest within South America to update aging fighter squadrons with expensive newer units, and the August 1997 lifting by the Clinton administration of a 20-year "presumption of denial" of U.S. exports of high-tech weapons to the region seemed to prefigure a surge in sales. By late 1998, however, no surge had occurred. No country ordered qualitatively different equipment (Ecuador ordered a squadron of F-1 fighters from France, matching a squadron already on hand). Uncertainty created by ripple effects of the Asian and Brazilian currency crises may have caused a postponement in purchases. On the other hand, South American states' military needs, given those states' evolving democratic political relations, may no longer point toward such big-ticket items, or those needs may be satisfiable with token numbers.[10]

The Southern Africa Development Community (SADC), although struggling to cope with the long-burning civil war in Angola, the battle for control of the Democratic Republic of Congo, and the related issues of Great Lakes border security, is an institution with potential to construct a security community amongst at least its democratic members. SADC aspires to create a joint peacekeeping unit, the first step in the development of a potentially valuable operational capacity for a security complex still beset with troubles. In general, regional peacekeeping enterprises may offer opportunities for regional military cooperation, training, and exercises, not to mention tools for keeping the peace.[11] So far, the major example of close cooperation in this area has been Scandinavia, whose countries' joint peacekeeping units served in Bosnia and Macedonia for a number of years. Such cooperative ventures abroad give military forces from smaller and middle powers an explicit and productive purpose that expands their horizons beyond internal security and the patrolling of national borders. Service with regional or UN peacekeeping operations to implement peace agreements is, moreover, an avenue by which the military forces of democratizing states can contribute to the evolution of democracy abroad.

In other regional security complexes, arms restraint and related security initiatives await movement in the political arena. Tenuous efforts by India and Pakistan to improve their relations were set back by 1999 fighting in Kashmir, and militants on either side have an interest in sabotaging reconciliation efforts. The evolution of political relations in Southeast Asia hinges upon what emerges from Indonesia's political upheavals and upon the future zeal with which China pursues territorial claims in the South China Sea. Peace and stability in Northeast Asia also depend upon political evolution in China and its relations with Taiwan, together with progress in coaxing North Korea out of its cave and in coping with its often-feral political instincts. The necessary precursors to quicker progress eventually may turn out to have been the economic crises in both North and South that mandated acceptance of outside assistance by the North and facilitated the election of Kim Dae Jung in the South.

Supplier Support for Regional Initiatives

To date, supplier restraint initiatives have been undertaken in the interest of suppliers and without consultation with or support from recipient states. These initiatives include the existing MTCR and Wassenaar Arrangement, the proposals of the Presidential Advisory Board on Arms Proliferation Policy, and the efforts of proponents of an Arms Transfer Code of Conduct, all discussed in chapter 4. A 1997 accord signed by the members of the Organization for Economic Cooperation and Development (OECD) that bans the bribery of government officials by OECD-based companies seeking business overseas, including arms transfer business, may also have an impact on arms sales and one that is welcome from the author's perspective, yet it remains a seller's initiative.[12] The abortive Perm Five talks on the Middle East after the Gulf War, undertaken even as the discussants were trying very hard to sell as much as they could as fast as they could to the region, also would have dictated restraint to the region had they succeeded.

Without denying either the need for or the potential utility of such purely supply-side programs, there is further need for an arrangement whereby arms suppliers—especially the Security Council's permanent members, plus significant suppliers like Germany and Italy—go on record as being prepared not to undercut arms restraint initiatives that arise from among the members of developing states' regional security complexes. The initial suppliers' pledge could be a generic one. The major suppliers would preposition themselves politically, as it were, to support restraint accords worked out by local players at a later date.

Since coordinated multilateral policy responses always take time to negotiate, and since peace and reconciliation accords can be fragile and sensitive to the timing of outside support, the basic framework of supplier support for regional restraint should be set up well in advance of any specific application. That framework could then be adapted as necessary to support specific regional initiatives as they arise, and be implemented using the diplomatic contacts and skeleton machinery already established to implement it. Such machinery could be as simple as an office within each supplier's foreign ministry whose job it would be to monitor regional restraint initiatives, share that information with each counterpart office, and advise its own arms export bureaucracy that a restraint initiative arrangement was underway or, perhaps, imminent. Part of this preparatory framework might be a "no-undercut" principle designed to discourage last-minute arms orders.[13]

The proposed framework would not preclude supplier governments from working with their allies and clients within a security complex to help craft a restraint regime. It could even encourage such cooperative drafting in the knowledge that other regional member states' principal arms patrons would be brought into the regional regime by the preexisting framework of supplier support for restraint.

The precedent for this sort of outside agreement not to undercut regional arrangements is Latin America's Treaty of Tlatelolco, whose protocols extract pledges from non-regional powers not to violate either its letter or its intent to ban nuclear weapons from the region. Nuclear weapon states that signed the protocol also pledged "not to use or threaten to use nuclear weapons" against any of the parties to the treaty.[14] Similar pledges of non-aggression, at a minimum, should accompany a regime to support regional restraint in conventional arms.

In addition to supporting regional initiatives, the major industrial democracies (the Group of Seven—the United States, Japan, Germany, France, Britain, Canada, and Italy—or Eight, if Russia ever stabilizes as a democratic polity) should be actively engaged on an urgent basis in regional dispute settlement, especially where participants are equipped with weapons of mass destruction and the means of their delivery. Focusing on arms alone without dealing with the political claims and grievances that generate the demands for those arms is likely to yield no long-lasting benefits, while waiting for the participants in every regional standoff to find their own way out of conflict mazes like Kashmir is to court disaster on a grand scale.

Modest Initiatives II:
Continuing Need for Supply-Side Restraint

Although suppliers should be poised to support regional initiatives, should encourage the evolution of such measures by pledging their support in advance, and should bend every diplomatic effort toward their achievement, such restraint regimes and underlying conflict resolution will tend to emerge slowly, and most slowly in those security complexes where political tensions and turmoil are greatest. Thus there remains good reason for the United States and other arms suppliers to support focused initiatives aimed at keeping certain arms from falling into the wrong hands. Definitions of broad classes of technology à la Wassenaar notwithstanding, effective supply-side measures require better agreement among suppliers as to which arms to hold back and which hands to avoid.

Focus on Specific Arms

Among major weapon systems, transfers of longer-range ballistic and cruise missiles are voluntarily constrained under the MTCR, largely because of their utility as delivery vehicles for WMD. They are, however, potential targets of broader restraint measures because they are difficult to defend against except at great expense and may give but five minutes radar warning of their approach, both features useful to an offensive military strategy. Although the average longer-range ballistic missile is likely to become more accurate with time, it will remain useful primarily as a terror weapon against civilian populations if equipped with conventional warheads and as a delivery vehicle for WMD, nuclear weapons in particular. These are good grounds for taking more definitive steps to halt and reverse the spread of such missiles, to which we will return in a moment.

Over the past two decades, incrementally minded supporters of conventional arms trade control have also advocated limits on so-called weapons of ill repute, a category that at various times has included shoulder-fired surface-to-air missiles, incendiaries, antipersonnel land mines, and blinding lasers, as discussed in chapter 4. Perhaps because they do not involve technologies that can, in relatively small numbers, directly threaten the survival of a state, governments have in fact been willing to sign accords to limit some weapons of ill repute, perhaps viewing the accords more as norm-building exercises than as instruments with immediate military impact. Thus the 1997 land mines convention signed in Ottawa, Canada, will not itself cause explosives to be lifted out of Angolan or Cambodian farmland, nor does it provide for any direct enforcement of its strictures, yet it does create a presumption of perfidy on the part of any person, group, or state that uses or deals in antipersonnel mines in the future. The fact that three big powers accounting for three-tenths of humanity (Russia, China, the United States) are not yet parties to the Ottawa Convention may dismay some of its proponents, but the presumption of the community of nonprofit, non-governmental organizations (NGOs) that fostered the treaty is that eventually these states, too, will join.[15] Having succeeded in bringing a land mines convention to signature, NGOs have been turning their attention increasingly to small arms, which are considered to pose particular threats to stability in postconflict situations. While the small arms campaign may have the look and feel of the land mines campaign, it is likely to unfold in a quite different manner that does not as readily lend itself to a single international agreement, because mines and small arms have very different uses.[16]

Arms control agreements like the Conventional Forces in Europe (CFE) Treaty have specified constraints on types of conventional armaments, but with the objective of being inclusive of all the major military equipment deployed within the treaty area, not with the objective of singling out a particular type for special restraint or condemnation. CFE might be considered a prototype for a land forces agreement in South Asia or the Korean peninsula, but such an accord would be likely only in the context of a prior, general easing of tensions, one that has a life span of more than three or four months. Moreover, India and Pakistan, in particular, must grapple with the additional factors of internal challenges to governmental authority that arise both from religious extremist groups and from regional secessionist movements, which undermine the stability of both countries.

Focus on Specific Actors

When actions are taken to limit a state's or a group's access to conventional weapons, they are usually taken after the fact—that is, only after costly breaches of international law or behavioral norms have occurred. Sanctions and embargoes are imposed, in other words, as punishments rather than preventive measures. This is the case in large part because, as related in chapter 4, there are no agreed upon and reliable predictors by which preventive action might be taken and no operational definitions of "excessive" or "destabilizing" accumulations of armaments that might tip off the international community to shut down the arms pipeline before a would-be aggressor accumulates a critical mass of firepower. As a result, embargoes on countries like Iraq, Libya, Yugoslavia, Liberia, or Somalia are imposed only after aggression, terrorism, or vicious internal warfare is evident to all. Usually, such measures are enacted by a body like the UN Security Council without enforcement measures to back them up. Where enforcement measures are available, a UN arms embargo like the one imposed on Iraq in 1990 can sometimes be effective, but even Iraq has been able to maintain a flow of replacement parts, and most UN arms embargoes have been even less effective. The embargo imposed on the former Yugoslavia, although the subject of some enforcement by NATO, was consistently violated on behalf of the breakaway states of Bosnia and Croatia. Embargoes imposed on African states and groups embroiled in civil wars have gone largely unenforced and have been roundly ineffective.

Voluntary supply-side measures like the MTCR and Wassenaar rely on enlightened self-interest on the part of major arms supplying states (augmented, perhaps, by the latent threat of U.S. sanctions for non-compliance) to halt trade in a specific set of missile-related technologies and to induce restraint in sales of a broad spectrum of advanced military and dual-use technologies. Both regimes attempt to be proactive by specifying technologies not to be transferred at all or only with care and restraint. Wassenaar tries to increase transparency within the club of high technology exporters and to focus export restraints on states whose governments violate international behavioral or legal norms. Its current, tacit rogues' gallery is, however, much smaller than the United States would like. As a voluntary arrangement lacking formal sanctions for non-compliance, Wassenaar depends completely on the cooperation of its members, case by case. The global buyers' market in major arms makes such cooperation difficult to induce, and the temptation to defect from the regime may well be overwhelming unless sellers' costs of defection are higher than the costs of cooperation, that is, the costs of sales foregone.

Because the United States has greater interest than most of the arrangement's other members in blocking technology to certain destinations, Wassenaar's effectiveness will depend in large part on the success of the United States in persuading those others to support American views. Doing so will require balance and tact, as heavy pressure and threats of sanctions could be counterproductive, generating, for example, reciprocal sanctions in any of a number of trade areas. While the MTCR may help to keep rogue nations from building missiles that might eventually reach the technology sellers' home turf, Wassenaar offers little that is tangible in return for suppliers' restraint, and neither regime holds out anything to would-be recipients. The targets of these regimes gain nothing in return for accepting either regime's strictures. It is not too surprising, then, that reports on missile developments in North Korea and Iran, and on Iraqi military technology smuggling, suggest neither regime has been especially effective against its prime targets, states that also have managed in several instances to establish supply relationships among themselves.[17]

More Difficult Initiatives

If we think in terms of politically or technically more difficult measures that might yield significant payoffs for regional peace and stability, several measures tend to stand out, although prospects for their implementation are unclear. In terms of measures focused on weapons, one might seek to formalize the constraints of the MTCR and make them universally applicable and reciprocal in nature, with a shorter-range limitation that aims at the eventual elimination of ballistic missiles as weapons of war. Other initiatives might include the addition of enforcement mechanisms to the Wassenaar Arrangement and the devising of a global system for monitoring end use of advanced technologies transferred internationally—although some suggest that energy might better be devoted to constraining the transfer of complete weapon systems, instead. If it is the case that how states are governed and how they plan to use their weaponry are at least as important for regional peace as their weapon inventories per se, then ambitious stability measures ought to focus on particular states, ought to implement arms transfer decision filters like the Code of Conduct, and ought to energize diplomacy backed by specific economic incentives to end the regional standoffs that can ratchet up demand for arms. Implementing such efforts will require closer coordination, even partnership, among the industrial democracies.

Moving Toward a Ban on Ballistic Missiles

At various points since Ronald Reagan proposed the idea to Mikhail Gorbachev at their Reykjavik, Iceland, summit in 1986, arms control advocates have argued that to help stem the proliferation of ballistic missiles and to roll back current arsenals, the United States ought to consider some variant of a ban on ballistic missiles, not just in the context of the Strategic Arms Reduction Talks with the states of the former Soviet Union but also globally. Some have advocated globalizing the Intermediate Nuclear Forces Treaty, which retired U.S. and Soviet ground-launched cruise missiles and ballistic missiles with ranges between 500 and 5,500 km. Critics of making that treaty multilateral note that it would leave short-range and intercontinental-range missiles unfettered.

Another approach would seek to eliminate ballistic missiles that can carry a 500 kg payload at least 100 kilometers. This "zero ballistic missiles" approach would tie together, in a sequence of steps, the process of U.S. and Russian strategic force reductions, missile cuts by other nuclear powers, and rollbacks by regional ballistic missile owners, recognizing that cuts in regional missile arsenals would have to be tied into a process of conflict resolution.[18] The 100 km range threshold would encompass the ubiquitous Scud and Scud-derivative missiles in several states' inventories, as well as such nuclear-capable missiles as the Soviet/Russian SS-21 (120 km) in the inventory of Syria and the U.S.-built Lance (110 km), which is held in store by Israel.[19] Perhaps equally important, the range threshold would capture an operational American weapon system, the U.S. Army's ATACMS, which has a range of up to 135 km.[20] Giving up ATACMS would be the U.S. price for a regime designed to rid other regions of Scuds, SS-21s, Al Husayns, Jerichos, Prithvis, Hatfs, Ghauris, Taepo Dongs, Shehabs, and other similar missiles.

Up to now, control of the trade in ballistic missiles has involved a discriminatory regime that allows the missiles that it prohibits to others to remain in the inventories of regime participants. A globally applicable agreement that captured operational U.S. armaments would substantially reduce both the perception and the reality of discrimination.

What about broadening such a ban to include other weapon systems capable of delivering WMD at long range? In the early 1990s, Stanford analyst John Harvey conducted a detailed comparative analysis of the destructive potentials of ballistic missiles and long-range, high-performance strike aircraft such as the Russian Su–24, U.S. F-111, and UK/German/Italian Tornado. Harvey concluded that such aircraft could be as dangerous as missiles for delivery of WMD and, in a number of mission scenarios, more effective militarily.[21] But attempting to fold such aircraft into a missile ban could undo the enterprise, precisely *because,* in one of those paradoxes of arms control, aircraft are in many instances substitute delivery platforms. If an Israel, India, or Pakistan knew that it could still deliver its knockout punch by air, it might be more willing to part with its missiles. Moreover, singling out strike aircraft would not necessarily accomplish much militarily. Various sorts of multirole combat aircraft can be configured for ground attack, including WMD delivery, while remaining capable of contributing to air defense and other defensive missions and strategies, as did allied air power in the Gulf War and NATO air power in the 1999 Kosovo crisis.

Going after ballistic missiles, on the other hand, would eliminate a weapon type that is very difficult and costly to defend against and may encourage first-strike strategies. The threat of accurate missile attack may confront military planners with a "use or lose" dilemma, while the threat of inaccurate attack implies targeting of easy-to-hit concentrations of population. Either prospect is likely to make military planners search for options to pre-empt such a threat before it can be launched.

Alton Frye, at the Council on Foreign Relations, points out that Russia's continuing military weakness has led it to lean on its ballistic missile forces as the last remaining vestige of former Soviet strength. Thus Moscow may be reluctant to participate in a ballistic missile ban. However, Russia is potentially threatened by the proliferation of missile/nuclear capabilities to countries like Iran or Pakistan. Moreover, Russia is projected by the next decade to have great difficulty maintaining a force with even 1,500 deployed strategic nuclear warheads, one-quarter of the number deployed in 1998.[22] While an immediate missile ban would very likely be rejected by Moscow, a phased prohibition consistent with or even faster than the reduction rates negotiated in the second Strategic Arms Reduction Treaty (START II) and anticipated for START III might be more acceptable.[23] At the end of that road, Russia might be sufficiently recovered economically to cover its security needs with other, more useful types of forces.

No prohibition on ballistic missiles could be enacted quickly, nor could one be crafted in isolation from the political situations unique to each security complex where missiles and/or WMD have become elements of the respective standoffs—in the Middle East, South Asia, and Korea. Rolling back missile deployments requires addressing the sources of demand for missiles, an issue to which we return in the section on democratic states and regional security below.

Enforcing the Wassenaar Arrangement

Since the nearly three dozen arms-supplying states that belong to the Wassenaar Arrangement have taken great pains to work out elaborate lists of technologies whose transfer is reportable and restraint of which is desirable, at least for some applications and to some destinations, why not take the next logical step and attempt to turn what is now a voluntary system into a legally enforceable treaty regime? The answer would seem to be that a voluntary system is all that the mid-1990s political traffic would bear. Wassenaar's predecessor, the Coordinating Committee on

Multilateral Export Controls (COCOM, discussed in chapter 4) was developed as an instrument of collective defense against the Soviet bloc. The requirement that transfers of sensitive technology to targeted countries be acceptable to all COCOM members fit a period of political-military struggle and a regime whose objective was to minimize technology transfer from West to East. At present, however, and looking further into the twenty-first century, the objectives of technology export controls are hazier, and both the targets and the criteria of restraint are matters of disagreement. As noted above and in chapter 4, Wassenaar's member states have not been able to agree on a list of "states of concern," or to define "destabilizing" or "excessive" arms transfers. What they do have is a list of commodities and a voluntary reporting arrangement that has been violated in precisely those circumstances in which the arrangement was intended to work, such as the 1998–1999 Eritrean-Ethiopian border conflict. Can Wassenaar member states that violate the current norms when it is convenient and lucrative to do so be expected to agree to transform such norms into legal obligations? That may well depend upon whether the members of the Wassenaar Arrangement can see their way clear to adopt criteria for arms and technology transfer decisions that are more readily made operational. Such criteria might be based upon elements of the arms transfer Code of Conduct, as discussed shortly.

Global Monitoring of Advanced Technology End-Use

The 1996 report of the Presidential Advisory Board on Arms Proliferation Policy noted the difficulty of constraining the spread of dual-use high technology. Limits become ever more difficult to impose with the increasing volume and rate at which technology and information diffuse across borders, pushed by the economic interests of the commercial world and the social interests of civil society organizations. Such technology proliferation is self-reinforcing in the sense that it makes international movement of goods, services, ideas, and values across oceans and cultures increasingly rapid and cheap.[24]

Rather than try to keep a lid on the movement of high technology with military applications (except for certain dangerous weapon systems and components), the Advisory Board emphasized tracking the distribution and use of technology with a global, cradle-to-grave system of registration and end-use monitoring. Built, perhaps, on the Wassenaar Arrangement, such a monitoring system would be universal in nature and not applicable just to a few "states of concern." Similar to the monitoring protocols to be used in verifying compliance with the global Chemical Weapons Convention but covering a much broad broader array of technologies, such a system would be designed to alert supplier governments to the diversion of technology from originally declared destinations and/or applications. Such a system would level the playing field between technology producers and governments, according to former Brookings analyst Wolfgang Reinecke: producers know what they make and usually to whom they sell it, whereas government often has to play catch up, despite official licensing systems. A global technology monitoring system—complex, expensive, and slightly Orwellian in the perspicacity of its observation requirements—would not necessarily slow the spread of technology, but it could give governments a better idea of who has access to what and thus what military capabilities might be building where. If companies were allowed to sell conventional dual-use technologies (that is, technology with application to con-

ventional weaponry but not to WMD) once they subscribed to the end-use monitoring system, complying with the system could be much more profitable and much less
stressful than trying to generate profits under the table. It would be much less costly, as
well, if the penalties for commercial non-compliance with the monitoring system were
made sufficiently onerous.[25]

The United States, Germany, and other Western states already have reasonably strict
systems of accounting for dual-use technology exports. However, a number of transitional economies with advanced technological know-how are still struggling to gain
a modicum of control over customs in general, let alone the disposition of high-tech
exports after they leave the country.[26] Control systems meeting U.S. standards may be
a long way off in such countries as Ukraine or the Russian Federation, which poses a
problem, because an end-use tracking system would require participation by all producers of comparable or substitutable technology to function as intended. Its effectiveness could be significantly compromised if one or more high-technology
producers, especially large countries difficult to penalize effectively (a Russia, China,
or India, for example), proved unable or unwilling to participate in the regime. International technical assistance might improve the capacity to monitor, but only a sense
of shared national interests could impart the willingness to do so.

The need to police the proliferation of all technologies with military potential may
be overstated, however. While market-driven innovation has accelerated the diffusion
of technologies applicable to both civilian and military use, it has not necessarily accelerated the ability of governments or militaries to *exploit* such technology effectively
by molding it into functioning military systems. In addressing this issue, defense industry analyst (and later undersecretary of defense) Jacques Gansler noted that fighting ability stems from the capacity to (a) integrate individual technology components
into a weapon; (b) train its operators and integrate them into a fighting force; and (c)
create doctrinal, command, control, communications, intelligence, and logistics systems to utilize and support the force effectively. Thus, Gansler argued, it is the major
weapons and systems integration end of the arms trade spectrum, not the technology
and subcomponent end, that should be the principal focus of export policy.[27] Superior U.S. ability to integrate components into systems and to create effectively organized and trained fighting forces, in other words, would allow the United States to
maintain sufficient operational superiority as to undertake necessary operations at acceptable levels of risk in overseas theaters more or less independently of the rate at
which militarily relevant technology proliferates around the world. If Gansler's arguments are correct, then controlling the proliferation of complete weapon systems is
more efficacious than controlling conventional dual-use technology.

If, in addition, Stephen Biddle's arguments in chapter 2 are correct, that overall
force structure and strategy have more impact on regional stability than the presence
or absence of particular weapon systems, then international peace and security would
be even better served by focusing U.S. and like-minded states' policies on particular
states and their political structures, their relationships with their people and within
their security complex, and how they plan to use their militaries.

Bringing Democracy to Bear on Regional Security

One implication of democratic peace theory is that the world would be a less conflictual place if more states were governed according to the principles of democracy

and respect for individual human rights.[28] A further implication is that democratic states have a mutual stake in one another's security and in the continuing spread of democratic governance around the globe. By extension, democracies also have an interest in preventing the rollback of democratic governance.

In theory, cooperating democracies would have a wide array of potential influence tools at their disposal to further these objectives, rooted in three basic types of power: threat power (the ability to coerce or to destroy); economic power (the ability to produce, to exchange, and to bribe); and psychological power (the ability to persuade or to integrate).[29] Military force, international embargoes, and other restrictions on communications and trade are some of the instruments of threat power. Trade itself, foreign direct investment, military or economic assistance, and "cooperative threat reduction" (whereby country A pays to dismantle weapons in country B) are some of the instruments of economic power. Effective psychiatry, missionary work, teaching, negotiation, mediation, and advertising are some of the instruments of psychological power.

These three types of power are often used in combination. Negotiators may offer side payments or level implicit threats to persuade their counterparts across the table to reach agreement. Mediators may threaten to withdraw their services to encourage disputing parties to converge on a settlement. European missionaries could at one time rely upon the power of colonial empires to help them convert the heathen, while teachers use both promises (of gold stars) and threats (of bad grades) to encourage attention to learning.

The different neighborhoods of the democratic community have tended to favor different combinations of power tools to advance their international interests, including interests in regional peace. Japan, constrained by its constitution from projecting military power abroad, has emphasized economic tools—aid and trade. The countries of greater Europe have more collective military power but little ability or desire to project it at a distance, stressing instead economic and integrative tools in their dealings with developing regions, not only aid and trade but also tangible support for post-conflict peacekeeping operations and for democratization, via official action and through the activities of NGOs. Canada's emphasis is similar. The democracies of Latin America have until fairly recently been fully engaged sorting out their own problems of conflict, stability, and legitimacy, but several among them have contributed peacekeepers to international operations and economic integration is increasing via common markets like MERCOSUR. The United States, with the most powerful and mobile military in the world and the largest and most dynamic economy, has a self-conscious need to remain number one in both of those realms. While it distributes modest amounts of developmental assistance, stumps for freer trade, and funds the National Endowment for Democracy, Washington also reaches for threat-based foreign policy tools more readily than any other democracy, in the form of military action, unilateral sanctions, and penalties leveled against governments and firms that fail to support such sanctions. Although Washington received help in dealing forcefully with Saddam Hussein (initially from many countries and continuously from Great Britain) and although, together with Britain, it convinced NATO to bring force to bear against Milosevic's Yugoslavia, the United States is far more likely to be found policing regions outside its own backyard than are other industrial democracies.

Phasing in a Code of Conduct

Are the democracies sufficiently like-minded and interested in building new neighborhoods of democratic states as to adopt as common policy something like the Arias

Code of Conduct for arms transfer decision making? The code, as discussed in chapter 4, aims to restrict arms transfers to states that are democratically governed, that respect human rights internally, and that meet their obligations internationally. U.S. and European Union (EU) actions to date suggest that neither entity is quite ready to support such a code. The EU has adopted a much weaker set of guidelines and the U.S. Congress has directed the executive branch to talk about such a code with other countries and specified arms transfer eligibility criteria that the president is free to waive on national security grounds.

The code, as presented by its supporters, is an all-or-nothing proposition that forbids arms sales if prospective recipients fail any one of its litmus tests. As such, it would seem to hand the arms market among states that fail one or more litmus tests to nondemocratic or otherwise non-participating arms suppliers—hungry quasi-democracies like Russia and non-democracies like China—thereby furthering a demimonde in which tyranny is, in effect, reinforced. Such an outcome would not seem to be what code supporters had in mind.

How might such an outcome be avoided while still applying a Code of Conduct standard to arms transfers? One way would be to phase in the code by degrees, beginning with those elements that are most important to international peace and security, rather than applying it as an all-or-nothing standard. Some of these elements may be relatively easy for most arms recipient states to comply with and some may be more difficult, requiring varying degrees of policy, behavioral, or institutional changes, or even major alterations in a country's internal power relations. Both factors—ease and importance—should be considered in designing a phased implementation of the code. Rapid, early adherence by suppliers and recipients alike to the *principle* of a Code of Conduct should be a primary objective, achieving a kind of political-military beachhead that can be further exploited later.

The elements of the Code of Conduct as delineated in chapter 4 are listed in table 6.2, ranked roughly according to their importance (1 is assessed to be most important, 18 least) and ranked also according to the relative ease with which they could be implemented in most countries (1 being easiest, 18 hardest).[30] Ease may be a function of the degree to which internal politics and power relations have to be altered to implement an element of the code, or it may be a function of the difficulty in defining the meaning of an element. The latter criterion applies, for example, to the meaning of armaments "beyond levels needed for legitimate self-defense," and to the meaning of "new technology." Problems with both of these terms were discussed at length in chapter 4.

The illustrative rankings in table 6.2 suggest that what may be easy for states to comply with in the code may not be its most important elements. However, in several instances, ease and importance do line up. This is easier to see in figure 6.1, which compares the two sets of rankings. Elements in the upper right quadrant of figure 6.1 are estimated to be both important and relatively easy to implement without major internal upheavals in governance. Elements in that quadrant include not fighting an aggressive war, respecting international borders, not engaging in genocide, supporting the Geneva Conventions on the treatment of prisoners of war and populations in war zones, supporting international conventions on terrorism, agreeing to apprehend and extradite terrorist suspects, and not inciting nationalist, racial, or religious hatred—all fairly fundamental stuff and all potential elements of a first phase implementation of the Code of Conduct. Not torturing or murdering one's populace, in whole or in

Table 6.2 Ranking the Elements of the Arias Code of Conduct by Importance and Ease of Implementation

Rank 1 (importance)	Rank 2 (ease of implementation)	Principle or Behavior
1	8	Commitment to promote regional peace, security, and stability (a) no arms to countries at war except in self-defense or in a UN-mandated military operation.
2	1	Commitment to promote regional peace, security, and stability (d) recognition of other states' borders.
3	4	Compliance with international human rights standards (a) banning genocide.
4	15	Compliance with international human rights standards (b) banning torture and extralegal/ arbitrary execution.
5	6	Opposition to terrorism (a) ratification of relevant conventions.
6	7	Opposition to terrorism (b) cooperation in apprehending/extraditing terrorists.
7	3	Compliance with international humanitarian law (a) Geneva Conventions and Protocols.
8	9	Commitment to promote regional peace, security, and stability (e) no advocacy of national, racial, religious hatred.
9	13	Respect for democratic rights (b) freedom of expression.
10	12	Respect for democratic rights (c) civilian control of the military.
11	11	Respect for democratic rights (a) free/fair elections with secret ballots.
12	17	Commitment to promote regional peace, security, and stability (b) no arms beyond legit. self-defense.
13	5	Respect for international arms embargoes and military sanctions.
14	14	Compliance with international humanitarian law (b) cooperation with international tribunals.
15	10	Compliance with international humanitarian law (c) granting access to non-governmental humanitarian organizations in time of emergency.
16	16	Promotion of human development (health/ education expenditures exceed military expend.)
17	2	Participation in UN Register of Conventional Arms
18	18	Commitment to promote regional peace, security, and stability (c) no new technologies.

Note: All rankings are author's estimates, for illustrative purposes.

part, is equally fundamental from the standpoint of human rights but more difficult to verify and to root out from the outside. It would nonetheless be a highly desirable element of phase one, arguably more important than, and a precursor to, freedom of expression and true democratic governance, inasmuch as no opinion can be free and no vote fair if the price for such expression is pain, imprisonment, or death.

The elements of the code in the upper left quadrant of figure 6.1 might be incorporated into phase one because they should be relatively easy for most countries to implement. Most already support the UN Register and UN sanctions regimes. Although access for NGO humanitarian relief in times of emergency is never simple, it is usually granted, with conditions that are situation-dependent.

Free expression, fair elections, and civilian control of the military fall in the mid-range of the importance rankings and in the harder-to-implement half. Civilian control is ranked lower in importance than code elements already discussed, not because it isn't important but because a military regime that meets those higher-ranked tests would not be an immediate threat either to neighboring states or to its own populace—that is, it would not be engaged in aggression, would respect borders, support the laws of war, extradite terrorists, and hew to the genocide convention, would not engage in torture or foster intergroup hatred, and might even allow public debate and political dissent. Arms transfers and military training would already have been denied under a phase one Code of Conduct if a government failed any of these latter tests, regardless of who ran the government and whether they wore a uniform. Tests of free expression, fair elections, and civilian control would therefore be incorporated into the second phase implementation of the code.

The elements of the code found in the lower left of figure 6.1 (support for international war crimes tribunals, human development spending in excess of military spending, "no excess arms" and "no new technologies"). Support for the work of war crimes tribunals may soon be considered as much a part of states' basic behavior under international law as support for the Geneva Conventions, but the United States, for example, supports the tribunals' work largely to the extent that there is zero risk of an American citizen being hauled before the bar, as there could be in the case of the International Criminal Court, indicating that implementation will be neither easy nor automatic.[31] The difficulty with defining excess arms and new technologies

Figure 6.1   Cross-Ranking the Elements of the Code of Conduct

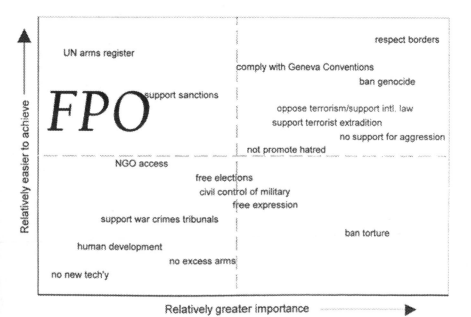

has already been noted, and the fact that a technology is "new" may not be good grounds for denying its transfer, even if "new" could be reliably defined. Nor is newness per se likely to pose that much of a threat if, as would be the case here, a would-be recipient already meets the other tests by which good international citizenship and good governance would be measured.

Finally, a country's ratio of spending on defense to spending on human development may be a useful criterion for judging arms eligibility in some cases, but it would not be a good criterion in circumstances in which a country that otherwise met the criteria of the code was under military pressure from hostile forces. Such a state should not be denied means for self-defense in precisely those circumstances in which its needs are, for the time being, most acute.

If the Code of Conduct were phased in as described, and with the sorts of exceptions described, a stronger case could be made to all arms suppliers, and not just the major democracies, to join in its implementation, at least for phase one. Depending on the time scales involved (which would be a matter of negotiation), by the time phase two and the elements of the code related to democratic governance were due to be implemented, the remaining major suppliers would either have made a transition to democracy themselves (making phase two easier to implement) or the world would be looking at a new Cold War and the rules of the game—for much more than arms transfers—would need to be redrawn.

Democratic Security Partnerships

RAND Corporation executive David Gompert has offered a model for protecting the interests of democratic countries and extending the writ of democracy and open markets that he calls the "democratic security partnership."[32] The Western industrial democracies, Gompert argues, are already well on their way to forming one large transatlantic security community with a raft of common interests—the "core" or nucleus of a potentially much larger community. Gompert would build the initial partnership on a sort of merger of NATO and EU that would share security tasks in places like the Persian Gulf that are now borne largely by the United States with some assistance from Britain. But the partnership's highest priority goals would also include "completing the democratic transformation of Eastern Europe, including the Balkans, the Baltics, Ukraine, and Russia itself"; achieving peace between the Arab states and Israel; and "providing security from nuclear, chemical, and biological weapons in the hands of hostile regimes."[33] The initial Atlantic partnership could grow as the number of willing and able democratic states grows—an optimistic but not entirely ungrounded hope—with the initial democratic security partnership supporting the evolution of democracy and security communities in Latin America, southern Africa, East and Southeast Asia, and wherever else that democracy and open markets have gained more than a toehold.

By implication, such a partnership would offer support to those democracies challenged by antidemocratic forces, whether internal or external in origin, extending the principles of the basic collective defense bargain underlying the NATO alliance to pledges like the Organization of American States' 1991 Santiago Commitment to Democracy. Such support could take many forms, from preventive measures such as military-to-military programs that help to professionalize military institutions and parallel aid programs to strengthen responsible civilian governance, to deterrent measures such as international security guarantees and remedial/reconstructive measures like electoral monitoring and postconflict peace implementation operations. Although

NATO has shown no collective inclination to conduct military operations outside of Europe, neither had it shown any inclination for such operations within post–Cold War Europe before sustained air operations against Yugoslavia in 1999 and the rehearsal four years earlier over neighboring Bosnia. The peacekeeping forces deployed after those enforcement actions in Bosnia and Kosovo may be emblematic of stabilization requirements that the democracies will face in coming decades in other security complexes teetering between chaotic and productive futures.

If the world's democracies choose to shoulder responsibility for supporting the global peace that makes their accustomed lifestyles sustainable in an economically intertwined world, they will want to share that responsibility as rapidly and as widely as possible. Thus the democracies have a deep and abiding interest in cloning their governing principles and their military capacities in other regional security complexes, so that the initial "core" need play only an indirect or supporting role in those regions' security affairs. Transfers of military technology and expertise to democratic and democratizing countries will play a particularly important role in creating such capacity, in parallel with economic and political development.

The spread of technology within the communities of democratic states ought not to be viewed as proliferation but as exchange within a "circle of friends," a concept discussed in chapter 4. Although the idea was criticized for creating a system of "ins" and "outs" for purposes of regulating exports of military arms and technology, discrimination among recipients is what arms export policy has always been about.[34] In this case, it would be mostly about building and defending democracy.

Arms exports outside the democratic community might best be managed by recourse to the phased Code of Conduct discussed previously. Non-democratic states that otherwise meet phase one criteria of the code could be eligible to receive certain types of armaments or certain classes of technology. The applicable engine for deciding on the types and classes, and monitoring their shipment and end-use, might be a formalized or enforceable Wassenaar system of the sort also discussed above.

The democratic community should be actively engaged, moreover, in dispute settlement, not just in the Middle East but in South Asia, where the half-century-long, zero-sum game played by India and Pakistan over Kashmir threatens to bathe those countries and much of their security complex in the soft blue glow of residual radioactivity. Nor can the democratic community afford to ignore central Africa, where a vacuum of governance threatens to suck in all the countries that border it, turning back the progress that the southern part of the continent has only recently begun to make toward democracy, prosperity, and peace. Withholding arms is not enough: active, urgent engagement in peacemaking is the order of the day.

Implications for U.S. Policy

Arms transfers have been a significant element of U.S. foreign and defense policy for more than a half-century. They served distinct political-military purposes for most of that period and provided economic support for a downsizing defense industry toward the period's end. As we move into the twenty-first century, what role should arms and military technology transfers play in U.S. policy? How should the United States plan to deal with the spread of military technologies and with the leveling effect (direct or "asymmetric") that broad availability of technology may have on its ability to use military power effectively and at acceptable risk?

As the world's principal supplier of armaments, the United States should be ready to lead other suppliers in support of regional arms restraint initiatives, should be ready to help enforce the OECD's anti-bribery convention, and should be ready to revisit how it views technology proliferation, especially conventional dual-use technologies. The point of limiting access to technology is not, or should not be, simply to keep it from reaching any non-U.S. destinations at all. First, there are already a number of producers of dual-use technology of all sorts in the world, so this goal is already largely unattainable except for a handful of special military techniques like radar signature reduction ("stealth") in which the United States holds a clear lead. Second, the growth of private sector industry in the twenty-first century will be cross-national in nature, buoyed by open trade and investment regimes, which will increase the potential for intrafirm "transfers" that may be difficult for governments to track. Third, destination makes a difference. The implications of technology transfer to the United Kingdom and to Iraq are quite different from the standpoint of U.S. interests. The basic U.S. approach within the Wassenaar Arrangement is correct—that certain things should not reach certain destinations—but this is a political decision that could be much more solidly grounded, using criteria from the phased Code of Conduct discussed above, in lieu of a basically unattainable definition of "excessive or destabilizing accumulations" of arms. All of the current members of Washington's "rogues gallery," for example, would fail even the phase one transfer criteria of such a technology transfer code.

The eventual diffusion of much military-related technology may be largely inevitable, but as noted earlier, unless that technology is effectively integrated into military forces and strategies abroad, U.S. policy and practice need not change radically to meet future international political-military challenges. If the rate of integration should increase, however, among countries unfriendly to democracy, then U.S. policymakers would face some tough choices. They could:

a. *accept* that local military capabilities in some places where the United States and its collaborators contemplate military intervention may make any action other than dropping smart bombs on fixed targets too risky, politically, to contemplate; or

b. *redefine* the interest-adjusted risk so that intervention with ground troops in some hostile situations is deemed worth it despite anticipated casualties; and/or

c. *invest* heavily in research and development to limit risk against all potential intervention targets, banking on superior U.S. technology integration, training, command and control, intelligence data, and logistics, while working assiduously to keep military expertise away from undesirable destinations, in part by tracking the peregrinations of military and dual-use technologies through the international system on a continuing basis to avoid surprises.

Course (a) would acknowledge that risks to interventionary forces may well increase as military-related technology spreads and that the costs of exerting dominant military influence in such circumstances may be more than the *political* traffic in the United States and other democracies will bear. Some uses of air power (as in Kosovo) may produce entré for ground forces, while others (as in Iraq since the end of the Gulf War) will not. Decisive uses of force may in any case be few and far between even if the United States is capable of besting any adversary that chooses to contest its military power directly.

Course (b) notes that with sufficiently important interests engaged, the prospect of greater casualties becomes more acceptable politically. Political leaders would, however, need a clear sense of those goals and objectives worth sending troops to die for and would need to have communicated them effectively to their publics in advance of such use.

Course (c) suggests that the risks of intervention may be manageable, as technology spreads, with continual effort on the part of U.S. (and allied) industry to stay a jump ahead of overseas competitors. This last course presumes that there are high-leverage military technologies that globalizing defense firms could be made to reserve for the exclusive use of the United States and its close collaborators. It presumes government willingness to compensate industry for its forbearance in high technology sales to restricted destinations, if only in the form of much higher unit costs of equipment for U.S. forces, as well as a fine-grained ability to keep track of exported technologies and their end use. The latter task would require the close cooperation of other countries' industry and governments to prevent circumvention of destination and end-use restraints. If adopted, course (c) can and should be refined to be consistent with U.S. interests in having more like-minded, militarily competent states to share responsibility for building and maintaining a peaceful and democratically governed world; that is, "undesirable destinations" in course (c) should be not be taken to mean "all foreign destinations" or even "all destinations in the developing world."

If the United States is serious about limiting the spread of certain conventional armaments, ballistic missiles in particular, then over the longer term it must be willing to give something to get something. There must be clear reciprocity in conventional arms restraint arrangements, or the political stresses created by its lack will eventually pull such arrangements apart. Thus, in the interest of achieving global restraint, the United States must be ready to give up a weapon system that an American military service actually considers useful, like ATACMS, in the interest of promoting elimination of a larger class of dangerous conventional weapons that is closely linked in many regions to the proliferation of WMD and that poses dangers to U.S. forward-deployed forces that are costly and difficult to counter.

Finally, the United States should take the lead in engaging the community of democracies in dispute settlement initiatives as a matter of standing national security policy. It should be as involved in resolving conflict in other places as it has been in the Middle East. U.S. policymakers should work hard at coordinating governmental and non-governmental initiatives in these areas, discreetly supporting in particular the work of NGOs that provide technical training in conflict resolution and political organizing skills, as well as those involved in relief and economic development work and population planning.

The globe has not always had an international political structure or a distribution of power like that of the late twentieth century. Indeed, that structure and distribution are unique in history, and although the structure has been extensively institutionalized, there is no inexorable reason why the twenty-first century's eventual political structure and power distribution need look the same.[35] International relations in the new century will be affected by many factors in ways that are hard to predict. These include the spread of technology of all sorts, the continuing evolution of dense international information networks and the computing power that drives them, the growth of human population in the world's poorest places, the radicalization of communities of belief, the institutionalization of democratic governance, the spread of open markets and, one hopes, growing concern by governments that basic human rights be respected and enforced everywhere. We cannot know the contours of the future, but with all of these factors in play we can readily guess that they will not be quite what we expect. Thus, in military affairs as in other facets of public policy, the basic U.S. stance must be adaptive and its basic orientation one of trying to shape the future in line with American interests in peace, freedom, and individual dignity. If they fail to do that, Americans and like-minded peoples risk finding themselves living in a future that has been shaped for them in ways that they do not like.

APPENDIX A

Mapping Security Complexes

A security complex is defined in terms of physical proximity, patterns of amity and enmity, and an established distribution of power.[1] A number of states on the boundaries between nominal security complexes face at least two ways in their security planning, and they were assigned here to whichever security complex seemed to require the greater attention of their governments (see table A.1).

The *Middle East and North Africa* security complex includes countries from Morocco eastward to Syria and Jordan. The Arab-Israeli conflict is the defining contest within this complex, but there are a number of localized enmities as well (Morocco-Algeria, over borders and unresolved claims to the Western Sahara; Libya-Egypt, deriving from Egypt's rapprochement with Israel; and Syrian aspirations to hegemony over war-torn Lebanon). Important relations across security complex "boundaries" include Syria's rocky relationship with Iraq, Saudi Arabia's role as banker to those states actively confronting Israel, Libya's meddling in both Chad and Sudan's civil wars, and its efforts to annex a bit of Chad (the uranium-bearing Aouzou Strip).

The *Persian Gulf* security complex includes Iran, Iraq, Kuwait, Saudi Arabia, and the rest of the states on the Arabian peninsula (Bahrain, Qatar, the United Arab Emirates, Oman, and one or two Yemens, depending on the time period). The Gulf Arabs fear their larger, northern neighbors' hegemonic ambitions (Iran's, played out with imperial grandeur and hubris under the late shah, and with true-believing zeal by his Shia Islamist successors; and Iraq's, reflected in its initiation of the Iran–Iraq War and the Gulf War). Subsidiary tensions include Saudi-Yemeni bickering over their mutual border (how it is drawn determines who has control over several interior oil fields). Cross-boundary issues (aside from the Arab-Israel conflict) include the ongoing question of the Kurds, who populate northern Iraq, far northwestern Iran and southeastern Turkey; the post-Soviet war on Iran's northern border between Armenia and Azerbaijan; the 16-year meltdown in Afghanistan; and ongoing unrest in southern Pakistan.

Since the breakup of the Soviet Union, Central Asia has gained five new republics, but their brand-new status and continued close security ties to Moscow left them out of the analysis, leaving *Afghanistan* in its own "complex" or, more correctly, making it a transitional state at the juncture of three other complexes (the Persian Gulf, the post–Soviet Commonwealth of Independent States, and South Asia).

South Asia comprises Pakistan, India, Sri Lanka, Bangladesh, and the small Himalayan states of Nepal and Bhutan, all of which were once part of British India and whose principal security concerns focus on present-day India. India's concerns focus in turn on Pakistan and, to a lesser extent, China. Burma (dubbed Myanmar by its illegitimate military regime in 1988), although a former British protectorate as well, is as Buzan notes a transitional or buffer state between South and Southeast Asia.[2] Because it is more closely linked to the latter—racially, culturally, religiously, and by the drug trade of the Golden Triangle—that is where it sits in this volume.

Table A.1 Regional Security Complexes

Super-region	Security Complex	Country	
Middle East	Persian Gulf and Arabian Peninsula	Bahrain	Qatar
		Iran	Saudi Arabia
		Iraq	United Arb Emirates
		Kuwait	Yemen
		Oman	Yemen, South
	Middle East (Levant and North Africa)	Algeria	Libya
		Egypt	Morocco
		Israel	Syria
		Jordan	Tunisia
		Lebanon	
Latin America	Meso-America, Central America, and the Caribbean	Cuba	Guatemala
		Belize	Haiti
		Costa Rica	Honduras
		Dominican Republic	Mexico
			Nicaragua
		El Salvador	Panama
	South America	Argentina	Guyana
		Bolivia	Paraguay
		Brazil	Peru
		Chile	Surinam
		Colombia	Uruguay
		Ecuador	Venezuela
Asia	Northeast Asia	Korea, North	Taiwan
		Korea, South	
	Southeast Asia	Brunei	Papua New Guinea
		Burma (Myanmar)	Philippines
		Cambodia	Singapore
		Indonesia	Thailand
		Laos	Vietnam, N. and all
		Malaysia	Vietnam, S.
	South Asia	Afghanistan	Nepal
		Bangladesh	Pakistan
		India	Sri Lanka
Sub-Saharan Africa	Southern Africa	Angola	Namibia
		Botswana	South Africa
		Lesotho	Swaziland
		Madagascar	Tanzania
		Malawi	Zambia
		Mozambique	Zimbabwe
	Central Africa	Burundi	Equatorial Guinea
		Cameroon	Gabon
		Central African Republic	Kenya
			Rwanda
		Chad	Uganda
		Congo (Brazzaville)	
		Congo, Democratic Republic of	

(continues)

The other states of *Southeast Asia* include Thailand, Laos, Vietnam, Cambodia, Malaysia, Singapore, Indonesia, and the Philippines. During the Cold War, the major security dynamic in this complex was ideological, with a global Cold War overlay (communist North Vietnam, later communist Indochina, and its Chinese and Soviet supporters versus the rest of the region and the United States). Since the end of the Cold War, the Association of Southeast Asian Nations (ASEAN), the region's political consultative group, has admitted Vietnam as a member, and its members tend to look toward China as a common problem. (China, due to its size, borders at

Table A.1 *(continued)*

Super-region	Security Complex	Country	
	Western Africa	Benin (Dahomey)	Liberia
		Burkina Faso	Mali
		Cote d'Ivoire	Mauritania
		Gambia	Niger
		Ghana	Senegal
		Guinea	Sierra Leone
		Guinea-Bissau	Togo
	Horn of Africa	Djibouti	Somalia
		Ethiopia	Sudan

least five regional security complexes and is in a category by itself, as it has been by choice, historically. In this study it is treated as a supplier, not a recipient.)

Northeast Asia includes Taiwan and North and South Korea. Since the amity-enmity relations in this security complex were and to some extent remain a joint result of the Pacific War, the Chinese Civil War, the Cold War, and U.S. political and/or military involvement in each, the United States might also be thought of as a member of this complex, with military forces based in the region (in South Korea and Japan) and ongoing interests there driving its engagement. While the U.S. presence could be thought of in terms of the "overlay" of the higher global security complex on Northeast Asia during the Cold War, in the 1990s the reasons for that presence are more regionally focused.

In terms of local enmity, the two Koreas have focused mainly on one another while Taiwan has focused on China. In terms of amity, South Korea's and Taiwan's principal ally throughout the Cold War was the United States, while North Korea had the support of China and the USSR. As the Cold War ebbed and outside support dwindled, North Korea pulled further into itself and reinforced its international reputation as a rogue state, a reputation diminished only somewhat by a 1994 agreement to trade its nuclear weapons program for enhanced commercial nuclear power, food, and fuel.

Skipping across the Pacific, the *South American* complex includes all states from Colombia southward. All but two of these states (Chile and Ecuador) share a border with Brazil, linking Andean and Amazonian politics together.

The *Central American* security complex starts with Panama and continues northward to Guatemala.[3] Several Central American states were wracked by civil war in the 1980s, and Guatemala was so cursed more or less continuously from the 1950s through 1996. The conflicts of Central America have had primarily economic roots, which translated into struggles for power and the right to redistribute wealth, although Guatemala's struggle also had heavy ethnic overtones. These struggles were, however, seen in Washington to mirror the global capitalist-communist struggle, an impression encouraged by the parties in power in those states throughout the Cold War and reinforced by Soviet and Cuban assistance to the Sandinista regime in Nicaragua and to rebel forces in El Salvador.

Mexico's and Canada's relationships with the United States are largely amicable, the migration and drug trafficking issues with Mexico notwithstanding, and the three are members of the North American Free Trade Agreement. In the Caribbean, Cuba has been locked into a hostile relationship with the United States since Fidel Castro took power in 1959, while other states have had a neutral relationship to U.S. power with periodic, short-term exceptions (the Dominican Republic, subject of U.S. intervention in 1965; Grenada, the subject of a brief U.S. invasion-cum-rescue-mission in 1983; and Haiti, disgorger of economic refugees and subject of American occupation under UN auspices in the mid-1990s). The foursome of Canada, the United States, Mexico, and Cuba really should be considered as a single North American security complex. Canada and the United States are outside the scope of this study, making Mexico and Cuba orphans that I have tried to track without bundling into Central America, as Cuba's import patterns and technology levels, in particular, are historically quite different from those of Central America.

I distinguished four subregional security complexes in sub-Saharan Africa, a region in which the flow of major weaponry is not great but conflicts and turmoil abound. The *Horn of Africa* includes Sudan, Ethiopia, Eritrea, and Somalia, linked by climate and by mutual refugee flows from their respective long-running civil wars. Somali territorial claims led to a brief war with Ethiopia (1977–1978), and an ill-defined border led to war (1998–1999) between Ethiopia and newly independent Eritrea, which tore away from Ethiopia in 1993 after three decades of secessionist struggle.

West Africa includes countries from Senegal to Nigeria. Its countries' security concerns are largely self-contained but heavily influenced by Nigerian politics, which descended into civil war once (1967–1970) and may yet do so again despite the best efforts of newly elected president Olusegun Obasanjo. French military presence among the Francophone states of the region has helped stabilize (some might say ossify) local politics, but civil war in Liberia (1990–1995) and Sierra Leone (1992–1999) seriously unsettled the area and displaced millions of people.[4] By and large its states are too poor to acquire major arms except through political largesse.

Central Africa incorporates states from the Cameroon to Kenya, centered around the great mass of the Democratic Republic of Congo (DRC, formerly Zaire). The DRC's vast, impoverished expanse borders both east-central states like Uganda, Rwanda, and Burundi that have been the site of periodic bloodshed, and the southwestern state of Angola, site of a 25-year tribal-cum-ideological military conflagration. Kenya is a buffer state that also borders Sudan, Ethiopia, and Somalia, that has hosted refugees from all three of those states' civil wars, and that supplies tons of the quickly perishable, mountain-grown stimulant *khat* to Somali bazaars.

By the late 1990s, *Southern Africa* seemed to be emerging at long last from a nightmare of racial and ideological conflict that has kept the region in turmoil for decades. Portugal's abrupt abandonment of colonial holdings in Mozambique and Angola in 1974 exposed the flanks of the white minority regimes in Rhodesia (now Zimbabwe) and South Africa to attack by forces based in Angola, Zambia, and Mozambique, supplemented in the Angolan case by substantial numbers of Cuban troops. But the white regime in Rhodesia acceded to a British Commonwealth-supervised transition to black majority rule in 1980; American diplomacy arranged for a Cuban (and South African) exit from Angola in 1989; Namibia made the transition to independence shortly thereafter under United Nations supervision; and Mozambique's war ended in 1992 as South Africa pulled support from a favored insurgent group. The UN oversaw the return of several million displaced persons and national elections in Mozambique in 1994. South Africa made its own transition to black majority rule in 1994 with more grace and less trauma than anyone would have predicted a decade earlier.

Regrettably, as of mid-1999, Angola had lapsed once again into major warfare as its 1994 peace agreement fell apart and the DRC was a battleground where states of the region contributed troops to opposing sides of a sprawling contest for control of the country and its resources. Moreover, southern Africa's multiple, long-lasting, and overlapping conflicts flooded the region with light arms and land mines, while its peace accords flooded markets with ex-soldiers looking for work and armed for alternative employment.

APPENDIX B

The Arms Transfer Data Set

The data on transfers of major conventional weapons to developing countries that served as the basis for the analysis in chapters 1 and 2 were derived from the arms transfer registers of the *SIPRI Yearbook*. Although the SIPRI data set is imperfect, it is the best and most consistent time series that tracks major weapon transfers. SIPRI analysts comb the trade press and other relevant public data sources to compile their annual assessments, and while SIPRI may not have access to the sort of inside data on countries' forces that is the bread and butter of intelligence agencies, it is nonetheless an invaluable source of basic data. It is also self-correcting, in the sense that entries from prior years are changed in subsequent editions of the *SIPRI Yearbook* as new information comes to light. In this analysis, the most recent delivery estimate was used for any given transfer.

SIPRI has also published dollar-value time series data that it calls "trend indicator values." Brzoska and Ohlson (1987) based their analyses on trend-indicator values running from 1950 through 1985, and they published individual supplier and recipient totals for that period.[1] Rather than report the actual value of an arms sale/transfer, SIPRI calculates what the transfer ought to have cost buyers if the prices of the items in question were not deeply discounted for political purposes, paid for in kind rather than cash, or paid for in non-convertible currencies (Russian rubles or Indian rupees, for example). SIPRI's pricing also attempts to account for the fact that technology has enhanced the performance of armaments over time. But SIPRI tracks only part of the arms trade, that involving combat aircraft, fighting ships of greater than 100 tons displacement, artillery of 100 mm or greater caliber, armored fighting vehicles, guided missiles and their launchers, and radar equipment directly associated with target tracking or weapon guidance. Its arms trade registers do not count the cost of military construction or support services, the cost of weapon components or upgrades, or the cost of transport vehicles, small arms, or ammunition.[2]

As a result, SIPRI tends to report a lower valuation for the international arms trade than does the U.S. Arms Control and Disarmament Agency (ACDA), which also attempts to estimate the value of the arms trade rather than just the out-of-pocket costs to recipients, but which incorporates a wider range of items in its annual tallies, including the cost of support services and the value of direct commercial sales. Periodically, ACDA would revalue prior-year arms sales based upon late-breaking data or evolving methodology. In its 1988 edition, for example, the agency substantially increased Soviet arms trade totals for the period 1977–1986. In 1992, it revalued U.S. arms exports upward, and in 1998 it changed its method of estimating the value of exports made by commercial firms under licenses issued by the State Department. The licenses create a three- or four-year window in which the companies holding the license may try to complete a deal with a specific country for specific armaments; a significant fraction of these licenses never actually result in a completed transfer. The new methodology increased the reported value of U.S. commercial arms exports as much as tenfold; whether the reported increase is actually justified is not clear.[3]

To check for the impact of the new commercial sales calculations on U.S. export totals, ACDA's data were compared to those reported annually by Richard F. Grimmett of the U.S. Congressional Research Service (CRS).[4] Like ACDA, Grimmett counts a broader range of materials and services in his calculations than does SIPRI, but unlike ACDA, he does not include the value of U.S. commercial arms exports.

The differences between SIPRI, ACDA, and CRS numbers can be seen in figure B.1, which tracks the aggregate value of all suppliers' arms transfers to developing countries between 1950 and 1996, expressed in constant 1995 U.S. dollars, using inflation adjustments published by the U.S. Department of Defense.[5] As expected, the data from the three sources do not match exactly, yet the shapes of the curves that they draw are not that different. Relative to the late 1970s and 1980s, the value of the trade through the mid-1960s is relatively modest, and that value comes down once again in the 1990s, so that all three data sets nearly converge by 1996.

Figure B.2 compares just the ACDA and CRS data sets, showing Grimmett's numbers as a percentage of ACDA's, and breaking them out by supplier. Positive percentages indicate that Grimmett's reported delivery numbers were higher than ACDA's; negative percentages indicate the obverse. From the late 1980s through 1995, Grimmett reported substantially lower values for U.S. deliveries than did ACDA, a difference that could be accounted for by the growing role of commercial exports in the American arms trade. Grimmett also tends to report lower Russian delivery numbers than does ACDA, but much higher numbers for small suppliers. Since these suppliers account for a tiny fraction of the total arms market (about 4 percent), the difference this makes in his aggregate totals as compared to ACDA's is minimal.

SIPRI arms delivery data from 1950 through 1985, and from 1987, were compiled in machine-readable form by Prof. William Baugh and Dr. Michael Squires, who graciously granted permission to use their compilation for the analyses in this volume.[6] Originally designed in an 80-column format for manipulation by Fortran-based custom programs, the data set was translated into standard Dbase (.dbf) format at the Henry L. Stimson Center, exported to a Foxpro 2.0 data base, and augmented with data from delivery years 1986 and 1988–1993. The data set

Figure B.1   Valuations of Arms Transfers to Developing Countries

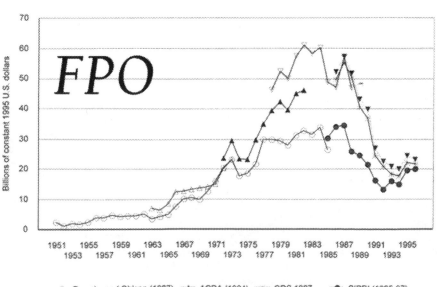

Figure B.2   Comparing CRS and ACDA Arms Delivery Data Values

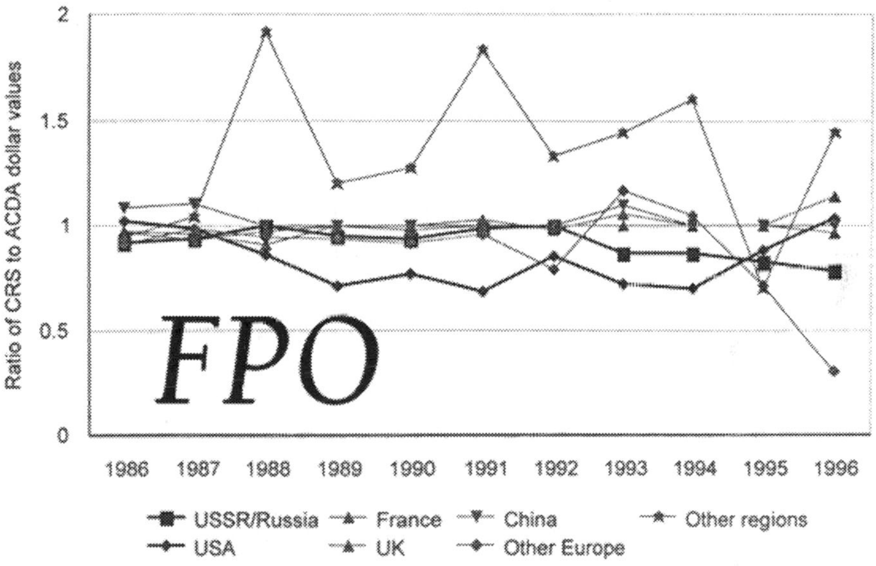

All values adjusted to 1995 dollars.

was later moved again to Windows-based Lotus Approach, and data for deliveries in 1994 and 1995 were added from the 1995 and 1996 *SIPRI Yearbook*.

Considerable care was taken to weed out duplicate data entries as well as data entries representing undelivered arms orders. Because the dates associated with weapon deliveries are generally firmer than the dates ascribed to arms orders, the study used SIPRI's arms delivery data and not its data on orders. Leaving the latter out of the picture works well for a retrospective study such as this one, but the ability to *predict* arms acquisitions would require working knowledge of orders as well, as neighboring states may well react to news of an order and not wait until they see new hardware before placing orders of their own.

In general, SIPRI denotes dates or quantities that are less certain by using parentheses. Where such entries seemed anomalous, they were cross-checked against tables of country holdings in *The Military Balance*, published annually by the International Institute for Strategic Studies (IISS) in London. If IISS listings showed no inventory of the specified system within two or three years of SIPRI's record of its transfer to a given country, the SIPRI record was marked spurious and deleted from the data base used for this study. (The time lag of two to three years accounted for the possibility that new arms inventory data might come to the attention of IISS only after equipment had been operational for a few years. The lag also allowed for data that may have been caught by SIPRI in a given year after the publication deadline for *The Military Balance*.)

Altogether, the data set used for this study contains roughly 13,000 records. A record is one year's deliveries of a particular weapon system to a particular country as a result of a particular sales agreement. Most agreements generate deliveries over several years. In its earlier years, and especially for sales whose details had to be estimated in part, SIPRI provided the total quantity of weapon systems delivered and a range of years over which they were believed to have been delivered, updating quantities and times in successive yearbooks. Here, multiyear deliveries were annualized by averaging them out over the indicated span of years, there being no better estimate at hand than the average (that is, the statistical "expected value") for quantities delivered in each year. Fractional quantities were avoided by rounding down in the earlier years of a series and rounding up in the later years to match the reported total delivered, mimicking slightly the ramping up in deliveries that might be expected as an agreement matures.

Deliveries were coded according to a standard list of weapon system descriptions that differed from those used by SIPRI (whose descriptions varied somewhat over time). A two-letter weapon-type coding scheme developed by Baugh and Squires was expanded into three letters to facilitate data sorting. The first letter in each type code describes the major category: Aircraft, Land forces equipment, Missiles, and Naval vessels and equipment. The second and third letters are a mnemonic for the specific weapon type within the major category, for example, MAS stands for "Missile, Air to Surface," and MTG stands for "Missile, anti-Tank, Guided." Table B.1 lists weapon system types within each major category, indicating the number of different weapon systems represented within each type. Types listed in bold were included in the analysis.

Weapon systems were differentiated in the data base by model as well as basic platform (thus, for example, deliveries of F-16A and F-16C jet fighters could be extracted and analyzed separately). Order dates, size of order, delivery dates, and size of delivery were logged for each record, which also includes the supplier and a "transfer type" variable indicating whether transfers came from a Western country, Eastern bloc country (for Cold War years), or a developing state arms producer (and whether it was licensed or indigenous production), or whether it was known to be secondhand equipment.

Because weaponry has become more sophisticated over time, particularly in the case of high-performance combat aircraft, missiles, and heavy armor, these categories were sorted by "generation," a rough estimate of technological sophistication and performance. "Transfer delay" measured the elapsed time between a weapon system's initial operational capability (IOC) with the original supplier state's armed forces and its delivery to a developing country recipient. This measure helped to define the rate at which new weapons have entered the world market and specific regional markets. Looking at transfer delay over time tells us whether weapons proliferation is accelerating, as is sometimes contended, and if so where and in what major weapon categories. Measuring generation and transfer delay together indicates whether weapons have been proliferating more quickly as their sophistication has increased.

Defining Weapon System Generations

Although the military trade press often declares that a weapon system represents "a new generation" of armaments, the term is used loosely, and the definition of *generation* is harder to pin down.

The performance characteristics of aircraft and other weapon systems can be subjected to far more rigorous quantitative scrutiny than was done here, involving not just the mechanical characteristics (such as, for aircraft, combat radius, top speed, weapon payload, thrust-to-weight ratio, and wing loading) but also the capabilities of aircraft electronics (or "avionics," radars, night attack systems, navigation, and the like). In the late 1970s, Lewis Snider used a statistical technique known as factor analysis to sort through more than a dozen aircraft characteristics in an effort to differentiate aircraft according to their aerial combat and ground attack capabilities. Ronald Sherwin and Edward Laurance published a trenchant critique of this application of factor analysis and recommended instead the use of "multi-attribute utility" (MAU) analysis. MAU uses panels of judges to assign relative importance to a set of weapon system characteristics and to evaluate different systems capabilities with respect to those characteristics, to develop an index number that can be compared across systems. The U.S. Army used a variation of MAU to create Weapon Effectiveness Indices and compile estimates of combat power of U.S., other Western, and Soviet weapon systems and fighting units.[7]

Although Sherwin's and Laurance's critiques were taken to heart, a lack of resources required to empanel the expert judges needed to evaluate dozens of models of aircraft, tanks, and missiles suggested a simpler approach. The generations developed here for aircraft and tanks are broad cat-

Table B.1 Weapon Types

Type	System Description[a]	No. Models	Totals
Fixed-wing aircraft			
AAL	**attack, light (weapon payload <6,000 lb.)**	37	
AAM	**attack, medium**	21	
ABB	**bomber (internal weapons carriage)**	32	
ACH	cargo, heavy (200+ passengers, 40+ tons of cargo)	5	
ACL	cargo/utility (6–25 passengers)	132	
ACM	cargo, medium (26–199 passengers, <40 tons of cargo)	112	
ACU	cargo, light utility (5 or fewer passengers)	97	
ADO	Drone, other	2	
AEI	electronic intelligence/electronic warfare	7	
AEW	early warning/control	6	
AFB	**fighter/bomber**	162	
AIR	inflight refueling tanker	8	
AMP	maritime patrol	21	
ARA	reconnaissance	31	
ASR	search & rescue/ambulance	4	
ASW	antisubmarine warfare	18	
ATA	**trainer and light attack**	43	
ATF	**trainer variant of fighter/bomber**	35	
ATR	trainer (unarmed)	130	903/**330**
Helicoptors			
AHM	**missile-armed**	27	
AHR	search and rescue	4	
AHT	**transport**	170	
AHW	antisubmarine	25	226/**197**
Land forces systems			
LAA	**antiaircraft gun, non-self-propelled**	1	
LAS	**antiaircraft gun, self-propelled**	14	
LAC	**armored car, reconnaissance**	47	
LCC	command personnel carrier	4	
LGC	**ground-launched cruise missile launcher**	1	
LGM	**ground-launched ballistic missile launcher**	10	
LGS	**field gun, self-propelled**	6	
LGT	**field gun, towed**	9	
LHS	**howitzer, self-propelled**	14	
LHT	**howitzer, towed**	10	
LIC	**infantry combat vehicle, mechanized**	10	
LMC	**mortar carrier**	3	
LPC	**personnel carrier, armored**	62	
LRM	**rocket launcher, mobile (surface-to-surface)**	15	
LRV	recovery/engineering vehicle, armored	6	
LSA	**surface-to-air missile launcher, fixed site**	5	

(continues)

egories based on an educated eyeball assessment of a range of system capabilities. Clearly, each generation contains weapon systems of varying capabilities, but variation within generations should be less than variation between, and should appear so to observers familiar with the systems. In the case of missiles, generational assignments were used to distinguish step-function improvements in performance within a particular military mission area.

Table B.1 *(continued)*

Type	System Description[a]	No. Models	Totals
LSM	**surface-to-air missile launcher, mobile**	**20**	
LTD	**tank destroyer (gun-armed vehicle)**	**12**	
LTG	**anti-tank guided missile launching vehicle**	**2**	
LTH	**tank, heavy (36+ tons, 90+ mm main gun)**	**38**	
LTL	**tank, light (<20 tons, tracked, 76+ mm main gun)**	**18**	
LTM	**tank, medium (20–35 tons, tracked, 76+ mm main gun)**	**17**	
LVB	bridging vehicle, armored	2	
LVS	specialized vehicle (misc. category)	3	329/**314**
Missiles			
MAA	**air-to-air missile**	**32**	
MAN	**air-launched anti-ship cruise missile**	**13**	
MAR	**air-to-surface anti-radiation missile**	**3**	
MAS	**air-to-surface missile, general purpose**	**12**	
MGC	**around-launched cruise missile**	**3**	
MGM	**surface-to-surface ballistic missile**	**7**	
MGN	**ship-launched anti-ship cruise missile**	**18**	
MRM	multiple-launch rockets	2	
MSA	**surface-to-air missile, fixed-site**	**7**	
MSM	**surface-to-air missile, mobile**	**25**	
MSN	**surface-to-air missile, naval**	**13**	
MSP	**surface-to-air missile, man-portable**	**11**	
MTG	**antitank missile, mobile ground-launched**	**18**	
MTH	**antitank missile, helicoptor-launched**	**7**	
MTP	**antitank missile, portable**	**8**	179/**177**
Naval vessels			
NAC	amphibious warfare craft	12	
NAS	amphibious warfare ships	27	
NAX	auxiliary ships	50	
NCA	**cruiser, gun-armed**	**7**	
NCG	**cruiser, missile-armed**	**1**	
NCM	**corvette, missile-armed (300–1000 tons full load displacement)**	**20**	
NCO	**corvette, gun-armed**	**25**	
NCV	**aircraft carriers**	**3**	
NDD	**destroyer, gun-armed (>2000 tons full load displacement, >30 knots speed)**	**17**	
NDG	**destroyer, missile-armed**	**3**	
NFF	**frigate, gun-armed (1–2000 tons full load displacement, <30 knots speed)**	**32**	
NFG	**frigate, missile-armed**	**19**	
NMW	mine warfare ships	32	
NPB	**patrol combatants conventional**	**122**	

(continues)

Heavy Combat Aircraft

Basic performance characteristics used to assign combat aircraft to generations are given in table B.2, along with representative samples of systems from each generation. Speed of course increases with each generation. Maximum takeoff weight climbs considerably in the jet-powered generations through the fifth; combat aircraft simply got bigger with time to accommodate

Table B.1 *(continued)*

Type	System Description[a]	No. Models	Totals
NPM	**patrol combatants missile**	**27**	
NSC	service craft	19	
NSN	**submarine, nuclear-powered**	**1**	
NSS	**submarine, conventionally powered**	**27**	444/**304**
			2132/**1322**

[a]**Boldface** indicates weapon type data used in chapters 1 and 2.

larger engines, more weapons, and much more avionics. A sixth-generation F-15 Eagle is three times heavier than a late-1940s F-80 Shooting Star, for example, but it has ten times the engine thrust, giving it a thrust-to-weight ratio that is three times better (making for better acceleration, faster climb and turn rates, and overall maneuverability). Finally, ground attack payloads in jets increased over time from about one ton to eight tons; a fifth- or sixth-generation fighter-bomber could carry nearly the same load of ordnance as a World War II heavy bomber like the Boeing B-17.

The first jets (the second generation in table B.2) were designed in the mid-1940s primarily as pursuit and interception aircraft. Their limited ordnance carrying ability was one reason why a first-generation, propeller-driven aircraft like the Skyraider was kept in service for 20 years.

Third-generation aircraft were built in large numbers and formed the backbone of the major powers' tactical air forces into the early 1960s. Turnover in the 1950s was quite rapid, however, as the technologies of jet-powered flight, airborne radar, and air-to-air weapons changed and matured rapidly under the impetus of the Cold War and growing fears that Armageddon might be just around the corner. This rapid evolution of combat jet technology, especially in engines but also in scientific knowledge of the aerodynamics of faster-than-sound flight, produced a generation of transonic aircraft able to exceed the speed of sound by 20 to 30 percent in high-altitude level flight and to carry greater payloads. This fourth generation includes aircraft like the U.S. F-100 Super Saber, Soviet MiG-19, and the Chinese F-6, which is based on the MiG-19.

These models were in turn rapidly eclipsed, however, by more capable fighter-bombers able to fly at twice the speed of sound that began to enter air forces in the early to mid-1960s (except for the Chinese air force). The fifth generation includes a number of highly successful and widely distributed aircraft such as the French Mirage III and V, American F-104 Starfighter and F-4 Phantom II, the Soviet MiG-21 (and its Chinese clone, the Shenyang F-7), and the MiG-23, the first Soviet fighter with a look-down/shoot-down capability (but not enough maneuverability to rate inclusion in generation six). Most of the aircraft in the fifth generation are multirole—that is, capable of both air-to-air and air-to-ground combat. That is true only for the latest versions of the MiG-21, new models of which continued to be developed through the mid-1970s.

The sixth generation is something of a catch-all category for heavy combat aircraft that have entered service since the mid-1970s. The sixth generation's inclusiveness reflects, in part, the maturation of combat jet technology after several decades of rapid change, and entails a distinct slowdown in the proliferation of aircraft models, as table B.2 indicates. The U.S. Air Combat Command, for example, is built around substantial numbers of C-model F-15s and F-16s that first rolled out of the factory in 1979 and 1980 respectively. The F-16C/D has subsequently gone through a number of "block" component upgrades that are not reflected in the data base.

If aircraft incorporating substantial radar deflecting/absorbing ("stealth") technology were to be sold to buyers in developing countries, then a seventh generation would be in order. The U.S. Air Force operates one stealthy attack aircraft, the F-117, and has a stealthy air superiority fighter,

Table B.2 Generations of Heavy Combat Aircraft: Basic Characteristics

Generation	Sample Aircraft (US)	Year Initially Operational	Max Level Speed (mph)	Combat Radius (miles)	Maximum Takeoff Weight (lbs.)	Main Armament	Ground Attack Payload (lbs.)	Engine (lbs. static thrust)	Avionics
First	F-4U5 Corsair	1943	470	450	14,100	4 × 20 mm cannon	2,000	[2,300 hp]	
Second	F-80C Shooting Star	1949	580	550	16,900	6 × 0.5 in. mach. guns	2,200	4,600	
Third	F-86F Saber	1952	687	370	20,600	2 AAM 6 × 0.5 in. mach. guns	2,000	6,000	
Fourth	F-100C Super Saber	1954	864	600	34,800	4 × 20 mm cannon	7,500	16,000	
Fifth	F-4B Phantom II	1961	1,500	600	54,600	4–8 AAM	16,000	34,000	APQ-72 radar, IR detector
Sixth	F-15A Eagle	1975	1,650	700 est.	56,000	8 AAM + 20 mm Gatling gun	16,000	47,600	APG-63 radar, ECM

Sources: Michael J. H. Taylor, ed., Jane's Encyclopedia of Aviation (New York: Crescent Books, 1989); Bill Gunston, An Illustrated Guide to Modern Fighters and Attack Aircraft (London: Salamander Books, Ltd., 1980).

the F-22 Raptor, under development, but neither is likely to be exported to the markets that are the focus of this volume.

Other open sources also discuss combat aircraft generations, in general terms. The Congressional Budget Office (CBO), for example, addresses the issue and also refers to generational categories used by the Office of Naval Intelligence (ONI), and to a generational assessment done by Robert Berman at Brookings in 1978. The generational concept is used in unclassified ONI publications to position other countries' aircraft in a rough technological hierarchy, but the performance measures underlying the categories are not further defined.[8] Table B.3 offers a comparison of CBO's discussion of aircraft generations and the assignments made in this volume.

Guided Missiles

Postwar American missile designers initially built upon late World War II German surface-to-surface and surface-to-air designs. The looming threat of Soviet nuclear arms and the belief that Soviet power needed to be both deterred and defended against stimulated research in offensive and defensive missiles. By the mid-1950s, the United States had fielded a variety of battlefield ballistic missiles, ground-launched cruise missiles, surface-to-air, air-to-air, and air-to-surface missiles.

The Soviet Union did the same, applying its artillery tradition to the development of extensive families of guided and unguided battlefield rockets. It was also the first to field short-range naval cruise missiles. Its shorter-range ballistic missiles eventually found their way into the arsenals of selected Third World friends and allies, were copied by China and North Korea, and played a significant political-psychological role in the Iran-Iraq War as urban terror weapons. That role was reprised in the Gulf War when Iraqi missiles shelled Israeli and Saudi cities, further stimulating wide interest in effective theater ballistic missile defenses.

Radar-guided surface-to-air missiles (SAMs) are sizable and costly, their tracking and guidance radars vulnerable to air-launched countermeasures (as demonstrated by NATO over Bosnia in the fall of 1995 and Kosovo in the spring of 1999). These limitations have led some countries to seek air defense missile systems with infrared (IR) homing capability. Such missiles

Table B.3 Comparing Generations of Aircraft

Sample Aircraft	Congressional Budget Office Generations[a]	Generations Used in This Volume
DeHavilland Mosquito	n/a	First
Douglas A-1 Skyraider	n/a	First
Lockheed F-80 Shooting Star	n/a	Second
Gloster F.8 Meteor	n/a	Second
MiG-15/17, F-86 Saber	First	Third
MiG-19, F-100 Super Saber	First	Fourth
Mig-21, F-4B Phantom II (1961)	Second	Fifth
Mig-23, F-4E Phantom II (1968)	Third	Fifth
Mirage 3, Mirage 5	Second/Third	Fifth
Panavia Tornado, Mirage F-1	Second/Third	Sixth
F-14 Tomcat, F-15 Eagle	Fourth	Sixth
Su-27, MiG-29, Mirage 2000	Fourth	Sixth
F-22 Raptor, Eurofighter 2000	Fifth	Not ranked (not in arms trade)

[a]U.S. Congressional Budget Office, *A Look at Tomorrow's Tactical Air Forces* (Washington, D.C.: U.S. GPO, January 1997), pp. 14–15.

detect and fly toward the heat emitted by their targets but emit nothing themselves before launch, giving their users an exploitable element of surprise, especially against relatively slow and low-flying cargo aircraft and helicopters. Many IR-guided SAMs can be carried by one or two soldiers; many thousand of these "shoulder-fired" weapons have made their way into the inventories of developing states and insurgent movements. In the 1980s, they proved to be great battlefield equalizers for such movements in Afghanistan, Angola, and El Salvador.

Compact air-to-air (or "air intercept") missiles have been available for use on some fighter aircraft since the 1950s. As with SAMs, air-to-air missiles come in two basic flavors, IR and radar homing. The United States deployed the AIM-9 Sidewinder IR-homing missile in 1956 but not until 1976 had missile seeker technology evolved to the point at which a Sidewinder could attack a target from any angle and not just from the direction of the hot engine exhaust. This new "all aspect" capability permitted pilots to engage without first maneuvering behind their target, a powerful air combat advantage that has been given to fewer than ten states outside NATO Europe.

Because infrared missiles lock onto their target before launch and engage it independently thereafter, they are true "fire and forget" systems. For the same reason, however, they can be used only against targets that their seeker can identify, which is generally no further than the unaided human eye can see. To engage targets beyond visual range, fighters need radar-guided missiles.

Fighters cannot carry the large radars that support long-range SAM systems, much less the two-ton missiles that such systems fired in the 1950s, forcing design and capability compromises that have been rendered steadily less onerous over time with progress in propulsion, electronic miniaturization, and computing power. Most radar-guided air-to-air missiles used "semi-active" homing, a compromise in which the fighter's radar illuminated the target and the intercept missile homed on that reflected radiation. The launching fighter had to keep the target illuminated until its missile struck home, leaving it vulnerable in the meantime to counterattack. An example of such a missile is the U.S. AIM-7 Sparrow. Through the 1980s, active onboard radar was confined to a few large (300 to 400 kg class) missiles like the U.S. Navy Phoenix, in service with the Navy's F-14 Tomcat fighter since 1974, or the Soviet AA-10 Alamo, which entered service in 1985. Used for long-range intercepts, the Phoenix was sold to just one customer, Imperial Iran. The Alamo did not reach a developing state until 1995, when it became part of aircraft packages offered to Malaysia and Vietnam.

The United States' newest air intercept missile, the Advanced Medium Range Air to Air Missile (AMRAAM), is lighter than the Phoenix (at 150 kg) and has 10 to 20 percent better range than the Sparrow, plus "fire and forget" active-radar guidance—something of a miracle of miniaturization. Operational since 1992, it has been sold so far to half of NATO and to Finland, Sweden, and Switzerland, but to few countries outside Europe.

Precision-guided munitions have extended the ability of fighter aircraft to attack ground targets. Some of these are bombs with strap-on, semi-active laser guidance and maneuvering kits, while others are fully integrated missiles with standoff ranges from 5 to 1,500 kilometers. Autonomous weapons that follow an aerodynamic path (as opposed to a ballistic trajectory) to their targets are known generically as "cruise" missiles. In this volume, lest every nonballistic air-to-ground and ground-to-ground missile be swept into that category, only autonomous weapons with ranges greater than 20 kilometers were defined as cruise missiles, thus excluding, for example, antitank weapons from the category.

The competition between battle tanks and tank-killing technology produced the first antitank guided missiles in the mid-1950s in Europe, where NATO countries faced massed Soviet armor across the inner German and Czech borders. Relatively slow-flying (much slower than a tank round) and guided to their targets by means of a trailing wire, these missiles gave the foot soldier or infantry squad tremendous new firepower. However, the soldier responsible for keeping a bead on the target to guide the missile was exposed to counterfire, spawning adaptations to protect the operators, like armored launch vehicles and elevated platforms for missiles and optics. Later generations of missiles dispensed with the wire, homing on reflected laser light, but somebody still had to remain within line of sight of the target to shine the laser on it. Systems just coming into operation with U.S. forces in the late 1990s have autonomous homing capabilities. One shoulder-fired missile uses imaging infrared sensors that lock onto a target before launch. Another, helicopter-carried missile uses extremely high frequency radar to illuminate and track its target. Once these missiles have been fired, the launcher/operator can either duck and cover or turn attention to other business.

Military missiles have, in short, developed considerably in 40 years, sufficient to divide that development into distinct generations of equipment. Table 1.10 presents examples of four missile types and up to four generations of capability for each by the mid-1990s. Because missiles intended for different applications use different technologies, it is not easy to make direct generational comparisons across missile types. Generational assignments within types are based upon the evolution of performance characteristics but are inevitably subjective, in part. Table B.4 presents performance characteristics for selected missiles of different generations in each of the four main missile types.

Heavy Armor

Like other weapon systems, armored vehicles have evolved over the last four decades in the three basic categories of mobility, firepower, and protection.[9] Engine horsepower has roughly tripled in 50 years and suspensions have improved, combining to give the latest tanks nearly double the road speed and much higher overland speed and maneuverability than their predecessors. Offensive firepower has been enhanced over time by larger-caliber guns, night vision and sighting systems, laser rangefinders, and stabilized sights and gun aiming systems. Stabilizers keep a tank's main gun aimed at a target even while the vehicle is moving at speed across rough terrain. Armor has become harder to penetrate and has evolved into elaborate composite laminates using ceramics, depleted uranium, and other materials in addition to the traditional high-strength steel. Add-on armor even incorporates its own explosive charges: so-called reactive armor is designed to detonate outward and disrupt the blast of an antitank missile warhead.[10]

Using these mobility, firepower, and protection advances, one can notionally assign tanks, at least, to vehicular generations, much as was done above with heavy combat aircraft and missiles (see table B.5). First-generation tanks are mostly of World War II design, are relatively lightly armed and armored, and lack stabilizers, night vision, and laser ranging technology. Second-generation tanks include either stabilized guns or night vision, and gun calibers of at least 100 mm. The U.S. M-47/48 series lacked these firepower characteristics but had much better armor and cross-country capacity (as measured by horsepower-per-metric ton) than most first-generation units and so merited inclusion in generation two. Third-generation tanks incorporate two out of three firepower improvements (stability, night vision, laser range finding). Fourth-generation tanks incorporate all three firepower improvements plus advanced laminate armor.

Armored vehicles remain in service longer than combat aircraft, and while the SIPRI arms trade registers do differentiate to a degree among models of tanks, many makers' vehicles have undergone upgrades over the course of their long production lives that the SIPRI data do not pick up. So in reviewing the generational assignments in table B.5, the reader should bear in mind that they refer to the initial production versions of each model listed.

Table B.4 Examples of Missile Characteristics by Type and Generation

	1, 2, 3, 4 = Generation (IOC)	Range (km)	Speed (mach)	Guidance Type	Warhead (weight, type)	Other
Air Defense	*IR-guided AAMs (US)*					
	1. AIM-9A Sidewinder (1956)	3.2	2.0	Lock-on-before-launch (LOBL) self-guidance in all models.	4.5 kg blast/frag.	Vacuum tube electronics
	2. AIM-9H " (1970)	to	to		9.0 kg continuous rod	Solid state electronics
	3. AIM-9L " (1976)	18.5	2.5		9.5 kg blast/frag.	All-aspect seeker, laser fuse
	Radar-guided AAMs (US)					
	1. AIM-7A Sparrow (1956)	8	3.7	Radar beam rider	30 kg blast/frag.	Vacuum tube electronics
	2. AIM-7C/E " (1958/69)	25	to	Semi-active radar homing	30 kg blast/frag.	Solid state electronics
	3. AIM-7F/M " (1981/83)	44	4.0	Semi-active radar homing	39 kg blast/frag.	Chooses among multiple targets
	4. AIM120 AMRAAM (1991)	74	4.0	Command, inertial, active radar homing	22 kg directed frag. "smart fuse"	
Anti-armor	1. AT-1 Snapper (USSR) (1960)	2.3	0.15	Wire, CLOS	5.3 kg hollow charge	Penetrate 350 mm armor
	2. BGM-71A TOW (US) (1970)	3.8	0.51	Wire, SACLOS	3.5 kg shaped charge	Penetrate 500 mm armor
	3. AT-5 Spandrel (USSR) (1977)	4.0	0.74	Wire, CLOS	n/a	
	4. AGM-114A Hellfire (US) (1985)	7.0	1.10	Laser beam rider	Tandem twin shaped charges	
Cruise	*Anti-ship*					
	1. AS-1 Kennel (USSR) (1961)	90	0.90	Cmd. link, radar homing	n/a	
	2. SS-N-9 Siren (USSR) (1969)	110	1.4	Cmd. link, radar homing	500 kg blast	
	3. AGM-84 Harpoon (US) (1977)	92	0.85	Inertial plus active radar homing	222 kg blast	

(continues)

Table B.4 *(continued)*

1, 2, 3, 4 = Generation (IOC)	Range (km)	Speed (mach)	Guidance Type	Warhead (weight, type)	Other
Other Air-to-ground					
1. AGM-12B Bullpup (US) (1959)	10	2.0	Cmd. link, track-by-flare	113 kg blast	
2. AGM-65A Maverick (US) (1972)	13	1–2	Daytime TV guidance link	59 kg shaped charge	
3. AS-14 Kedge (USSR) (1980)	40	1.0	Cmd. link, IIR	100+ kg	
4. AGM-65G (US) (1991)	30	1–2	IIR	136 kg blast/frag.	

Notes: CLOS = Command Line of Sight (missile guided manually to target by operator); SACLOS = Semi-automatic CLOS (operator tracks target, missile alters course to match); IIR = imaging infrared.

Sources: R. T. Pretty and D. H. R. Archer, eds., *Jane's Weapon Systems, 1973–74* (New York: McGraw-Hill, 1973); R. T. Pretty, ed., *Jane's Weapon Systems, 1986–87* (London: Jane's Publishing Co., 1986); Forecast International, *Missile Forecast* (1996). *Flight International,* March 16–22, 1994, pp. 34–36; *Aviation Week and Space Technology,* March 16, 1992, pp. 80–118; Ray Bonds, ed., *The Soviet War Machine* (London: Salamander Books, Ltd., 1977); Ray Bonds, ed., *The US War Machine* (London: Salamander Books, Ltd., 1978); Col. Timothy M. Law and Steven L. Llanso, *Encyclopedia of Modern U.S. Military Weapons* (New York: Berkeley Books, 1995).

Table B.5 Tank Characteristics

Name	Made In	Age (IOC)	Mobility				Protection		Firepower				Gen.
			Gross Weight (tonne)	Engine (hp)	HP per tonne	Max Road Speed (kph)	Max Glacis Armor (mm)	Advanced Armor?	Gun Caliber (mm)	Gun Stable?	Night Vision?	Laser Ranging?	
Comet	UK	44	33	600	18.2	52	32	0	77	0	0	0	1
Centurion	UK	45	50	650	13.0	35	76	0	77	0	0	0	1
Centurion Mk3	UK	47	50	650	13.0	35	76	0	83	0	0	0	1
M-4	USA	42	31	450	14.5	42	61	0	75	0	0	0	1
M-36	USA	44	28	450	16.1	44	51	0	90	0	0	0	1
T-34	USR	40	32	500	15.6	55	41	0	76	0	0	0	1
T-34 85	USR	43	32	500	15.6	55	41	0	85	0	0	0	1
JS-2	USR	44	46	520	11.3	37	91	0	122	0	0	0	1
JS-3	USR	45	46	520	11.3	37	104	0	122	0	0	0	1
TYPE 59	CHN	59	36	520	14.4	50	87	0	100	0	1	0	2
AMX-30	FRA	66	36	720	20.0	65	79	0	105	0	1	0	2
AMX-30S	FRA	77	37	620	16.8	65	79	0	105	0	1	0	2
Vickers Mk 1	UK	65	39	650	16.7	48	60	0	105	1	0	0	2
M-47	USA	50	46	810	17.6	48	101	0	90	0	0	0	2
M-48	USA	52	45	810	18.0	42	101	0	90	0	0	0	2
M-48A2C	USA	58	47	825	17.6	48	101	0	90	0	0	0	2
M-60	USA	60	50	750	15.0	48	n/a	0	105	0	1	0	2
M-60A1	USA	62	53	750	14.2	48	n/a	0	105	0	1	0	2
M-48A5	USA	75	49	750	15.3	48	101	0	105	0	1	0	2

(continues)

Table B.5 *(continued)*

Name	Made In	Age (IOC)	Gross Weight (tonne)	Engine (hp)	HP per tonne	Max Road Speed (kph)	Max Glacis Armor (mm)	Advanced Armor?	Gun Caliber (mm)	Gun Stable?	Night Vision?	Laser Ranging?	Gen.
T-54	USR	47	36	520	14.4	48	87	0	100	0	1	0	2
T-10	USR	55	50	400	8.0	42	230	0	122	1	0	0	2
T-55	USR	59	36	580	16.1	50	87	0	100	1	1	0	2
TYPE 69	CHN	82	37	580	15.7	50	87	0	100	1	1	1	3
AMX-30B	FRA	76	37	700	18.9	65	79	0	105	0	1	1	3
TAM	GMW	78	30	720	24.0	75	n/a	0	105	1	1	0	3
OF-40	ITA	82	43	830	19.3	60	61	0	105	0	1	1	3
Centurion Mk9	UK	59	52	650	12.5	35	118	0	105	1	1	0	3
Chieftain	UK	67	54	730	13.5	48	n/a	0	120	1	1	1	3
Chieftain Mk3	UK	71	54	730	13.5	48	n/a	0	120	1	1	1	3
Chieftain Mk5	UK	71	55	750	13.6	48	n/a	0	120	1	1	1	3
Vickers Mk3	UK	79	40	720	18.0	50	60	0	105	1	1	1	3
Khalid	UK	79	58	1200	20.7	48	n/a	0	120	1	1	1	3
M-60A3	USA	78	53	750	14.2	48	n/a	0	105	1	1	1	3
T-62	USR	62	40	580	14.5	50	87	0	115	1	1	0	3
LeClerc	FRA	93	50	1500	30.0	72	n/a	1	120	1	1	1	4
MERKAVA Mk2	ISR	83	56	900	16.1	46	n/a	1	105	1	1	1	4
Challenger	UK	83	62	1200	19.4	52	n/a	1	120	1	1	1	4
M-1	USA	80	54	1500	27.8	72	n/a	1	105	1	1	1	4

(continues)

Table B.5 *(continued)*

Name	Made In	Age (IOC)	Mobility				Protection			Firepower			
			Gross Weight (tonne)	Engine (hp)	HP per tonne	Max Road Speed (kph)	Max Glacis Armor (mm)	Advanced Armor?	Gun Caliber (mm)	Gun Stable?	Night Vision?	Laser Ranging?	Gen.
M-1A1	USA	86	57	1500	26.3	67	n/a	1	120	1	1	1	4
M-1A2	USA	90	57	1500	26.3	67	n/a	1	120	1	1	1	4
T-72	USR	73	41	780	19.0	80	200	1	125	1	1	1	4
T-80	USR	84	42	985	23.5	75	n/a	1	125	1	1	1	4

Sources: Jane's Armor and Artillery, 1982–83, 1989–90 (London: Jane's Publications).

APPENDIX C

Conflict Scaling and Path Analysis

Scaling Conflict Data

To be able to analyze quantitatively the relationship between arms transfers and conflict, one needs a measure of conflict. A single scale for both internal and external conflict is not adequate. If internal conflict is coded at the "low" end and interstate conflict at the "high" end of such a scale, the impact of verbal threats or minor clashes between states may be overstated, while the virulence of internal conflicts may be understated. Angola's long civil war, for example, caused at least 200 times more casualties than the Falklands/Malvinas War between Argentina and Great Britain. Several authors' approaches to conflict scaling were evaluated for use in this study; ultimately, two new scales were created for external and internal conflict respectively.

Alternative Conflict Scales

James Dunnigan and William Martel use a single scale that begins with minor terrorist activity and moves up in severity through sustained internal guerrilla warfare before encompassing interstate wars at levels up to and including world war; they provide data on 200 years of conflict, including 41 conflicts that occurred within the time frame of interest here. However, a single scale for both internal and external conflict really does not allow one to effectively code minor armed clashes between states and may understate the virulence of some internal conflicts. The Falklands, as a low-scale conventional interstate war, would seem to rank higher than the vicious and bloody Angolan civil war, for example, on the Dunnigan-Martel scale.[1] (See tables C.1 and C.2.)

Joshua Goldstein recruited seven fellow faculty members from the University of Southern California to join him in creating a conflict-cooperation scale based on the 61 event-types used by the World Events Interaction Survey (WEIS) data collection. Goldstein took some pains to reinforce the linear and interval nature of the results by means of paired comparisons of items in which the second member of the pair had been assigned a scale value twice as large as the first member (for example, comparing event-types scaled at 3 with those scaled at 6). This was done "to ensure that [panelists agreed that] events in the second box were twice as conflictual as the first." On the resulting scale, a +10 was assigned to full cooperation and a −10 to "military attack, clash, [or] assault."[2]

The elements of Goldstein's scale that are relevant to this analysis are those involving threats or the use of military force, ranging from −4.4 to −10. They are listed in table C.1. A simple transformation was used to re-anchor Goldstein's scale at +2, opposite the "threat" category in the Militarized Interstate Dispute (MID) data set, while retaining the original numerical ratios. Thus transformed, Goldstein's numbers match the MID hostility levels fairly closely, suggesting that the MID levels—and similar scales—can be used for analytic purposes as quasi-interval measures.

Table C.1 Scales for External Conflict

Dunnigan and Martel, How to Stop a War	Militarized Interstate Disputes, v. 2.10	Joshua Goldstein (original scale value in parentheses)	Durch Composite Coding Scale
	1 no militarized action		0 no hostile activity
	2 threats to use force (verbal)	2.0 threat without specific negative sanction (–4.4)	1 fighting or turmoil next door
		2.6 threat with specific negative non-military sanction (–5.8)	
	3 display of force (alert, mobilize)	3.2 threat with force specified; ultimatum (–7.0)	2 cross-border air duels, other minor coercive uses of military force; up to battalion-level engagements
		3.5 armed forces mobilization, display, exercise, buildup (–7.6)	
4 sustained high-intensity guerrilla war or low-level conventional war	4 use of force (raid, blockade, occupy territory, declare war)	4.1 seize position or possession (–8.8, –9.2)	3 large cross-border operation (brigade or larger; Israeli Operation Litani)
5 state of war; continuous operations using primarily conventional forces	5 interstate war	4.6 military attack, clash, assault (–10)	4 low-scale conventional war (Falklands)
6 low-scale conventional war (e.g., Falklands)			6 medium-scale conventional war (Iran-Iraq)
7 medium-size conventional war (Iran-Iraq)			
8 multitheater conventional war ("US, Chinese civil wars")			
9 world war			

Sources: James F. Dunnigan and William Martel, *How to Stop a War: The Lessons of Two Hundred Years of War and Peace* (New York: Doubleday, 1987), p. 181; Stuart A. Bremer, "Format of the Militarized Interstate Dispute Data," January 1, 1996, file FORMAT.TXT, with version 2.10 of the MID dataset maintained by the Peace Science Society International, Internet: http://pss.la.psu.edu/MID_DATA.HTM, last accessed March 4, 2000. See also Daniel M. Jones, Stuart A. Bremer and J. David Singer (1996) "Militarized Interstate Disputes, 1816–1992: Rationale, Coding Rules, and Empirical Patterns." *Conflict Management and Peace Science* 15(2), pp. 163–213; Joshua Goldstein, "A Conflict-Cooperation Scale for WEIS Events Data," *Journal of Conflict Resolution* 36(2), pp. 369–85.

Table C.2 Scales for Internal Conflict

Ted Robert Gurr, Why Minorities Rebel (Rebellion Scale)	Dunnigan and Martel, How to Stop a War	Durch Composite Coding Scale
0 no activity	0 terrorism handled by police	0 no reportable activity
1 political banditry, sporadic terrorism, failed coups	1 terror requiring paramilitary action or low-level guerrilla war involving regular armed forces	1 coup attempt, scattered terrorism, political rioting
2 campaign of terror, successful coup	2 deal with terrorist activities involves most of the armed forces; medium-level guerrilla war	2 successful coup, terror campaign, death squads
3 small-scale guerrilla activity	3 high-intensity guerrilla war, high-level terrorist activities	3 state of emergency, "dirty war," guerrilla war in part of country
4 large-scale guerrilla activity (>1,000 fighters, frequent attacks over a substantial area)	4 sustained guerrilla war or low-level conventional war (overlaps table C.1)	4 nationwide insurgency
5 protracted (>15 years) civil war fought by military units with base areas		5 sustained insurgency with cross-border bases
		6 large-scale civil war on a par with Angola, Lebanon

Sources: Ted Robert Gurr, *Why Minorities Rebel: A Global View of Ethnopolitical Conflicts* (Washington, D.C.: US Institute of Peace, 1993), p. 95; Dunnigan and Martel (see table C.1).

Ted Robert Gurr calculates separate scales for measuring two kinds of internal upheaval, one related to "protest" and the other related to "rebellion." Gurr's scales, while clearly interpretable as to severity of event measured, apply only to internal conflict. Table C.2 presents his scale for "rebellion," defined as action that "aims directly at . . . fundamental changes in government and in power relations among groups."[3]

The well-known MID data set from the long-running Correlates of War (COW) project uses a very simple five-point scale: (1) peace, (2) verbal threat of force, (3) display of force, (4) use of force, and (5) war (defined by the usual COW standard as conflict involving at least 1,000 battle fatalities between regular armed forces of two or more states). Thus, MID lumps all uses of force short of war, of whatever severity, into a single category that includes actions ranging from minor clashes to fighting that causes up to 999 battle fatalities. Nor can it discriminate by size of engagement. The five-part scale is used mostly in default of any other agreed approach to ranking the 23 different specific actions that are coded in the data set. Wayman and Singer, for example, criticized one attempt to scale the 23 actions, in part because it imputed greater seriousness to mobilization of forces (a "display" level action) than it did to several different uses of force. It also gave less weight to military alerts (also a "display" level action) than it did to some verbal threats that involved no other military activities at all.[4]

Some of the difficulty in interpreting and scaling the specific actions within the MID data may derive from the historical nature of the data set (1816–1992) and the need (or the impulse) to code events consistently over long stretches of time in which military doctrine, military technology, and communications technology have each undergone revolutionary change more than once. Since nineteenth-century states needed to call up their forces in order to fight on any substantial scale, mobilization was indeed a more serious and commitment-laden action than any minor clash of standing forces along a common border or seizure of assets on the high seas.

In the present age of large standing forces, fast and accurate delivery vehicles, and electronic communications and intelligence, the United States needs to "mobilize" nothing before it launches cruise missiles at targets like Iraq or Afghanistan, although it certainly increased the alert levels of the specific forces involved in the launch. Indeed, the power of modern military technology can make it dangerous to advertise in advance what one is about to do, particularly if the political support for one's move is likely to hinge importantly on the casualties incurred in making it. Such factors argue for quick moves that combine the display and use of force in one tidy strike package that may fulfill a threat made some time back and that has been "left on the table" to await an appropriate triggering event. U.S. attacks on Libya in the 1980s and Israeli attacks on Palestinian and Hezbollah targets in Lebanon from the late 1970s onward would be actions of that sort.

Composite Conflict Scales

None of these conflict scales was completely satisfactory for use in testing the impact of actual military clashes of varying severity on states' arms acquisition behavior. Needed here was a measure that captured the size of the forces involved in an altercation rather than the number of deaths it may have caused, because the former type of measure seemed more relevant to an assessment of the relationship of conflict and arms imports. Most developing countries' major weapons are imported, so the larger the engagement, the more arms may have been imported to make it possible. Similarly, the larger the engagement, regardless of fatalities involved, presumably the greater the stimulus for reactive/protective purchases of arms to guard against the next encounter.

The conflict scales used in this study are discussed in chapter 2 and listed in the last column of tables C.1 and C.2. As noted in chapter 2, for external conflict, a scale value of 1 was assigned to states that were neighbors of states involved in some level of fighting, on the assumption that such states would experience greater anxiety than countries in completely peaceful security complexes. Such anxiety might anticipate the spillover of conflict and might translate into hedging behavior involving arms acquisitions. The assignment of states to this category was to some extent a matter of judgment, so a validity test appeared to be in order. In the test, all country entries coded as 1 were recoded as 0—that is, as "no hostile activity." In no instance did regressions using the recoded conflict data do a better job accounting for arms transfers than regressions using conflict data that included the "tension" category, and in several instances the results were much worse with the recoded data. So for that reason, the "tension" category was retained as useful.

Selecting and Coding Conflict Data: Augmenting "Militarized Interstate Disputes"

As noted in chapter 2, the external conflict data used in the analysis of conflict and arms transfers was a composite of a simple data set of my own devising and relevant portions of the MID data set created by the Correlates of War Project.[5]

The highest hostility level in the MID data for each state in each year between 1970 (the first year of the first five-year period of the quantitative analysis) and 1992 (the last year covered by the data set) was extracted for use in this study. The highest hostility level attained by a state within each time period defined in chapter 2 was used as that state's conflict score for that time period.[6] In the resulting subset of the MID data, states using force tend to far outnumber states that just employed threats or displays of force, possibly because in this coding scheme, states could engage in any number of threats or displays within a time period without that being reflected in this data subset, if the same state also *used* force at least once in the same period. The logic for this coding rule is that a state's worst behavior is likely to have more of an impact on its neighbors than is its average behavior.

Figure C.1 graphs out the MID data subset extracted for this analysis. It displays as a single line MID hostility levels two and three ("threats" to use force and "displays" of armed forces). A second line traces "uses of force" short of war, and a third line traces patterns of interstate war. While the apparent prevalence of uses of force in the MID data may be partly an artifact of my coding procedure that records the worst instances of conflict, the *shape* of the use of force curve is not an artifact. The number of states using force at least once in a year, according to MID, but not involved in interstate war as MID defines it, rose sharply through the 1970s to include more than 30 states in 1976, subsided a bit and then rose again in the mid-1980s before tailing off to early 1970s levels by 1992, the last recorded year. The number of states participating in interstate wars moved in the opposite direction, with numbers peaking in 1973 (associated with the Yom Kippur War) and tailing off to zero by 1976 but bouncing back to meander through the 1980s and reach zero once again in 1992.

To supplement the MID data and better differentiate among uses of force, a very simple data collection was created that focused on conflictual actions by states, as opposed to verbal threats and displays, using the coding scale in the last column of table C.1 and drawing on secondary sources of conflict data.[7] The data record peak annual conflict levels for each state. The highest annual conflict level in each half-decade time period became a state's conflict score for that time period, just as was done with the MID data.

The coding in this supplementary data set was compared with the MID data, state by state, for each of the five time periods. The two sets track closely at higher levels of conflict, while MID proved more comprehensive in its record of lesser uses of force. That part of the MID data was combined with the author's data collection in a fashion that corrected for MID's undifferentiated coding of most uses of force. MID's coding was tracked back to individual cases and recorded as a 2 on the external conflict scale used here (i.e., as a minor use of coercive military force) unless at least a brigade-equivalent force was known to be involved in the incident, in

Figure C.1   Militarized Interstate Disputes, Developing Countries, 1970–1992

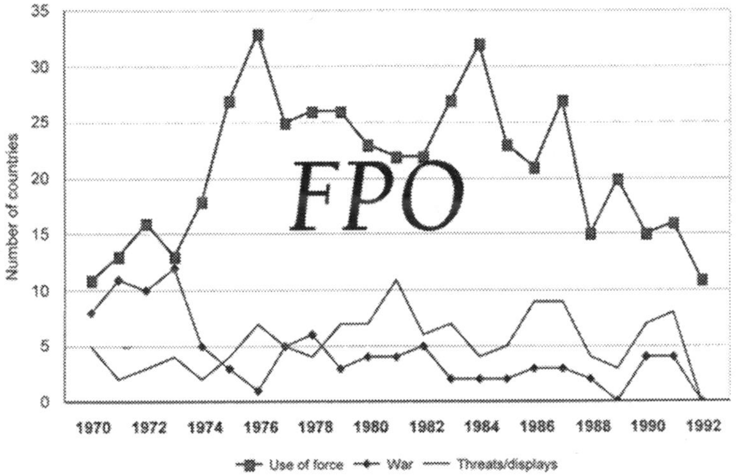

Source: Peace Science Society International, *Militarized Interstate Disputes*, V. 2.10. Internet: *http://pss.la.psu.edu/MID DATA.HTM*. Last accessed March 4, 2000.

which case the MID data point was coded as 3. Threats and displays, as opposed to uses, of force were excluded from the analysis. The graph of this composite external conflict data was presented as figure 2.1 in chapter 2.

There was no ready-made internal conflict data appropriate for this project when the analysis began in the mid-1990s. The data set that Ted Gurr and his colleagues compiled for their *Minorities at Risk* project, while detailed and comprehensive with respect to ethnic and communal violence, does not address political or ideologically based conflict. Thus, Gurr's data does not notice the civil wars in El Salvador or Cambodia, or the simmering guerrilla wars or drug cartel violence aimed at the state in countries like Colombia. These are important omissions for a study interested in all kinds of internal violence and not just ethnic or religious conflict, important as those may be.

The internal conflict data set used in this study was drawn from the same basic sources as the external data. As such, it may somewhat underreport conflict at the lower end of the scale (failed coups, minor terrorist actions). But some of this potential underreporting is mitigated by the fact that the data actually used in the regression analyses discussed below are based on five-year reporting periods. One would have much less confidence that the data set accurately reflected conditions in each state if the reporting period were much more fine-grained—let us say annual or quarterly. (The conflict and arms delivery data used in the analysis may be found in tables C.3 and C.4.)

Static Versus Dynamic Measurement

The static multivariate models presented in chapter 2 to evaluate relations between arms transfers and conflict do not measure something that might be useful to know: whether the military forces of various competitive states or groups of states within a security complex were in rough equilibrium in terms of fighting power or out of balance, and whether the imbalance righted itself or grew worse over time. Nor do the models examine whether various security complexes have exhibited signs of "arms race" behavior. A consensus explanatory (indeed, descriptive) measure of such competitive arms acquisitions has eluded social scientists, although that has not kept many from tackling the problem.[8]

Table C.3 External Conflict and Heavy Arms Transfers

	External Conflict Data					Heavy Weapons (log₁₀ of deliveries)				
	1970–1974	1975–1979	1980–1984	1985–1989	1990–1995	1970–1974	1975–1979	1980–1984	1985–1989	1990–1995
Afghanistan	0	4	4	4	2	1.4624	2.6274	3.0976	3.1746	2.9562
Algeria	1	1	1	1	1	0.0000	2.2330	2.5575	0.0000	2.3010
Angola	0	3	3	3	0	0.0000	1.9294	2.5786	2.9004	2.0755
Argentina	0	2	4	2	0	1.8921	1.8692	2.1038	0.4771	1.6812
Bahrain	0	2	2	1	4	0.0000	2.0000	2.0000	1.8976	2.4713
Bangladesh	0	2	2	2	1	1.1461	1.8692	1.9243	1.4150	2.0334
Belize	0	0	1	1	1	0.0000	0.0000	0.0000	0.0000	0.0000
Benin (Dahomey)	0	0	0	0	0	0.0000	0.0000	0.0000	0.0000	0.0000
Bolivia	0	0	0	0	0	1.2553	1.6021	1.0792	1.2041	0.0000
Botswana	0	2	2	2	0	0.0000	0.0000	1.4771	0.0000	1.2041
Brazil	1	1	2	1	1	1.6021	1.7160	1.6335	1.8865	0.0000
Brunei	0	0	0	0	0	0.0000	0.0000	0.0000	1.3802	0.0000
Burkina Faso	2	2	2	3	0	0.4771	0.0000	1.1767	0.0000	0.0000
Burma (Myanmar)	0	2	2	2	2	0.0000	1.2553	0.0000	0.0000	2.1584
Burundi	2	0	0	0	0	0.0000	0.0000	1.0414	0.0000	0.0000
Cambodia	4	4	4	4	1	1.9085	1.3979	0.3010	2.3892	1.3010
Cameroon	1	2	2	1	1	0.0000	0.0000	0.6021	0.0000	1.1461
Cent Afr Repub	0	2	1	1	1	0.0000	0.0000	0.6990	0.0000	0.0000
Chad	0	2	3	3	1	0.0000	0.0000	1.3010	2.5366	0.0000
Chile	0	2	2	0	0	1.4624	1.8325	1.9494	1.5185	1.4914
Colombia	0	0	2	2	0	1.2041	0.0000	1.0792	1.0414	0.0000
Congo	0	0	0	0	0	0.0000	0.0000	0.4771	0.0000	0.0000
Costa Rica	0	2	1	2	1	0.0000	0.0000	0.0030	0.4771	0.0000
Cote d'Ivoire	0	0	0	0	1	0.0000	0.0000	1.2788	0.0000	0.0000
Cuba	2	4	3	3	1	1.8129	1.3010	2.2810	2.4564	0.7782

(continues)

Table C.3 *(continued)*

	External Conflict Data					Heavy Weapons (log$_{10}$ of deliveries)				
	1970–1974	1975–1979	1980–1984	1985–1989	1990–1995	1970–1974	1975–1979	1980–1984	1985–1989	1990–1995
Cyprus	4	2	2	2	1	0.0000	0.0000	0.0000	0.0000	1.3222
Djibouti	1	1	1	1	1	0.0000	0.0000	0.0000	0.0000	1.5315
Dominican Repub	0	0	0	1	1	0.0000	0.0000	0.0000	0.0000	0.0000
Ecuador	2	2	2	1	2	0.4771	1.5185	1.2041	1.0792	0.4771
Egypt	6	3	2	2	4	3.4478	2.6425	3.3985	3.2378	2.7938
El Salvador	0	2	1	2	0	0.4471	1.2553	1.4150	0.6021	0.0000
Equit Guinea	2	0	0	2	0	0.0000	0.0000	0.0000	0.0000	1.5563
Ethiopia	2	4	2	2	1	1.9085	2.6848	2.1367	2.8645	1.4771
Gabon	0	0	0	0	0	0.0000	0.4771	0.9542	1.0792	0.6021
Gambia	2	0	0	0	0	0.0000	0.0000	0.0000	0.0000	0.0000
Ghana	0	0	2	0	0	0.0000	0.0000	0.0000	0.0000	0.0000
Guatemala	2	1	1	1	1	1.1761	0.0000	0.0000	0.0000	0.0000
Guinea	2	0	0	0	1	0.0000	0.0000	1.6990	0.9031	0.0000
Guinea-Bissau	0	0	0	0	0	0.0000	0.0000	0.0000	0.0000	1.2553
Guyana	2	2	0	0	0	0.0000	0.0000	0.0000	0.0000	0.0000
Haiti	0	0	0	0	0	0.6990	0.0000	0.0000	0.0000	0.0000
Honduras	0	2	1	1	1	0.7782	1.0792	0.7782	1.0792	0.0000
India	6	2	2	3	2	3.0233	2.7853	3.1673	3.0039	2.6435
Indonesia	0	2	2	2	2	1.4771	2.0000	1.8633	1.4472	1.3802
Iran	2	2	6	6	2	2.9410	3.39808	3.1581	3.4567	2.9015
Iraq	2	2	6	6	6	2.5198	2.9899	3.7440	3.8678	0.0000
Israel	6	3	4	3	2	3.2230	3.1495	3.2480	2.7083	2.7127
Jordan	6	1	1	1	2	2.5611	2.8513	2.8692	2.5119	0.0000
Kenya	1	2	1	2	1	0.4711	1.9395	2.1399	0.0000	0.0000
Korea, North	2	2	2	2	2	2.4048	1.3424	2.3617	2.7993	1.9138

(continues)

Table C.3 *(continued)*

	External Conflict Data					Heavy Weapons (log₁₀ of deliveries)				
	1970–1974	1975–1979	1980–1984	1985–1989	1990–1995	1970–1974	1975–1979	1980–1984	1985–1989	1990–1995
Korea, South	4	2	2	2	1	2.3502	2.8215	2.1004	3.0022	3.3402
Kuwait	2	0	1	1	6	1.3979	2.6542	2.0899	2.3636	2.8949
Laos	4	4	2	2	1	1.1761	1.9345	1.5798	1.5798	0.3010
Lebanon	2	3	3	3	3	1.9031	1.5563	2.5038	2.0755	2.5011
Lesotho	0	0	2	1	1	0.0000	0.0000	0.0000	0.0000	0.7782
Liberia	0	0	0	0	3	0.0000	0.0000	0.0000	1.1461	0.0000
Libya	2	3	3	3	0	3.1532	3.4897	2.6812	2.2122	0.0000
Madagascar	0	0	0	2	2	0.0000	0.0000	0.0000	0.0000	0.0000
Malawi	2	1	1	1	1	0.0000	0.0000	0.0000	0.0000	0.0000
Malaysia	0	0	0	2	2	1.6990	1.2041	2.7973	1.6812	1.8513
Mali	2	2	0	2	0	0.4771	0.0000	0.0000	0.0000	0.0000
Mauritania	0	1	0	3	1	0.0000	0.0000	0.0000	0.0000	0.0000
Mexico	2	1	2	1	2	0.0000	0.0000	1.1139	0.0000	1.9956
Morocco	2	2	2	2	0	1.7559	2.6981	2.6739	2.3464	2.0792
Mozambique	0	3	1	2	0	0.0000	2.5855	1.8062	1.4771	0.0000
Namibia	0	0	0	1	1	0.0000	0.0000	0.0000	0.0000	0.0000
Nepal	0	0	0	0	0	0.0000	0.0000	0.0000	1.0000	2.1303
Nicaragua	0	2	2	2	2	0.0000	0.0000	1.8325	1.9685	0.9542
Niger	1	1	2	1	1	0.0000	0.0000	0.9031	0.0000	0.0000
Nigeria	1	1	2	1	1	0.0000	0.0000	0.0000	2.1239	2.2810
Oman	2	2	0	2	4	1.0792	1.6335	1.5682	1.5315	1.9868
Pakistan	6	2	2	2	4	2.9186	2.4814	2.9159	3.0056	2.9269
Panama	0	2	0	3	3	0.0000	1.0000	0.0000	0.0000	0.0000
Pap. New Guinea	0	0	2	0	0	0.0000	0.0000	0.0000	0.0000	0.0000
Paraguay	0	0	0	0	0	0.0000	0.0000	0.0000	0.0000	0.0000

(continues)

Table C.3 *(continued)*

	External Conflict Data					Heavy Weapons (log_{10} of deliveries)				
	1970–1974	1975–1979	1980–1984	1985–1989	1990–1995	1970–1974	1975–1979	1980–1984	1985–1989	1990–1995
Peru	2	2	2	1	2	2.3304	2.4150	1.8451	1.8751	2.1139
Philippines	3	1	2	1	1	1.1761	1.2304	1.7993	0.6021	1.6902
Qatar	0	0	1	2	4	1.1761	1.6232	2.1584	2.2480	1.2304
Rwanda	2	0	0	0	2	0.0000	0.0000	0.0000	0.0000	0.0000
Saudi Arabia	6	1	2	1	6	1.9956	3.1529	2.8000	3.0220	3.2135
Senegal	2	0	0	2	2	0.0000	0.0000	0.0000	0.7782	0.0000
Sierra Leone	0	0	0	2	2	0.0000	0.0000	0.0000	0.0000	0.3010
Singapore	0	0	0	0	0	1.8261	2.3541	2.0969	1.8129	1.7709
Somalia	2	4	2	2	3	2.2148	0.0000	2.0792	2.2068	1.5052
South Africa	2	3	2	3	1	1.8451	2.7135	0.0000	0.0000	0.0000
Sri Lanka	0	0	2	2	2	0.6990	0.0000	0.0000	0.7782	1.6990
Sudan	2	1	2	2	2	2.2856	1.3802	2.3444	1.8129	1.1461
Surinam	0	2	0	0	0	0.0000	0.0000	0.0000	0.0000	0.0000
Swaziland	0	0	0	2	0	0.0000	0.0000	0.0000	0.0000	0.0000
Syria	6	3	3	3	4	3.0366	3.0603	3.5139	3.3397	2.3729
Taiwan	1	1	1	2	1	1.8451	1.7482	2.5391	2.5775	2.1987
Tanzania	2	4	1	1	1	1.8325	0.0000	0.0000	0.0000	0.0000
Thailand	4	2	2	2	2	0.0000	1.9031	2.4281	2.8293	2.8414
Togo	0	0	2	0	0	0.0000	0.0000	0.0000	0.0000	0.0000
Tunisia	0	0	3	2	1	1.6435	1.8129	2.2355	1.8573	0.0000
Uganda	2	4	0	2	2	1.3617	1.8751	0.0000	1.1139	0.0000
UAE	0	1	1	1	4	0.0000	1.6128	1.4314	1.8513	2.1335
Uruguay	0	1	1	0	0	0.0000	1.0792	0.0000	0.0000	0.0000
Venezuela	2	1	2	2	1	2.2945	0.4771	1.6435	0.7782	0.0000
Vietnam, N. & All	6	6	4	4	1	2.5378	3.0334	1.1761	1.5185	0.0000

(continues)

Table C.3 *(continued)*

	External Conflict Data					Heavy Weapons (log₁₀ of deliveries)				
	1970–1974	1975–1979	1980–1984	1985–1989	1990–1995	1970–1974	1975–1979	1980–1984	1985–1989	1990–1995
Vietnam, S	6	6	0	0	0	3.2114	0.0000		0.0000	0.0000
Yemen, N. & All	2	4	2	2	1	0.0000	2.5740	2.2788	1.6628	0.0000
Yemen, S.	2	4	0	2	0	1.3222	2.6675	2.8531	0.6021	0.0000
Zaire	2	2	2	2	2	0.0000	1.5911	0.0000	0.6021	1.3979
Zambia	2	2	2	2	0	1.0000	0.9931	1.6628	0.0000	0.0000
Zimbabwe	0	0	1	3	3	0.0000	0.0000	1.4150	1.9085	0.0000

Source: See appendix C, endnote 7.

Table C.4 Internal Conflict and Medium Arms Transfers

	Internal Conflict Data					Medium Weapons (log$_{10}$ of deliveries)				
	1970–1974	1975–1979	1980–1984	1985–1989	1990–1995	1970–1974	1975–1979	1980–1984	1985–1989	1990–1995
Afghanistan	2	6	6	6	6	0.0000	1.9912	2.3464	2.4409	2.0000
Algeria	0	0	0	0	4	0.3010	1.2304	2.0253	1.6990	1.7482
Angola	0	6	6	6	6	0.0000	0.0000	1.7076	2.7451	2.1072
Argentina	3	3	3	3	0	2.1959	1.5441	2.4133	1.3222	1.2553
Bahrain	0	0	0	0	0	1.2553	0.0000	0.3010	0.0000	0.0000
Bangladesh	6	3	3	3	3	0.7782	1.0000	0.8451	0.0000	0.9031
Belize	0	0	0	0	0	0.0000	0.0000	0.0000	0.0000	0.0000
Benin (Dahomey)	2	0	0	1	0	0.0000	0.0000	0.0000	2.1038	0.0000
Bolivia	0	0	0	0	0	0.9542	1.8195	1.6532	0.4771	0.0000
Botswana	0	0	0	0	0	0.0000	0.0000	1.3979	1.1761	1.5798
Brazil	0	0	0	0	0	1.7076	1.3617	1.1461	2.0374	1.6435
Brunei	0	0	0	0	0	0.0000	0.0000	1.2553	0.3010	0.0000
Burkina Faso	2	0	2	1	0	0.0000	0.0000	1.3010	0.7782	0.0000
Burma (Myanmar)	3	3	3	3	3	1.0792	2.0792	0.0000	1.3979	1.6902
Burundi	4	2	0	4	3	0.0000	0.0000	0.0000	0.0000	0.0000
Cambodia	6	6	6	6	6	2.0828	0.0000	0.0000	1.1761	1.6628
Cameroon	0	0	1	0	1	0.6021	0.0000	1.5563	1.4771	0.4771
Cent Afr Repub	0	2	2	0	0	1.0000	0.0000	0.0000	0.0000	0.0000
Chad	4	4	4	4	3	1.3010	0.6021	0.8451	2.2380	0.3010
Chile	3	2	2	2	0	1.1461	2.2765	1.8451	1.5441	1.7559
Colombia	3	3	3	3	4	0.0000	1.6435	2.0294	1.4624	1.4314
Congo	0	2	0	0	2	0.0000	0.0000	0.0000	0.0000	0.0000
Costa Rica	0	0	0	0	0	0.0000	0.0000	0.0000	0.4771	0.0000
Cote d'Ivoire	0	0	0	0	0	0.8451	0.0000	0.8451	0.0000	0.0000
Cuba	0	1	0	0	0	0.0000	0.0000	0.0000	2.3802	0.0000

(continues)

Table C.4 *(continued)*

	Internal Conflict Data					Medium Weapons (log_{10} of deliveries)				
	1970–1974	1975–1979	1980–1984	1985–1989	1990–1995	1970–1974	1975–1979	1980–1984	1985–1989	1990–1995
Cyprus	1	0	1	0	0	0.0000	0.0000	0.0000	0.0000	0.0000
Djibouti	0	0	0	0	0		0.0000	2.0000	0.8451	0.0000
Dominican Repub	0	0	0	1	0	0.8451	0.0000	1.3979	0.0000	0.0000
Ecuador	0	1	2	2	2	2.0043	1.7924	1.2304	1.3010	1.2553
Egypt	0	2	5	5	2	2.9238	1.0792	1.5911	2.3032	0.6021
El Salvador	3	3	5	5	5	1.0792	0.0000	1.5441	1.2788	0.0000
Equit Guinea	3	3	0	0	0	0.0000	0.0000	0.0000	0.0000	0.0000
Ethiopia	3	3	3	3	6	1.7482	0.6021	2.3010	2.9079	1.3010
Gabon	0	0	0	0	0	0.0000	0.0000	1.7993	1.7482	0.0000
Gambia	2	2	0	0	0	0.0000	0.0000	0.0000	0.0000	0.0000
Ghana	2	0	0	0	0	0.7782	1.0792	0.0000	0.0000	0.6021
Guatemala	4	4	4	4	4	1.4472	0.0000	0.7782	0.0000	0.0000
Guinea	0	0	2	0	0	1.0792	0.0000	0.0000	0.0000	0.0000
Guinea-Bissau	0	0	2	1	0	0.0000	0.0000	1.3010	0.0000	0.0000
Guyana	0	0	0	0	0	0.6021	0.0000	1.4771	0.0000	0.0000
Haiti	0	0	0	2	3	0.6021	0.0000	0.0000	0.0000	0.0000
Honduras	0	3	3	3	0	0.0000	1.1461	1.5798	1.0000	0.9031
India	3	3	3	3	3	1.2553	0.0000	0.0000	1.4914	0.0000
Indonesia	4	3	3	3	3	1.0000	2.3997	1.9191	0.0000	0.9031
Iran	3	4	3	3	3	2.7803	2.9562	0.0000	2.2041	1.6021
Iraq	3	3	3	3	3	2.1173	2.5353	3.3088	2.9586	0.7782
Israel	1	2	0	0	3	1.8261	1.5052	0.0000	0.0000	1.9590
Jordan	3	0	0	0	0	1.3302	0.6021	1.3802	2.3856	0.0000
Kenya	0	1	0	0	2	1.3617	1.2041	1.3222	1.8261	0.0000
Korea, North	0	0	0	0	0	2.1584	0.0000	0.0000	0.0000	0.0000

(continues)

Table C.4 *(continued)*

	Internal Conflict Data					Medium Weapons (log$_{10}$ of deliveries)				
	1970–1974	1975–1979	1980–1984	1985–1989	1990–1995	1970–1974	1975–1979	1980–1984	1985–1989	1990–1995
Korea, South	0	2	1	1	0	0.3010	1.5563	1.5441	1.9138	1.9731
Kuwait	0	0	0	0	0	1.5441	1.0000	1.5315	1.9868	0.0000
Laos	6	6	0	0	3	1.6128	0.0000	0.0000	0.3010	0.0000
Lebanon	0	6	6	6	4	1.8976	0.0000	2.0000	1.4771	0.0000
Lesotho	3	2	3	3	2	0.0000	0.0000	0.4771	0.0000	0.0000
Liberia	0	0	3	3	6	0.0000	0.0000	0.0000	1.1461	0.0000
Libya	0	0	0	0	0	1.1139	2.5539	2.2967	0.0000	0.0000
Madagascar	2	0	0	0	0	0.0000	0.0000	0.0000	0.0000	0.0000
Malawi	0	0	0	0	1	0.9542	0.3010	0.0000	0.3010	0.0000
Malaysia	0	0	0	0	0	2.0253	2.0969	2.1271	2.0492	1.4624
Mali	0	0	0	0	2	0.0000	0.0000	0.9031	0.0000	0.0000
Mauritania	0	1	2	0	1	0.0000	0.9031	0.0000	0.0000	0.0000
Mexico	0	0	0	0	2	1.0414	1.2553	1.6628	1.9823	0.0000
Morocco	0	2	3	3	3	1.0792	1.8129	2.3181	2.3711	0.0000
Mozambique	0	5	5	5	5	0.0000	1.3010	1.4771	1.3010	0.0000
Namibia	0	0	0	0	1	0.0000	0.0000	0.0000	0.0000	0.6021
Nepal	0	0	0	0	0	0.0000	0.3010	0.0000	0.0000	0.0000
Nicaragua	0	3	5	5	0	0.6021	0.6990	2.4014	2.3502	0.0000
Niger	2	0	0	0	2	0.0000	0.0000	0.0000	0.4771	0.0000
Nigeria	4	2	2	2	3	0.6990	1.7634	2.2148	1.9956	2.2625
Oman	3	3	0	2	0	1.8325	0.9542	1.4150	0.4771	1.4150
Pakistan	3	3	3	3	3	1.4150	1.5682	0.6021	1.0414	1.0792
Panama	0	0	0	1	0	0.6990	1.0414	1.2304	0.3010	0.0000
Pap. New Guinea	0	0	0	3	0	0.0000	0.0000	0.0000	0.0000	0.0000
Paraguay	0	0	0	2	0	1.0792	1.3424	1.2304	0.3010	0.0000

(continues)

Table C.4 *(continued)*

	Internal Conflict Data					Medium Weapons (log$_{10}$ of deliveries)				
	1970–1974	1975–1979	1980–1984	1985–1989	1990–1995	1970–1974	1975–1979	1980–1984	1985–1989	1990–1995
Peru	0	2	3	3	4	2.5211	1.5315	1.6902	1.3010	1.3802
Philippines	3	3	3	3	3	1.8573	1.7634	1.9294	2.0899	1.5563
Qatar	0	0	0	0	0	1.3979	1.3802	1.0792	0.0000	1.3222
Rwanda	2	0	0	0	6	0.3010	0.0000	0.0000	0.9542	0.0000
Saudi Arabia	0	2	0	1	0	2.1553	2.7235	2.6355	2.3927	2.9978
Senegal	0	0	1	3	1	0.0000	0.0000	0.8451	0.0000	0.0000
Sierra Leone	0	0	0	1	4	0.4771	0.0000	0.0000	0.0000	0.0000
Singapore	0	0	0	0	0	1.6232	1.3010	2.2504	1.3424	2.0492
Somalia	0	0	0	3	6	2.4472	0.0000	0.0000	0.0000	0.0000
South Africa	0	2	2	3	2	2.3502	0.0000	0.3010	1.0000	0.0000
Sri Lanka	3	3	1	4	4	1.9590	1.0792	0.7782	1.8261	0.6021
Sudan	3	1	3	3	5	0.0000	1.6990	2.0569	1.5563	0.0000
Surinam	0	0	0	0	0		0.0000	1.0000	0.0000	0.0000
Swaziland	2	2	3	0	0	0.0000	0.0000	0.0000	0.0000	0.0000
Syria	2	0	0	0	0	0.3010	1.5911	0.0000	2.3010	0.0000
Taiwan	0	0	0	0	0	1.7482	1.6990	0.3000	1.7924	0.3010
Tanzania	0	2	0	0	0	0.9542	0.9031	0.3010	0.0000	0.0000
Thailand	0	0	0	0	2	2.1614	2.2227	2.7267	2.3075	1.9823
Togo	0	0	0	0	2	0.0000	0.4771	0.0000	1.1761	0.0000
Tunisia	0	0	0	2	0	1.1461	1.8325	2.2058	0.0000	0.0000
Uganda	3	3	4	4	2	2.2878	1.8129	1.7150	1.1139	0.0000
UAE	0	0	0	0	0	0.0000	1.8451	2.1106	1.9823	1.1461
Uruguay	4	2	0	0	0	0.3010	0.0000	1.2041	0.3010	0.3010
Venezuela	0	0	0	1	2	1.5441	1.0792	1.7782	2.2900	2.4133
Vietnam, N & All	0	0	0	0	0	2.0253	3.3222	0.0000	0.0000	0.0000

(continues)

Table C.4 *(continued)*

	Internal Conflict Data					Medium Weapons (log₁₀ of deliveries)				
	1970–1974	*1975–1979*	*1980–1984*	*1985–1989*	*1990–1995*	*1970–1974*	*1975–1979*	*1980–1984*	*1985–1989*	*1990–1995*
Vietnam, S	6	6	0	0	0	3.0660	0.0000	0.0000	0.0000	0.0000
Yemen, N. & All	4	2	0	0	6	0.0000	0.0000	1.2553	1.0792	0.0000
Yemen, S.	0	2	0	4	0	0.0000	0.0000	0.0000	0.0000	0.0000
Zaire	0	3	0	0	3	2.3139	1.9912	0.9031	0.0000	0.0000
Zambia	0	0	0	0	0	1.8062	1.8388	2.2430	1.2041	0.0000
Zimbabwe	0	0	3	3	0	0	0	2.09691	1.20412	0

Source: See appendix C, endnote 7.

A realistic comparison of military capabilities requires a dynamic, interactive model capable of simulating complex military engagements. Acquiring or developing such a model does not get one out of the woods, however, as there are many more opinions of each dynamic model's worth than there are models.[9] While the static models used in this study to investigate conflict and arms transfers are not as sophisticated as dynamic models, they have the virtue of being far more transparent to readers, a quality that dynamic models and simulations lose in their attempt to mimic human events more closely.

Quantitative Analysis Issues

Path analysis, used to create the research results depicted in figures 2.12 and 2.14, is a statistical technique for analyzing and comparing alternative assumptions about causal relations between variables. Thus, to construct a path diagram or path model is usually to posit certain causal hypotheses for testing. Path analysis can also be used, however, to test weaker claims about relationships among variables, "forecasting" the shape of those relationships without necessarily asserting causality—if only because all possible influential variables have not been specified in the model. In the analyses undertaken for chapter 2, the objective was to forecast (or, if you will, "retrocast") relations between levels of arms deliveries and levels of conflict, controlled for regional security complex membership.[10] The model was built up using ordinary least squares (OLS) regression techniques.

Addressing the Foibles of Ordinary Least Squares

OLS regression formally requires that the data used for the independent and dependent variables behave in certain ways. When those requirements are not met, problems arise in interpreting the regression results. Two of the more serious potential problems relate to the distribution of prediction error ("residuals") and autocorrelation in time series data.

Distribution of Residuals

Valid predictions from OLS require that residuals—the differences between actual values of the dependent variable and regression estimates of those values—do not show a pattern of variation ("heteroscedasticity") as the values of the independent variable (X) change. Residuals should not be substantially greater at one end of X's range than at the other, for example. If they do show a pattern, the estimators produced by the regression model (in our case, the path coefficients) are unaffected, but their statistical significance—whether they should be taken seriously or not—is difficult to determine. Statistical significance tests carry specific distributional assumptions as part of their own standard baggage, preferring that residuals be normally distributed and have a mean value of zero. Significance tests based on a badly skewed distribution of residuals can therefore be misleading.[11] Plots of the residuals of the various regression equations used to generate the path models in chapter 2 and measurements of the residuals' distribution using the Shapiro-Wilk "W" test for normality indicated that heteroscedasticity is a problem with some of the equations. It is a greater problem in the equations that predict to conflict than in those predicting to arms deliveries. The W statistic indicated non-normal distributions of residuals for the models for external conflict in time periods 1975–1979, 1980–1984, and 1985–1989, while the same statistic indicated normally distributed residuals for arms deliveries in those same time periods and non-normally distributed residuals for arms deliveries in 1990–1995 (see table C.5). Non-normal distributions of residuals were a much greater problem for the equations attempting to measure relations between arms deliveries and internal conflict.

Although one might mistakenly believe that an independent variable is a statistically significant predictor when residuals show heteroscedasticity, the problem is less daunting when one is mostly interested in pruning rather than adding variables to an equation. One could assume, for example, that variables labeled statistically *in*significant in any given regression run really were insignificant and could justifiably be dropped from the model. Borderline variables with t-statistics having probabilities just below the standard significance threshold of 5 percent could likewise be considered weak candidates for retention in the model.

Some comfort may also be derived from the fact that the data being analyzed do not represent a sample of conflict and arms transfer data but a population. Although other, more refined measures of that population might have been devised, the fact that the distribution of data does not always hew to statistical norms reflects not bad or biased sampling but the fact that states' actual behavior is not normally distributed.

Autocorrelation

Another important requirement of OLS is that residuals not be highly correlated with one another. Such "autocorrelation" is a particular problem in cross-time analyses and can distort an equation's path coefficients.[12] The Durbin-Watson statistic is a commonly used, "two-tailed" measure of autocorrelation. A value of 2.0 indicates zero autocorrelation, whereas perfect *positive* autocorrelation of residuals reduces the statistic to zero and perfect *negative* autocorrelation inflates it to 4.0. In evaluating its statistical significance, common threshold values of .05 and .95 were used, leaving a 5 percent or lower probability of being wrong in asserting the presence of significant autocorrelation.

The Durbin-Watson statistic is not so useful when the regression equation uses an earlier ("lagged") measurement of the dependent variable to predict a later measurement, as is done here. For example, conflict levels in 1970–1974 were used to predict conflict levels in 1975–1979. SAS, Inc.'s JMP v.3.2.2, the statistical package used to create the path models, also generates a second measure of autocorrelation independent of Durbin-Watson. Where an equation in table C.5 or C.6 contains only a lagged version of the dependent variable as a predictor, only this second measure is reported. In all but two instances in table C.5 (Conflict 1980–1984 and Arms Deliveries 1985–1989), calculated autocorrelation was very low. In table C.6, autocorrelation was a potential problem in one instance (Conflict 1980–1984).

Measuring "Lack of Fit"

JMP calculates a "lack of fit" parameter that suggests to the analyst that the model being measured could be better specified.[13] Five of the eight equations in table C.5 and two of eight in table C.6 had a "lack of fit" problem. Other variables need to be included in the equation to make it a good predictor. So even where variance explained (adjusted R-squared) is reasonably high, as in the equation for heavy arms deliveries in 1975–1979, the pattern of error in the prediction suggests that we are not doing such a great job of explaining the dependent variable.

Resolving Data Problems

Some of the problems just discussed can be alleviated with sophisticated data transformations and use of more esoteric regression techniques. But each new level of sophistication brings limitations of its own, not the least of which is the growing difficulty of interpreting the statistical results in real-world terms. Moreover, rough, simple data of the sort used here are probably best treated with the simplest techniques that will do the job without producing distorted results.[14]

Regression analysis can be distorted somewhat if the variables in the equation have ranges that differ by orders of magnitude. In the raw data used here, the conflict scores rose no higher than 6 but raw numbers of arms transferred to one country within a five-year period were as high as 7,000. To make the conflict and arms data more comparable in scale, the common logarithm of arms deliveries was used in lieu of the raw numbers. A country receiving 100 heavy weapons in a five-year period thus received a transformed score of 2.0; a country receiving 1,000 received a score of 3.0; and so on. (Use of logarithms required a decision as to how to portray zero deliveries or deliveries of just one weapon per time period, because the common logarithm of one is zero, while the log of zero is undefined. With the real-world impact on conflict of one

Table C.5 Building the Path Model for Heavy Weapons and External Conflict

Variable Explained	Variance Explained			Variable Selection Test		Residuals: Autocorrelation and Distribution				Lack of Fit Test		Equation Parameters					
	A	B	C	p	Cp	D	E	F	G	H	I	J	K	L	M	N	O
1975–79 conflict	.32	.30	95	3	7.76	1.78	0.130	.09	.30	1.94	0.01	Intercept	.77	.17	4.60	.00	.00
												con. 70–74	.23	.09	2.74	.01	.30
												arms 70–74	.44	.15	3.00	.00	.33
1980–84 conflict	.39	.38	98	3	−1.71	1.48	0.004	.26	.02	1.52	0.094	Intercept	.34	.17	1.96	.05	.00
												con. 75–79	.34	.08	4.16	.00	.35
												arms 80–84	.46	.10	4.68	.00	.40
1985–89 conflict	.58	.54	94	10	10	2.21	0.844	−.12	.02	.46	0.995	Intercept	.79	.26	3.08	.00	.00
												con. 80–84	.56	.08	6.96	.00	.58
												arms (air)					
												80–84	.40	.14	2.80	.01	.26
												So. America	−1.18	.35	−3.40	.00	−.31
												Africa	−.18	.28	−.63	.53	−.07
												Middle East	−.07	.40	−1.74	.09	−.16
												Persian Gulf	−.03	.37	−.09	.93	−.01
												So. Asia	−.30	.44	−.68	.50	−.06
												SE Asia	−.33	.36	−.92	.36	−.08
												NE Asia	−.50	.58	−.86	.39	−.07
1990–95 conflict	.56	.53	96	9	9	1.99	0.469	−.01	.22	2.53	0.001	Intercept	.85	.27	3.34	.00	.00
												arms (gra)					
												85–89	.42	.10	4.01	.00	.35
												So. America	−.57	.38	−1.52	.13	−.13

(continues)

278

Table C.5 *(continued)*

Variable Explained	Variance Explained			Variable Selection Test		Residuals: Autocorrelation and Distribution				Lack of Fit Test		Equation Parameters						
	A	B	C	p	Cp	D	E	F	G	H	I	J	K	L	M	N	O
												Africa	-.19	.30	-.65	.52	-.07
												Middle East	.30	.45	.67	.50	.06
												Persian Gulf	2.39	.45	5.32	.00	.49
												So. Asia	.20	.48	.42	.68	.03
												SE Asia	-.20	.39	-.50	.61	-.04
												NE Asia	-.66	.64	-1.02	.31	-.08
1975–79 arms transfers	.65	.61	92	10	10	2.17	0.783	-.08	.51	3.32	0.000	Intercept	-.02	.23	-.09	.93	.00
												con. 75–79	.18	.06	3.01	.00	.22
												arms 70–74	.34	.09	3.65	.00	.31
												So. America	.49	.30	1.61	.11	.14
												Africa	.16	.24	.67	.50	.07
												Middle East	1.54	.35	4.41	.00	.39
												Persian Gulf	1.71	.32	5.38	.00	.45
												So. Asia	.83	.37	2.23	.03	.17
												SE Asia	.61	.31	1.95	.05	.17
												NE Asia	.95	.49	1.95	.05	.14
1980–84 arms transfers	.65	.61	94	10	10	2.12	0.695	-.07	.76	1.28	0.220	Intercept	.27	.21	1.26	.86	.00
												con. 80–84	.15	.06	2.28	.02	.17
												arms 75–79	.50	.09	5.31	.00	.50
												So. America	.05	.29	.18	.86	.02
												Africa	.04	.23	.18	.86	.02

(continues)

Table C.5 *(continued)*

Variable Explained	Variance Explained			Variable Selection Test		Residuals: Autocorrelation and Distribution				Lack of Fit Test		Equation Parameters						
	A	B	C	p	Cp	D	E	F	G	H	I	J	K	L	M	N	O
												Middle East	.92	.36	2.55	.01	.23
												Persian Gulf	.62	.35	1.79	.08	.16
												So. Asia	.43	.37	1.17	.25	.09
												SE Asia	.01	.31	.02	.99	.00
												NE Asia	.62	.47	1.32	.19	.09
1985–89 arms transfers	.70	.67	95	10	10	2.38	.97	−.20	.31	2.08	0.020	Intercept	−.02	.27	−.09	.93	.00
												con. 80–84	.21	.06	3.50	.00	.24
												arms 80–84	.55	.08	6.55	.00	.55
												So. America	−.03	.27	−.12	.91	−.01
												Africa	.02	.21	.11	.91	.01
												Middle East	.19	.34	.57	.57	.05
												Persian Gulf	.50	.31	1.62	.11	.13
												So. Asia	.64	.34	1.89	.06	.13
												SE Asia	.32	.28	1.16	.25	.09
												NE Asia	1.18	.44	2.66	.01	.17
1990–95 arms transfers	.52	.48	96	9	9	2.07	.62	−.04	.00	1.82	0.020	Intercept	.23	.22	1.05	.30	.00
												arms 85–89	.35	.08	4.16	.00	.38
												So. America	.10	.31	.31	.76	.03
												Africa	.01	.25	.02	.98	.00
												Middle East	.54	.37	1.74	.09	.17
												Persian Gulf	.86	.35	2.44	.02	.24

(continues)

Table C.5 *(continued)*

Variable Explained	Variance Explained			Variable Selection Test		Residuals: Autocorrelation and Distribution				Lack of Fit Test		Equation Parameters					
	A	B	C	p	Cp	D	E	F	G	H	I	J	K	L	M	N	O
So. Asia													1.55	.40	3.89	.00	.34
SE Asia													.64	.32	2.02	.05	.19
NE Asia													1.33	.53	2.52	.01	.21

Notes: A = Raw R²; B = Adj'd R²; C = Degrees of Freedom; D = Durbin-Watson; E = DW signif.; F = Autocorr. of resids.; G = Shapiro-Wilk signif.; H = F-ratio; I = Signif.; J = Equation Terms; K = Regression estimates; L = Std Error of estimate; M = t-ratio; N = Prob>|t|; O = Standardized estimates (path coefficients)

Table C.6 Building the Path Model for Medium Weapons and Internal Conflict

Variable Explained	Variance Explained			Variable Selection Test		Residuals: Autocorrelation and Distribution				Lack of Fit Test		J	Equation Parameters				
	A	B	C	p	Cp	D	E	F	G	H	I		K	L	M	N	O
1975–79 conflict	0.47	0.46	95	3	3	1.70	0.06	0.07	0.00	0.93	0.60	Intercept	.17	.20	.85	.39	.00
												con. 70–74	.59	.07	8.01	.00	.60
												arms 70–74	.55	.15	3.77	.00	.28
1980–84 conflict	0.54	0.54	99	2	2	2.32	0.95	−0.17	0.00	0.56	0.45	Intercept	.22	.15	1.41	.16	.00
												con. 75–79	.75	.07	10.80	.00	.74
1985–89 conflict	0.67	0.67	102	2	2	1.88	0.27	0.06	0.00	2.82	0.02	Intercept	.36	.13	2.88	.01	.00
												con. 80–84	.84	.06	14.36	.00	.82
1990–95 conflict	0.41	0.41	102	2	2	1.79	0.14	0.08	0.00	0.93	0.44	Intercept	.65	.19	3.44	.00	.00
												con. 85–89	.71	.08	8.45	.00	.64
1975–79 arms transfers	0.31	0.25	93	9	9	2.13	0.71	−0.08	0.08	4.70	0.00	Intercept	0.23	0.24	0.96	0.34	0.00
												arms 70–74	.25	.10	2.48	.02	.24
												So. America	.76	.34	2.23	.03	.26
												Africa	.03	.26	.10	.92	.01
												Middle East	.79	.36	2.21	.03	.25
												Persian Gulf	.78	.35	2.24	.03	.26
												So. Asia	.53	.40	1.34	.18	.14
												SE Asia	.73	.35	2.07	.04	.25
												NE Asia	.51	.51	.98	.33	.10
1980–84 arms transfers	0.27	0.25	99	3	4.70	2.19	0.83	−0.11	0.01	1.19	0.28	Intercept	0.51	0.12	4.37	0.00	0.00
												con. 80–84	.11	.04	2.42	.02	.21
												arms 75–79	.46	.09	5.24	.00	.45

(continues)

Table C.6 *(continued)*

Variable Explained	Variance Explained			Variable Selection Test		Residuals: Autocorrelation and Distribution				Lack of Fit Test		Equation Parameters						
	A	B	C	p	Cp	D	E	F	G	H	I	J	K	L	M	N	O
1985–89 arms transfers	0.29	0.28	104	2	2	2.19	0.84	−0.10	0.01	0.98	0.52	Intercept	0.39	0.11	3.43	0.00	0.00
												arms 80–84	.55	.08	6.52	.00	.54
1990–95 arms transfers	.40	.34	96	9	9	2.23	.87	−.12	.00	2.76	.00	Intercept	−0.20	0.18	−1.11	0.27	0.00
												arms 85–89	.36	.07	5.27	.00	.43
												So. America	.78	.25	3.15	.00	.33
												Africa	.16	.20	.82	.41	.10
												Middle East	.18	.27	.66	.51	.07
												Persian Gulf	.66	.26	2.49	.01	.25
												So. Asia	.56	.31	1.81	.07	.17
												SE Asia	.87	.26	3.41	.00	.35
												NE Asia	.52	.40	1.29	.20	.11

Notes: A = Raw R²; B = Adj'd R²; C = Degrees of Freedom; D = Durbin-Watson; E = DW signif.; F = Autocorr. of resids.; G = Shapiro-Wilk signif.; H = F-ratio; I = Signif.; J = Equation Terms; K = Regression estimates; L = Std Error of estimate; M = t-ratio; N = Prob>|t|; O = Standardized estimates (path coefficients) weapon delivery every five years being essentially nil, I assigned no deliveries and one delivery a working value of zero.)

Using Dummy Variables

Knowing that regional security complexes differed considerably in their arms importing habits, seven regional "dummy variables" were included in the regression equations, for South America, sub-Saharan Africa, the Middle East and North Africa, the Persian Gulf, South Asia, Southeast Asia, and Northeast Asia. The distinct weapon distribution patterns of the Middle East (defined in Appendix A as the Levant and North Africa) and the Persian Gulf suggested defining them as two separate security complexes, despite the political linkage between them created by the Arab-Israeli dispute.

A dummy variable has a value of either zero or one: Each country received a score of one for the regional complex variable to which it belonged, and a score of zero for all the others. If all regions were included, the dummy variables would derail the regression model.[15] Since countries in Central America and the Caribbean, with the sole exception of Cuba, have received the fewest heavy weapons of any region and would likely have the least impact on the analysis, that variable was the one left out of the equations.

The effect of using the dummy variables is to give each region its own regression line. The dummy variables for regions to which a country does *not* belong drop out of the regression equation. Since the value of the remaining dummy variable is always equal to 1.0, its value in the regression equation is the same for all members of a particular region and becomes in effect an add-on to the regression intercept, alpha (the value at which the regression line, when plotted, crosses the Y axis). The result is a regression line for each region that intercepts the Y axis above alpha, if the dummy variable for that region has a positive regression coefficient, and below alpha, if that coefficient is negative. Alpha itself, in effect, anchors the regression line for Central America and the Caribbean—the region without a dummy regional variable of its own.

In most of the regression equations used to build the path models, only some of the regional variables returned statistically significant results. However, dropping the non-significant dummies and using fewer than the complete set in the equation alters the interpretation of the regression results and creates in effect a new, smaller "set" of regional dummy variables (those regions with variables remaining in the equation, and all the other, unmeasured regions lumped together). Rather than tailor the dummy variables in this way, which would alter the interpretation of regional effects equation by equation, all dummy variables were retained in an equation if any of them proved statistically significant.[16]

Variable Selection

The multistage path model was built using separate regressions for conflict and for weapon deliveries in each time period. Initially included in each regression were all of the variables whose arrows pointed directly to the dependent variable. A combined stepwise regression technique was used that put the strongest predictors into an equation first but tested for continuing statistical significance of each variable as new variables entered the equation, on occasion bumping variables out as others entered. Such "variable selection" can be problematic and is worthwhile doing only if apparent increases in accuracy of prediction (that is, higher R-squared) are not purchased at the price of increased bias. One way to ascertain whether or not a model is improved by adding or dropping variables is to monitor the value of Mallow's C_p, a statistic calculated by the JMP stepwise regression routine. The preferred equations are those where C_p is equal to or less than p, the number of predictor variables in the equation.[17]

Building the Path Models

Each stage of the path model (for 1975–1979, 1980–1984, and so on) was built using separate regressions for conflict and for weapon deliveries, starting in the past and adding modules for each subsequent time period. Each regression initially included all of the independent (or explanatory) variables from which arrows pointed directly to the dependent (or explained) variable. Details of the regression results used to produce the final path diagrams of figures 2.12 and 2.14 can be found in tables C.5 (heavy weapons and external conflict) and C.6 (medium weapons and internal conflict).[18]

Each path in the diagrams has a number associated with it called a path coefficient, which is a standardized regression coefficient. The path coefficients are the numbers reported in the last column of tables C.5 and C.6. The path coefficient is interpreted as the fractional unit change in the dependent variable (to which the arrow points) that is associated with a unit change in the independent variable (from which the arrow originates). Positive coefficients indicate a direct relationship between variables (i.e., they tend to increase or decrease together), while negative coefficients indicate an inverse relationship (when one goes up, the other goes down). The larger the coefficient, the stronger the relationship.

The first step in evaluating each stage of the path model was to test to see which if any of the arrows could be eliminated and which if any arrows to retain between variables within the same time period. Initially, all relevant variables were included in the regression for each dependent variable. Thus the regression equation for Conflict 1975–1979 initially included the regional dummy variables, plus Conflict 1970–1974, Arms Deliveries 1970–1974, and Arms Deliveries 1975–1979. Conversely, the initial equation for Arms Deliveries 1975–1979 included the regional variables plus Conflict 1970–1974, Arms Deliveries 1970–1974, and Conflict 1975–1979. Variables generating statistically insignificant contributions to the model were dropped from the regression equation and the regression was rerun with the reduced set of variables. In some instances, contemporaneous variables (that is, variables from the same time period as the dependent variable) generated significant path coefficients for both conflict and arms deliveries. In such ambiguous cases in which influence seemed to run in two directions simultaneously, the stronger of the two path coefficients dictated the ultimate direction of the arrow that remained in the model.

Gordon Hilton notes that such seeming "non-recursiveness," or looping of causal relations, is a result of the time slice being used in the model: "Whether we have a recursive or non-recursive model depends entirely upon the time aggregation with which we operate. If we have smaller periods of measurement, we are more likely to get the true model, which will always be recursive."[19] In our case, within the five-year slice of data used to drive the models there is almost certainly a series of back-and-forth interactions between arms and conflict in at least some of the national cases in which arms stocks are built up, war breaks out, arms stocks are depleted and replenished, and war ends (or not). Our data records merely the outbreak of war (most significant conflict event of the time period) and the total number of weapons delivered, a trade-off made for reasons noted earlier.

Notes

Introduction

1. See, for example, Michael Brzoska and Frederic S. Pearson, *Arms and Warfare, Escalation, De-escalation, and Negotiation* (Columbia: University of South Carolina Press, 1994), who analyze the role of arms transfers in ten recent wars.
2. Following World War I, U.S. advocates of temperance succeeded in passing and ratifying the Eighteenth Amendment to the U.S. Constitution, which banned the production and consumption of all but the weakest of alcoholic beverages. Demand for alcohol found ways around the law, however, and criminal organizations thrived on servicing that demand. Prohibition was repealed by the Twenty-first Amendment to the Constitution in 1933. See Thomas M. Coffey, *The Long Thirst: Prohibition in America, 1920–1933* (New York: W. W. Norton, 1975).
3. Barry Buzan, *People, States, and Fear,* 2nd ed. (Boulder, Colo.: Lynne Rienner Publishers, 1991), ch. 5.
4. Kalevi J. Holsti, *The State, War, and the State of War* (Cambridge: Cambridge University Press, 1997), ch. 8.
5. Barry M. Blechman, William J. Durch, David F. Gordon, and Catherine Gwin, *The Partnership Imperative: Maintaining American Leadership in a New Era* (Washington, D.C.: The Henry L. Stimson Center and the Overseas Development Council, 1997), pp. 5–13.

Chapter 1

1. Both numbers are derived from data compiled by the Stockholm International Peace Research Institute (SIPRI). Because SIPRI, for reasons discussed shortly in the text, counts the values only of major weapon systems transferred, $750 billion (adjusted for inflation to 1995 dollars) is a minimum estimate for the cumulative value of the arms trade with developing countries. It does not account for after-sale support, training, military construction, minor weapons, ammunition, or basic equipment not directly associated with weapon systems, such as trucks or radios. U.S. government figures indicate that weapons, ammunition, and direct support services for them account for just about two-thirds of the total value of U.S. arms sales and assistance programs. While U.S. support programs may be more comprehensive than most, adding another one-third to SIPRI's valuations—for an even $1 trillion—would probably not exaggerate the value of arms-related goods and services provided developing countries through 1997. For a breakout of U.S. data, see U.S. Defense Security Assistance Agency, *Foreign Military Sales, Foreign Military Construction Sales and Military Assistance Facts,* as of September 30, 1996 (Washington, DC: DSAA).
2. For those who think the analogy inapt or one-dimensional, recall that, in successive films, the eponymous character both takes life without remorse and shields it without flinching, reflecting the Janus-like nature of the arms trade itself, the weapons in which it deals, and the militaries that it enables.

3. See *SIPRI Yearbook 1995: Armaments, Disarmament, and International Security* (New York: Oxford University Press, 1995), p. 493.

4. Sovietologist Jamie McConnell attributed this greater activism in part to the adoption by Moscow of a "diplomacy of force" vis-à-vis the developing world in 1966. James M. McConnell, "Doctrine and Capabilities," in Bradford Dismukes and James M. Mc-Connell, eds., *Soviet Naval Diplomacy* (New York: Pergamon Press, 1979), pp. 24 ff.

5. See Randall Forsberg and Jonathan Cohen, "Issues and Choices in Arms Production and Trade," in Randall Forsberg, ed., *The Arms Production Dilemma: Contraction and Restraint in the World Combat Aircraft Industry,* CSIA Studies in International Security No. 7 (Cambridge, Mass.: The MIT Press, 1994), p. 274.

6. Roughly 40 percent of U.S. FMS buys weapons; the other 60 percent goes toward logistical support, construction, training, and the like. Craig M. Brandt, "American Weapon Sales and the Defense Industrial Base: Can Arms Transfers Save Defense Production?" A paper presented at the annual meeting of the International Studies Association, San Diego, Calif., April 16–20, 1996, p. 12.

7. John Donnelly and Colin Clark, "Merger Mania Hits $53 Billion This Year—So Far," *Defense Week* 18, no. 27 (July 7, 1997), pp. 1, 13–14.

8. Brandt, p. 12.

9. Mikhail I. Gerasev and Viktor M. Surikov, "The Crisis in the Russian Defense Industry: Implications for Arms Exports," in Andrew J. Pierre and Dmitri V. Trenin, eds., *Russia in the World Arms Trade* (Washington, D.C.: The Carnegie Endowment for International Peace, 1997), p. 10.

10. Gerasev and Surikov, "The Crisis in the Russian Defense Industry," pp. 18, 22–24. The authors cast the alternatives to export-led restructuring as: just don't modernize Russian forces; modernize with imported weapons; or, produce for domestic consumption at much higher unit costs.

11. These are numbers based on the counting methods of the Stockholm International Peace Research Institute (SIPRI), which SIPRI calls "trend indicator values." They are estimates of the performance-adjusted value of modern weapons, not the actual prices paid for those weapons. Brzoska and Ohlson (1987) report data in constant 1985 U.S. dollars as calculated by SIPRI. For consistency with later tables and figures in this chapter, I used U.S. Department of Defense (DOD) GDP deflators of roughly .73 to convert their data to 1995 dollars (only a rough conversion because DOD deflators are based on U.S. inflation rates only). As SIPRI trend value data were not available for 1986–1990 and 1991–1995 for all countries, I substituted data from the U.S. Arms Control and Disarmament Agency (ACDA). ACDA's valuation of arms imports is much higher than SIPRI's (see discussion in appendix B). The ACDA valuation comparable to $100 million in SIPRI's trend indicators turned out to be $132 million, derived by counting the number of states receiving at least $100 million in any year between 1981 and 1985 in Brzoska and Ohlson's data, and then finding the threshold valuation in the ACDA data that resulted in an identical tally (53 countries) for the same period. I then used the higher threshold to produce tallies for 1986–1990 and 1991–1995, with reasonable confidence that they could be reproduced using SIPRI data. U.S. ACDA, *World Military Expenditures and Arms Transfers, 1991–1992* and *1996* (Washington, D.C.: U.S. ACDA, 1992, 1997).

12. U.S. ACDA, *World Military Expenditures and Arms Transfers, 1991–1992* (Washington, D.C.: ACDA, 1992), p. 153.

13. John Stremlau, *Sharpening International Sanctions: Toward a Stronger Role for the United Nations,* a report of the Carnegie Commission on Preventing Deadly Conflict (Washington, D.C.: Carnegie Endowment for International Peace, November 1996), p. 30.

14. For discussion, see U.S. Congress, Office of Technology Assessment, *Global Arms Trade* OTA-ISC-460 (Washington, DC: U.S. Government Printing Office, June 1991), esp.

pp. 123 ff. See also Andrew L. Ross, "Developing Countries," in Andrew J. Pierre and Dmitri V. Trenin, eds., *Russia in the World Arms Trade* (Washington, DC: Carnegie Endowment for International Peace, 1997), pp. 89–127.

15. For a discussion of SIPRI's selection criteria for weapon systems in its data base, see *SIPRI Yearbook 1995,* pp. 532–33. See also appendix B in this volume.

16. According to British arms analyst Chris Smith, this American-sponsored pipeline to Afghanistan was *the* most adverse development of the last decade for regional stability, hastening Pakistan's fragmentation. See his "Light Weapons and Ethnic Conflict in South Asia," in Jeffrey Boutwell, Michael T. Klare, and Laura W. Reed, eds., *Lethal Commerce: The Global Trade in Small Arms and Light Weapons* (Cambridge, Mass.: American Academy of Arts and Sciences, 1995), pp. 62–64, 75. For additional treatments and policy prescriptions, see Edward J. Laurance, *Light Weapons and Intrastate Conflict,* a report to the Carnegie Commission on Preventing Deadly Conflict (Washington, D.C.: July 1998), and Virginia Gamba, ed., *Society Under Siege: Licit Responses to Illicit Arms,* Toward Collaborative Peace Series: Vol II (Halfway House, South Africa: Institute for Security Studies, 1998).

17. Laurance, *Light Weapons,* pp. vi, 2–3.

18. For extensive discussion, see Nicole Ball, *Security and Economy in the Third World* (Princeton: Princeton University Press, 1988), pp. 266 ff.

19. Grant T. Hammond, "The Role of Offsets in Arms Collaboration," in Ethan B. Kapstein, ed., *Global Arms Production* (Lanham, Md.: University Press of America, 1992), pp. 205 ff. See also Ball, *Security and Economy,* pp. 288–89, n. 90.

20. "Russia Seals 'Deal of the Century' with Malaysia," *Defense Marketing International* (newsletter), June 24, 1994, pp. 6–7; "Malaysia Concludes MiG-29 purchase," *Flight International,* June 15–21, 1994, p. 20; Pavel Felgengauer, "An Uneasy Partnership: Sino-Russian Defense Cooperation and Arms Sales," in Pierre and Trenin, eds., *Russia in the World Arms Trade,* pp. 95–98.

21. Bill Keller, *Arm in Arm: The Political Economy of the Global Arms Trade* (New York: Basic Books, 1995), pp. 54, 130–32, 169–70. See also Robert H. Trice, "Transnational Industrial Cooperation in Defense Programs," in Kapstein, *Global Arms Production,* pp. 159 ff.

22. Keller, *Arm in Arm,* pp. 4–5. Michael T. Klare, "The Subterranean Arms Market: Black-Market Sales, Covert Operations, and Ethnic Warfare," in Andrew J. Pierre, ed., *Cascade of Arms: Managing Conventional Weapons Proliferation* (Washington, D.C.: Brookings Institution Press, 1997), p. 44.

23. Klare, "Subterranean Arms Trade," p. 45; Stephen Kinzer, "Croatia Reportedly Buying MiGs, Defying UN," *New York Times,* September 23, 1993, p. A9; John Pomfret, "Iran Ships Materials for Arms to Bosnians," *Washington Post,* May 13, 1994, p. 1; Daniel Williams and Thomas W. Lippman, "U.S. Is Allowing Iran to Arm Bosnia Muslims," *Washington Post,* April 14, 1995, p. A1; and Michael Dobbs, "Saudis Funded Weapons for Bosnia, Official Says," *Washington Post,* May 29, 1996, p. A1.

24. The Iran-contra conspirators contrived to sell anti-tank and other missiles and spare parts to Iran in exchange for the freedom of American hostages held in Lebanon by Islamic radicals, with any profits being directed toward arms for counterrevolution in Central America. Trading arms for hostages and arming Iran were both counter to public Reagan administration policy, and arming the "contras" had been prohibited by the Congress. See Aaron Karp, "The Rise of Black and Gray Markets," in *The Annals* 535 September 1994, pp. 187–89; and John Tower, Edmund Muskie, and Brent Scowcroft, *The Tower Commission Report* (New York: Bantam Books, 1987).

25. See Laurance, *The International Arms Trade* (New York: Lexington Books, 1992), table 5–15, p. 111; Clifford G. Gaddy and Melanie L. Allen, "Russian Arms Sales Abroad: Policy, Practice, and Prospects," Brookings Discussion Papers (Washington, D.C.: The Brookings Institution, Foreign Policy Studies Program, September 1993), p. 38, n. 5; and

Stephen Foye, "Russian Arms Exports after the Cold War," *RFE/RL Research Report* 2:13 (March 1993), p. 62.

26. See U.S. Defense Security Assistance Agency, *Foreign Military Sales, Foreign Military Construction Sales and Military Assistance Facts* (Washington, DC: DSAA, September 1992).

27. Laurance, *The International Arms Trade,* pp. 82–109, 113–20, 167–69.

28. David J. Louscher and James Sperling, "Arms Transfers and the Structure of International Power," in Norman A. Graham, ed., *Seeking Security and Development: The Impact of Military Spending and Arms Transfers* (Boulder, Colo. and London: Lynne Rienner Publishers, 1994), pp. 55–77.

29. Raymond Garthoff, *Détente and Confrontation: American-Soviet Relations from Nixon to Reagan* (Washington, D.C.: The Brookings Institution, 1985), pp. 74–75.

30. Readers will note the sharp difference in value between figure 1.3's final year (1985) and figure 1.4's initial year (1986). This jump is mostly an artifact of using a different data set, from the U.S. ACDA, in figure 1.4, as SIPRI has not published dollar value data on individual suppliers' arms deliveries to developing countries since 1994. For a comparison of the SIPRI and ACDA data sets, please see appendix B.

31. For discussion, see Garthoff, *Détente and Confrontation,* pp. 412–13, 431–37.

32. U.S. Defense Security Assistance Agency, *Foreign Military Sales, Foreign Military Construction Sales and Military Assistance Facts* (Washington, DC: DSAA, December 1978, September 1987, September 1992, and September 1996).

33. The total is understated, partly because it leaves out the value of "excess defense articles" (EDA), used weapon systems transferred for free or at low cost to other countries after they are retired from use by American forces. DSAA's reporting system values these items at their original cost, without any effort to adjust for inflation. Since an appreciable fraction of EDA has been warships that were transferred after decades of U.S. service, their original cost represents only a small fraction of what it might have cost to build a comparable item in the year of transfer. On the other hand, backtracking DSAA's aggregate figures to individual items, noting individual year of construction, and reflating original costs to 1990s dollars would require prodigious effort, beyond the scope of even this study. Readers should, in short, be mindful that American military aid during the Cold War was "worth" even more than reported here. SIPRI's item counts of weapon systems transferred do, however, include EDA items, so the item-count charts in this chapter, which are based on SIPRI's data, accurately track actual flows of arms.

34. The following sections make use of data drawn from published SIPRI arms trade registers, cross-checked against the annual *Military Balance* published by the International Institute for Strategic Studies in London, to measure how quickly new weapons reached Third World markets and the relative sophistication of the arms transferred. The chapter examines by region the patterns of proliferation of fixed-wing combat aircraft, helicopters, missiles, armored vehicles, and naval systems. Details on the data set and the tools used to analyze it may be found in Appendix C.

35. An exception was made in including the Chinese Q-5 Fantan as a heavy attack aircraft. Derived from China's copy of the Soviet MiG-19 fighter, it has a relatively poor payload capability of 4,400 pounds, but it still carries nearly twice the weight of ordnance of the most capable light attack aircraft, the British Aerospace Hawk 100. Data derived from Michael J. H. Taylor, *Jane's Encyclopedia of Aviation,* rev. ed. (New York: Crescent Books, 1993).

36. Operations that are offensive in nature tactically (that is, in terms of immediate or localized objectives of battle) may support a defensive strategy (the longer-term or generalized objectives of the war). Thus a NATO concept called "offensive counter air" had as a tactical objective the destruction of air bases from which the Warsaw Pact could

launch attacks against NATO—which served NATO's strategic objective of defending Western Europe.

37. Compare, for example, the entries for the McDonnell-Douglas F-4 Phantom II and the Boeing B-17 in Michael J. H. Taylor, ed., *Jane's Encyclopedia of Aviation* (New York: Crescent Books, 1993), pp. 175, 645.

38. Comparable generational assignments are discussed in U.S. Congressional Budget Office, *A Look at Tomorrow's Tactical Air Forces,* a CBO Study (Washington, D.C.: U.S. GPO, January 1997), pp. 14–15, and Robert P. Berman, *Soviet Air Power in Transition* (Washington, D.C.: The Brookings Institution, 1978), p. 52. CBO's first generation starts with the MiG-15, however, skipping World War II designs and the early jets. For a table of comparisons with the generational assignments used in this volume, please refer to Appendix B.

39. Artillery is not covered in this section. For reasons best known to SIPRI, its coverage of artillery transfers is deficient, particularly prior to the 1980s. For example, the 1976–1977 *Military Balance* indicates holdings of more than 2,600 artillery pieces by states in the Middle East and Persian Gulf. Through 1977, however, the SIPRI data catalog the transfer of just under 300 pieces of artillery. One would expect transfers to far exceed 1976 holdings, to account for war losses and other attrition. Thus the artillery category was dropped from this analysis. (Other spot-comparisons of SIPRI and *Military Balance* ground forces data, such as tanks, did not reveal such glaring disparities of numbers.)

40. *SIPRI Yearbook 1995,* Appendix 14B.

41. Data on tanks taken out of service for all causes are given in Randall Forsberg, Andrew Peach, and Judith Reppy, "U.S. Airpower and Aerospace Industries in Transition," in Forsberg, *The Arms Production Dilemma,* p. 117.

42. The chart uses nominal tonnages for each category of vessel: 150 tons for patrol craft, 1,000 tons for corvettes, 2,000 tons for frigates, 4,000 tons for destroyers. While a reasonable estimate of actual tonnage transferred, it may be higher or lower in specific categories.

43. There are, of course caveats: if the enemy was unsupported by combat aircraft, and if friendly aircraft transmitted crucial data as to the general direction and range of over-the-horizon targets. In the 1986 U.S. raids on Libya, for example, Libyan missile craft were destroyed by U.S. carrier aviation miles from their effective missile launch points. David C. Martin and John Walcott, *Best Laid Plans: The Inside Story of America's War Against Terrorism* (New York: Simon and Schuster, 1988), pp. 282–84.

44. Chile and Argentina were taking delivery on ships ordered before they resolved their dispute over the Beagle Channel, which lies at the very tip of South America. The dispute flared into fighting in 1978–1979; the ships were ordered in 1978–1981. The two countries signed a Treaty of Peace and Friendship in 1984 that resolved the dispute. See Michael A. Morris, "Confidence-Building Measures in the Maritime Domain," in Augusto Varas, et al., eds., *Confidence-Building Measures in Latin America: Central America and the Southern Cone,* Report No. 16, Stimson Center/FLACSO-Chile (Washington, D.C.: The Henry L. Stimson Center, February 1995), p. 55.

45. See Bernard Prézelin, ed., *Combat Fleets of the World 1993,* trans. A. D. Baker III (Annapolis, Md.: Naval Institute Press, 1993).

46. Tom Stefanick, *Strategic Antisubmarine Warfare and Naval Strategy* (Lexington, Mass.: Lexington Books, 1987), esp. pp. 44, 52, 333.

47. Hastings and Jenkins, pp. 147–56, 160.

48. International Institute for Strategic Studies, *The Military Balance, 1997–98* (Oxford: Oxford University Press, 1997), pp. 299–301.

49. Coit D. Blacker and Gloria Duffy, eds., *International Arms Control, Issues and Agreements,* 2nd ed. (Stanford, Calif.: Stanford University Press, 1984), p. 91.

50. On February 2, 1995, Sweden's Kockums shipyard launched the A-19-class boat, *Gotland,* the first non-nuclear, AIP-capable submarine that is not a test-bed. *Gotland* uses a four-ton Stirling cycle engine fueled by diesel and stored oxygen (from liquid oxygen tanks) that is expected to give it two weeks underwater endurance at five knots. A stretched version, the T-96, is said to be offered for export, and Thailand, Malaysia, and Singapore have been cited as potential customers. Australia and Japan are said to be testing engines "identical to those in *Gotland."* (Joris Janssen Lok, "Scandinavians study joint procurement," *Jane's Defence Weekly,* February 11, 1995, p. 2; "World's first production AIP submarine," *Despatches: The IDR New Product Review,* February 1995, p. 1; "Air Independent Propulsion," *U.S. Naval Institute Proceedings,* August 1990, p. 59.)

51. Other missiles (for example, strategic ballistic missiles and antiballistic missile systems) that did not reach export markets are not included in this table.

52. Fear of losing either of their expeditionary force's aircraft carriers to Argentine Exocets kept the Royal Navy operating eastward of the Falklands. See Max Hastings and Simon Jenkins, *The Battle for the Falklands* (New York: W.W. Norton, 1983), pp. 156, 161.

53. U.S. ACDA, *WMEAT 1997,* p. 4.

54. "Clinton Administration Releases Arms Transfer Policy," *Arms Trade News,* March 1995, p. 1.

55. Brandt, "American Weapon Sales," p. 7.

56. See, for example, Giovanni de Briganti, "French Move Belies Policy," *Defense News,* April 7, 1997, p. 1.

57. "No Cause for Celebration" (Editorial), *Defense News,* January 5, 1998, p. 14. Britain, France, Germany, and Italy have a joint armaments agency intended to take over management of all joint weapon programs. Called the Organisation Conjoint de Cooperation d'Armement (OCCA), it was set up in February 1997. See Giovanni de Briganti, "Policy Disputes Stymie Europe Consolidation," *Defense News,* May 12, 1997, p. 6. On lack of progress toward a consolidated European armaments industry, see Working Group on the Future of the Defence Industry, "Conclusions of the First Meeting," of June 19, 1998, released by the Centre for European Reform, London. Internet:

58. "Global Defence Industry," *The Economist,* June 14, 1997, p. 16.

59. Barbara Opall, "East Asia Allies Distrust Sino-Russian Strategic Pact," *Defense News,* May 19, 1997, p. 8.

60. Pyotr Yudin, "Russian Exports Fall Short of Expectations," *Defense Week,* December 8, 1997, p. 16.

61. See, for example, Francis Tusa, "'Any Old Iron' in Tank Market," *Armed Forces Journal International,* May 1993, p. 36, and Mark Wagner, "Fighting Over the Scraps," *Flight International,* June 1–7, 1994, pp. 23–28.

62. See, for example, Clifford Beal and Barbara Starr, "Fighter Upgrades: The Only Game in Town," *International Defense Review* 6/1993, pp. 445–50; Giovanni de Briganti and Stephen C. LeSueur, "As Procurement Sags, Upgrade Markets Boom," *Defense News,* July 4–10, 1994, p. 8; "Programme Guide: Combat Aircraft," *Jane's Defence System Modernisation,* July/August 1994, pp. 36–39; Sept/Oct 1994, pp. 31–35; and Nov/Dec 1994, pp. 30–35; Simon Michell, "Remake, Remodel, Retrofit, Remarket," *Jane's Defence Weekly,* December 17, 1994, p. 25; and Pyotr Yudin, "Russia Offers Upgrade Deals," *Defense News,* December 8–14, 1997.

63. See, for example, Clifford Beal and Barbara Starr, "Fighter Upgrades: the only game in town," *International Defense Review* 6/1993, pp. 445–50; Mark Wagner, "Fighting Over the Scraps," *Flight International,* June 1–7, 1994, pp. 23–28; Giovanni de Briganti and Stephen C. LeSueur, "As Procurement Sags, Upgrade Market Booms," *Defense News,* July 4–10, 1994, p. 8; and "Upgrade Programme Guide: F-16 Electronic Warfare Options," *Jane's Defence Systems Modernisation,* April 1995, pp. 28–32.

Chapter 2

1. Mohammed Ayoob, *The Third World Security Predicament: State Making, Regional Conflict, and the International System* (Boulder, Colo.: Lynne Rienner Publishers, 1995), pp. xiii, 22–23.
2. Charles Tilly, "Western State-making and Theories of Political Transformation," in Tilly, ed., *Formation of National States in Western Europe* (Princeton: Princeton University Press, 1975), p. 598. Cited in Ayoob, *Third World Security Predicament,* p. 31. Ibid., pp. 24–30.
3. Ibid., pp. 31, 15–16.
4. Edward Azar and Moon Chung-In, "Legitimacy, Integration, and Policy Capacity," in Azar and Moon, eds., *National Security in the Third World* (Cambridge: Cambridge University Press, 1988), pp. 77–82.
5. Robert E. Riggs and Jack C. Plano, *The United Nations: International Organization and World Politics,* 2nd ed. (Belmont, Calif.: Wadsworth Publishing Co., 1994), p. 195.
6. Ibid., p. 79.
7. Ayoob, *The Third World Security Predicament,* pp. 83–84.
8. Ibid., pp. 167–71, 179–80.
9. Associated Press, "President-Elect Vows to Revive Ailing Nigeria," *Washington Post,* March 3, 1999; p. A18.
10. Brian Job, "The Insecurity Dilemma: National, Regime, and State Securities in the Third World," in Job, ed., *The Insecurity Dilemma, National Security of Third World States* (Boulder, Colo.: Lynne Rienner Publishers, 1992), pp. 14–16.
11. Harvard's Liah Greenfeld characterizes these two approaches as "civic" and "ethnic" nationalism. The former is inclusive in nature, the latter exclusive: you can't join it, you must be born into it. See her *Nationalism: Five Roads to Modernity* (Cambridge, Mass.: Harvard University Press, 1992), p. 11.
12. Job, "The Insecurity Dilemma," pp. 17, 25.
13. For a taste of eighth-century thought via twenty-first-century technology, see the Taliban site on the World Wide Web. Internet:. Accessed April 3, 1999.
14. Barry Buzan, *People, States, and Fear,* 2nd ed. (Boulder, Colo.: Lynne Rienner Publishers, 1991), p. 187.
15. Buzan, *People, States, and Fear,* pp. 194–95.
16. Buzan, *People, States, and Fear,* p. 209.
17. Buzan, *People, States, and Fear,* pp. 211–14.
18. Joel Colton and R. R. Palmer, *A History of the Modern World,* 8th ed. (New York: Knopf, 1995), passim.
19. Buzan, *People, States, and Fear,* pp. 135–40.
20. See John Jacob Nutter, "An Analysis of Threat," in Norman A. Graham, ed., *Seeking Security and Development: The Impact of Military Spending and Arms Transfers* (Boulder, Colo.: Lynne Rienner Publishers, 1994), pp. 47–48.
21. John J. Johnson, "The Latin American Military as a Politically Competing Group in a Transitional Society," in Johnson, ed., *The Role of the Military in Underdeveloped Countries* (Princeton: Princeton University Press, 1962), pp. 93–95.
22. A. F. Mullins, *Born Arming: Development and Military Power in New States* (Stanford: Stanford University Press, 1987), pp. 2–5.
23. K. J. Holsti, "International Theory and Domestic War in the Third World: The Limits of Relevance," paper presented at the annual meeting of the International Studies Association, Toronto, Canada, February 1997.
24. The deterrence literature is voluminous, but largely repetitive of the same basic constellation of ideas that were most highly developed in the context of the Cold War and the East-West nuclear standoff. Some of the best perspectives on deterrence logic and its evolution may be found in Thomas C. Schelling, *The Strategy of Conflict* (New York:

Oxford University Press, 1960); Philip Green, *Deadly Logic: The Theory of Nuclear Deterrence* (New York: Schocken Books, 1966); Lawrence Freedman, *The Evolution of Nuclear Strategy* (New York: St. Martin's Press, 1983); and Charles L. Glaser, *Analyzing Strategic Nuclear Policy* (Princeton: Princeton University Press, 1990). For evaluation of the effectiveness of post–Cold War deterrent threats by the United States, see Barry M. Blechman and Tamara Wittes, "Defining Moment: The Threat and Use of Force in American Foreign Policy Since 1989," Foreign Policy Project Occasional Paper No. 6 (Washington, D.C.: The Henry L. Stimson Center and the Overseas Development Council, May 1998).

25. A well-known discussion of security dilemmas and the spiral model may be found in Robert Jervis, *Perception and Misperception in World Politics* (Princeton: Princeton University Press, 1976).

26. The original Chamberlain, of course, came away from a 1938 meeting with Adolph Hitler reassured that he had achieved "peace in our time," months before Hitler's *blitzkrieg* rolled into Poland.

27. George W. Downs, "Arms Races and War," in Philip E. Tetlock, et al., eds., *Behavior, Society, and Nuclear War,* 2 (New York: Oxford University Press, 1991), pp. 90–91.

28. Holsti, "International Theory," pp. 24–25.

29. Holsti, "International Theory," p. 13.

30. Job, "The Insecurity Dilemma," p. 17.

31. Barry R. Posen, "The Security Dilemma and Ethnic Conflict," in Michael E. Brown, ed., *Ethnic Conflict and International Security* (Princeton: Princeton University Press, 1993), pp. 104, 110–11.

32. Holsti, "International Theory," p. 18, citing Russell Hardin, *One for All: The Logic of Group Conflict* (Princeton: Princeton University Press, 1995).

33. For a concise history of events in Kosovo, see International Crisis Group, *Kosovo Spring,* parts 1 and 2, March 20, 1998. Internet: Click on Publications, then South Balkans, and scroll to the title. Accessed April 1, 1999. See also Raymond Bonner, "NATO Is Wary of Proposals to Help Arm Kosovo Rebels."

34. For background on the larger meltdown of Yugoslavia in the 1990s, see Lenard J. Cohen, *Broken Bonds: The Disintegration of Yugoslavia* (Boulder, Colo.: Westview Press, 1993); Brian Hall, *The Impossible Country: A Journey through the Last Days of Yugoslavia* (Boston: David R. Godine, 1994); Susan Woodward, *Balkan Tragedy: Chaos and Dissolution After the Cold War* (Washington, D.C.: The Brookings Institution, 1994). For analysis of international efforts to cope with that meltdown, see William J. Durch and James A. Schear, *Faultlines: UN Operations in the Former Yugoslavia,* in W. J. Durch, ed., *UN Peacekeeping, American Policy, and the Uncivil Wars of the 1990s* (New York: St. Martin's Press, 1996).

35. For those who think it strange to cling to the memory of military defeat, consider how many Americans (and Anglos, in particular, to give the analogy the appropriate ethnic twist) travel to view the Alamo, Custer's Last Stand, and Pearl Harbor.

36. Paul Lewis, "Cash Crunch: Arms buying has come to a halt in the troubled Southeast Asian economies," *Flight International,* January 21–27, 1998, pp. 60–62. John Haseman, "Indonesia cuts back on spending as crisis bites," *Jane's Defence Weekly,* January 21, 1998, p. 6. Robert Karniol, "South Korea postpones programmes amid crisis," *Jane's Defence Weekly,* January 21, 1998, p. 14. Barbara Opall, "Thai F-18s Cost U.S. Navy," *Defense News,* March 16–22, 1998, p. 1.

37. Transparency International, "Corruption Perception Index." Internet: Accessed March 1998 and April 1999. TI's ratings are based on at least three and as many as 12 surveys for each country rated.

38. Only 16 countries of interest were included in the 1995 TI corruption survey. Their ratings and the average value of their arms deliveries in 1996–1997 (based on data from

U.S. Department of State, Bureau of Arms Control, *World Military Expenditures and Arms Transfers, 1998* (Washington, D.C.: 1999), Table II), are listed below. Regression of CPI on AT value returns an R^2 of 0.01.

	CPI 1995	AT Value (U.S. mil.)
Argentina	5.24	75.0
Brazil	2.7	380.0
Chili	7.94	135.0
Colombia	3.44	95.0
India	2.78	405.0
Indonesia	1.94	605.0
Malaysia	5.28	462.5
Mexico	3.18	105.0
Pakistan	2.25	385.0
Philippines	2.77	145.0
South Africa	5.62	105.0
South Korea	4.29	1100.0
Taiwan	5.08	5600.0
Thailand	2.79	762.5
Venezuela	2.66	230.0
Singapore	9.26	415.0

39. Nicole Ball, *Security and Economy in the Third World* (Princeton: Princeton University Press, 1988), p. 391.
40. Norman A. Graham, "The Quest for Security and Development," in Graham, ed., *Seeking Security and Development: The Military Impact of Spending and Arms Transfers* (Boulder, Colo.: Lynne Rienner Publishers, 1994), p. 257.
41. United Nations Development Program, *Human Development Report, 1996* (New York: Oxford University Press, 1996), table 1.9, p. 29.
42. The classic justification for such policies is Jeanne J. Kirkpatrick, "Dictators and Double Standards," *Commentary* 68 (Nov. 1979), pp. 34–45.
43. Samuel P. Huntington, "Arms Races, Prerequisites and Results," *Public Policy* 8 (1958). The subsequent academic literature on competitive arms acquisitions is enormous. For a comprehensive bibliography through the late 1980s, see Craig Etcheson, *Arms Race Theory, Strategy and Structure of Behavior* (New York: Greenwood Press, 1989). Influential works from that period include Colin S. Gray, "The Arms Race Phenomenon," *World Politics* (Oct. 1971), pp. 39–59, and "The Urge to Compete: Rationales for Arms Racing," *World Politics* (Jan. 1974), pp. 207–33; Graham T. Allison, "Questions About the Arms Race: Who's Racing Whom? A Bureaucratic Perspective," in Robert L. Pfaltzgraff, Jr., ed., *Contrasting Approaches to Strategic Arms Control* (Lexington, Mass.: Lexington Books, 1974); George W. Downs, David M. Rocke, and Randolph M. Siverson, "Arms Races and Cooperation," *World Politics* 38 (1985), pp. 118–46; Paul F. Diehl, "Armaments Without War: An Analysis of Some Underlying Effects," *Journal of Peace Research* 22 (1985), pp. 249–59; Paul Diehl and Jean Kingston, "Messenger or Message? Military Buildups and the Initiation of Conflict," *Journal of Politics* 49 (1987), pp. 801–13; Downs, "Arms Races and War," pp. 73–109; and Michael Brzoska and Fredric Pearson, *Arms and Warfare: Escalation, Deescalation, and Negotiation* (Columbia: University of South Carolina Press, 1994).
44. In a study published three years before Gorbachev began to signal Soviet disengagement from the Cold War, University of Illinois professor Paul Diehl examined two dozen "enduring rivalries" between major powers that involved arms races but no war.

He found that 78 percent ended in capitulation by one party, about 14 percent produced political stalemate, and 7 percent produced compromise. However, most of the
capitulating in Diehl's data occurred in the 1930s and during the postwar period before the onset of mutual East-West nuclear deterrence. (Diehl, "Armaments Without
War.") A subsequent historical study of the impact of major-power military buildups on
war initiation found little statistical relationship between the two. (Diehl and Kingston,
"Messenger or Message?")

45. David Kinsella, "Arms Transfers, Dependence, and Regional Stability: Isolated Effects or
General Patterns?" paper presented at the annual meeting of the International Studies
Association, Washington, D.C., February 16–21, 1999. Mimeo. As his arms transfer indicator, Kinsella used arms transfer "programs"—line items in the arms trade registers
of the Stockholm International Peace Research Institute (SIPRI)—rather than quantities or types of weapons transferred. One tank or one thousand tanks transferred to a
country count as one "program" if they are all the result of one transfer agreement. A
lot of variability in supply relationships and behavior over time (1950–1991) was
lumped into single counts of American transfer programs, Soviet programs, and "conflictual events" initiated by each arms recipient in that time period. Copies of the paper
are available on the Internet at:

46. Ball, *Security and Economy,* p. 294.

47. Downs, "Arms Races and War," pp. 77–78.

48. Prominent among proponents of the war-avoiding value of the nuclear standoff is historian John Lewis Gaddis; see his "The Long Peace: Elements of Stability in the Post-
War International System," *International Security* (Spring 1986), esp. pp. 121–23. The
prospect of nuclear holocaust, Gaddis argues, induced caution in leaders on both sides,
forcing them "to confront the reality of what war is really like, indeed to confront the
prospect of their own mortality" (p. 123).

49. For discussion of Israeli decision-making, see John J. Mearsheimer, *Conventional Deterrence* (Ithaca: Cornell University Press, 1983), pp. 143–55.

50. See, for example, Andreas von Bulow, "Defensive Entanglement: An Alternative Strategy for NATO," in Andrew J. Pierre, ed., *The Conventional Defense of Europe* (New York:
Council on Foreign Relations, 1988), pp. 112–51. For a critique, see Stephen J. Flanagan, "Nonprovocative and Civilian-Based Defenses," in Joseph S. Nye, Jr., Graham T. Allison, and Albert Carnesale, eds., *Fateful Visions: Avoiding Nuclear Catastrophe* (Cambridge,
Mass.: Ballinger Publishing Co., 1988), pp. 93–110.

51. Ground- and space-based beacons can augment the civilian signal to produce near-mil-
itary-quality guidance without access to the encrypted "P-code" from the GPS constellation itself. For a comprehensive review of applications and policy issues, see Scott
Pace, et al., *The Global Positioning System: Assessing National Policies,* The RAND Critical Technologies Institute, MR-614-OSTP (Santa Monica: RAND, 1995). The executive summary is on the Internet at: (downloaded March 31, 1998). See also *A Policy
Direction for the Global Positioning System: Balancing National Security and Commercial Interests,* RAND Research Brief 1501, on the Internet at: http://www.rand.org/publica-
tions/RB/RB1501/RB1501.html.

52. Basic information on ballistic missile technology may be found online at the website of
the Center for Defence and International Security Studies, Lancaster University, United
Kingdom (Accessed March 4, 1998. Data on missile reentry velocities may be found in
Herbert Lin, *New Technologies and the ABM Treaty* (Washington: Pergamon-Brassey's,
Inc., in cooperation with the MIT Center for International Studies, 1988), p. 38.

53. James Roche, "Proliferation of Tactical Aircraft and Ballistic and Cruise Missiles in the
Developing World," in W. Thomas Wander, Eric H. Arnett, and Paul Bracken, eds., *The
Diffusion of Advanced Weaponry: Technologies, Regional Implications, and Responses* (Washing-
ton, D.C.: American Association for the Advancement of Science, 1994), pp. 75–76.

54. Jack S. Levy, "The Offensive/Defensive Balance of Military Technology: A Theoretical and Historical Analysis," *International Studies Quarterly* 28 (1984), p. 226. The debate about whether an offense-defense balance can even be usefully calculated was renewed in 1998–1999 in the pages of the journal *International Security*. See Stephen Van Evera, "Offense, Defense, and the Causes of War," *International Security* 22, no. 4 (Spring 1998), pp. 5–43; and Charles L. Glaser and Chaim Kaufmann, "What Is the Offense-Defense Balance and Can We Measure It?" Ibid., pp. 44–82. Critiques from James W. Davis, Jr., Bernard I. Finel, and Stacie E. Goddard, with replies from the original authors, may be found in "Correspondence: Taking Offense at Offense-Defense Theory," *International Security* 23, no. 3 (Winter 1998–1999), pp. 179–206.

55. Stephen D. Biddle, *The Determinants of Offensiveness and Defensiveness in Conventional Land Warfare* (Ph.D dissertation, Harvard University, May 1992), pp. 21–56. Also Biddle, "Recent Trends in Armor, Infantry, and Artillery Technology: Developments and Implications," in Wander, et al., *The Diffusion of Advanced Weaponry*, pp. 108–116.

56. On alliance-building behavior, especially "balancing" against regional hegemons, see Stephen M. Walt, *The Origins of Alliances* (Ithaca: Cornell University Press, 1987).

57. Mullins, *Born Arming*, p. 44.

58. U.S. Congress, Office of Technology Assessment, *Global Arms Trade* (Washington, D.C.: U.S. Govt. Printing Office, June 1991), esp. pp. 3–16.

59. The same larger system of relations preserved and protected tiny and weak states that might otherwise have succumbed to violent, Euro-style political competition but whose peoples might also have emerged from that competition, as did many European groups, as citizens of larger and more survivable countries. However, given the lengths to which European bloodletting went before the continent found peace, such a solution, in a nuclear age, to regional boundary and ethnic disputes clearly would have been unwise. (For discussion, see Mullins; Ayoob.)

60. Whether Bosnia-Herzegovina will survive as a tri-ethnic state after three-and-one-half years of brutal intercommunal conflict, despite the efforts of outsiders from NATO, the Organization for Security and Cooperation in Europe, and the United Nations, remains to be seen. For an account of the conflict and the interventions, see William J. Durch and James A. Schear, "Faultlines: UN Operations in the Former Yugoslavia," in Durch, ed., *UN Peacekeeping, American Policy, and the Uncivil Wars of the 1990s* (New York: St. Martin's Press, 1996).

61. See, for example, Leonard S. Spector and Mark G. McDonough with Even S. Medeiros, *Tracking Nuclear Proliferation: A Guide in Maps and Charts, 1995* (Washington, D.C.: The Carnegie Endowment for International Peace, 1995. Scott D. Sagan and Kenneth N. Waltz, *The Spread of Nuclear Weapons: A Debate* (New York: W.W. Norton and Company, 1995). Gen. Andrew Goodpaster, et al., *An American Legacy: Building a Nuclear-Free World*, final report of the steering committee, Project on Eliminating Weapons of Mass Destruction (Washington, D.C.: The Henry L. Stimson Center, 1997). Michael J. Mazaar, *Nuclear Weapons in a Transformed World: The Challenge of Virtual Nuclear Arsenals* (New York: St. Martin's Press, 1997). Amy E. Smithson, *Chemical Weapons Disarmament in Russia: Problems and Prospects,* Report no. 17 (Washington, D.C.: The Henry L. Stimson Center, 1995) and Smithson, *Separating Fact from Fiction: The Australia Group and the Chemical Weapons Convention,* Occasional paper no. 34 (Washington, D.C.: The Henry L. Stimson Center, 1997). Brad Roberts, ed., *Biological Weapons: Weapons of the Future?* (Washington, D.C.: Center for Strategic and International Studies, 1993) and Amy E. Smithson, *Biological Weapons Proliferation: Reasons for Concern, Courses of Action,* Report no. 24 (Washington, D.C.: The Henry L. Stimson Center, 1998).

62. See Martin S. Navias and E. R. Hooton, *Tanker Wars: The Assault on Merchant Shipping during the Iran-Iraq Conflict, 1980–1988* (New York: I. B. Tauris, 1996); Dilip Hiro, *The*

Longest War: The Iran-Iraq Military Conflict (New York: Routledge, 1991); and David Segal, "The Iran-Iraq War: A Military Analysis," *Foreign Affairs* 66 (Summer 1988).

63. I have opted to exclude from the arms/conflict analysis the rather large number of island microstates that either have no military forces or that imported no heavy weapon systems during the period of analysis. These states were rarely involved in measurable conflict, with the exception of the occasional coup d'etat (Fiji, Comoros) or invasion by the United States (Grenada).

64. Ted Robert Gurr, with Barbara Harff, Monty G. Marshall, and James R. Scarritt, *Why Minorities Rebel: A Global View of Ethnopolitical Conflicts* (Washington, D.C.: United States Institute of Peace, 1993), pp. 95ff.

65. Filling data gaps is the principal reason why the final period encompasses six rather than five years. The change in total arms deliveries for the time period is not great, but a few more countries take deliveries in 1995 that did not do so in 1994 or 1993, virtually all such orders having originated in 1990 or later. There are no new conflict peaks in 1995 that change any country's conflict score.

66. Ted Gurr distinguishes rebellion from protest and scales them separately. "Protest typically aims at persuading or intimidating officials to change their policies toward the group; rebellion aims directly at more fundamental changes in governments and in power relations among groups." Rebellion is of greater interest here. See Gurr, et al., *Why Minorities Rebel*, p. 93.

67. This is also consistent with Gurr's coding methodology for ethnopolitical conflict. See Gurr, *Why Minorities Rebel*, p. 94.

68. Ordinary Least Squares (OLS) regression, R-squared $= .62$; t-test is significant at .02 level.

69. Also OLS regression. Uses U.S. ACDA arms import value numbers rather than SIPRI trend indicators. ACDA reports individual data for all developing countries.

70. For a discussion of the strictures on using path analysis for purposes of causal influence, see Thomas D. Cook and Donald T. Campbell, *Quasi-Experimentation: Design and Analysis Issues for Field Settings* (Boston: Houghton Mifflin Co., 1979), pp. 296–97.

71. In the last quarter-century, arms have reached active combatants sometimes within days of need (as in the case of U.S. and Soviet air and sea lifts of arms to Israel and Egypt in October 1973), and certainly within a year. Many of the arms that Iran and Iraq ordered in the latter part of their eight-year war reached them in the year of order. See SIPRI Yearbook 1988, pp. 230–34.

72. Dichotomous, or dummy variables, are used to group records within data sets. In this instance, each record represented one country and its conflict, arms transfer, and other data in one time period. Eight regional variables were part of each record; seven of them would be assigned a value of zero, and the eighth, representing that country's regional security complex, would be assigned a value of one. So as not to "overdetermine" a model that includes "N" dummy variables, regressions are run using N-1 dummies. In this case, I omitted the variable for Central America and the Caribbean since, with the periodic exception of Cuba, states in the region have not acquired much in the way of heavy weapons. The effect of these states on the model is built into the regression equation's constant instead of their own regional regression coefficient. On the uses of dummy variables, see N. R. Draper and H. Smith, *Applied Regression Analysis* (New York: John Wiley and Sons, 1966), p. 134.

73. The time period 1970–1974 serves as the baseline for the analysis and neither conflict nor weapon deliveries for that period are explained here.

74. James L. Payne, *Why Nations Arm* (New York: Basil Blackwell, 1989), pp. 1–3, 24, 104–32. Having looked at and found wanting a number of geopolitical indicators (such as number of neighboring states, or lengths of common frontiers), Payne ultimately em-

phasized expansionist ideology and culture as the keys to the puzzle of militarism, that is, to the buildup of military forces substantially larger than "rationally" required by a state. He focused especially on Marxism and "the Islamic tradition" as sources of militarism, as states in either camp had force ratios (numbers of soldiers per thousand population) substantially higher than the average non-Marxist or non-Muslim country.

75. The force ratio used in each instance was that for the five-year period preceding the weapon deliveries. So for deliveries in 1975–1979, I used the force ratio for 1973, the hypothesis being that the strength of military influence led to contracts being signed that produced the deliveries in subsequent years.

76. Fighting in Zaire, now the Democratic Republic of Congo (DROC), flared once again in fall 1998, this time involving multiple interventions by regional states, both in support of and in opposition to the new DROC government. For further discussion, see chapter 5 on the Southern African Development Community.

77. The postcolonial wars for control of Mozambique and Angola, for example, lasted 18 and more than 25 years respectively. Cambodia has been embroiled in conflict that has spanned insurgency, intervention, invasion, and genocide, from the late 1960s through 1998. Afghanistan, similarly, has been snarled in civil war since 1979, Colombia has struggled to defeat various guerrilla armies for at least a decade longer than that, and Guatemala brought its four-decade civil war to an end only in late 1996. On the difficulties of rebuilding peace in war torn lands, see I. William Zartman, ed., *Collapsed States: The Disintegration and Restoration of Legitimate Authority* (Boulder, Colo.: Lynne Rienner Publishers, 1995).

Chapter 3

1. This chapter does not deal directly with the Arab-Israeli dispute. While the countries actively involved in it absorbed a major proportion of arms shipped to developing countries during the Cold War, its problems also have been studied extensively. Israel's opponents have lost their source of concessional arms supplies, and with the Israeli-Palestinian accords, the focus of the regional political contest has shifted largely to the microcosm of Israel, the West Bank, and Gaza. The critical balance in the region is as much one of political will in Jerusalem, Hebron, and hundreds of other towns and villages as it is the balance of heavy armaments in the region. Similarly, Iraq and the Gulf Arab states are treated in this chapter in the context of their dealings with Turkey and Iran rather than separately.

2. Alan Moorehead, *Gallipoli* (New York: Ballantine Books, 1958, 1982), pp. 113, 122–24, 239–40, 269–73.

3. L. S. Stavrianos, *The Balkans Since 1453* (New York: Holt, Rinehart, and Winston, 1958), pp. 580–89. Kemal did not import communism along with his armaments, however. "Emissaries from Russia who sought to propagate the new proletarian gospel mysteriously disappeared on their arrival in Turkey" (p. 584).

4. Duygu B. Sezer, *Turkey's Political and Security Interests and Policies in the New Geostrategic Environment of the Expanded Middle East,* Occasional Paper No. 19 (Washington, D.C.: The Henry L. Stimson Center, July 1994), p. 5. This paper was commissioned in support of this study, with the support of the Ploughshares Fund.

5. Sezer, *Turkey's Political and Security Interests and Policies,* p. 5.

6. Metin Heper, "The State, the Military, and Democracy in Turkey," *The Jerusalem Journal of International Relations* 9, no. 3 (1987), pp. 52–64. Heper notes that where the idea of the state is separate from that of civil society, as it is in Turkey, it is embodied in "state elites," which in Turkey have included the military, from Ottoman times onward. The military sees its mission as setting it apart from and above the country's "political elites."

Indeed, in its 1980 intervention, the military abolished all existing political parties and promulgated a new constitution. In reviewing a new political history of Turkey, Robert Kaplan notes:

Because the military's hierarchical command culture stifles civilian responsibility . . . Turkish parliamentarians have "never really developed the concept of teamwork central to party politics." That is why Turkish politics is essentially a passionate drama of personal rivalries, often unconnected to policy differences. Such irresponsibility both permits and forces the generals to act as paternalistic pashas.

Robert D. Kaplan, "After the Ottoman Empire, review of 'Turkey Unveiled: A History of Modern Turkey," *New York Times Book Review,* January 17, 1999, p. 8. Looking at the same situation, an economist might say that the military's role as stabilizer of last resort creates a "moral hazard" for Turkish politics, whose leaders know that utterly reckless behavior will not have ultimately disastrous consequences for the country. Turkey's politicians periodically get a "time out" but are soon back in the game.

7. Celestine Bohlen, "Turkish Army in New Battle in the Defense of Secularism," *New York Times,* March 30, 1996, p. 5; Human Rights Watch, "Turkey—Translator and Publisher of HRW Report Charged in Court," press release, October 16, 1996. Available online at gopher://gopher.igc.apc.org:5000/00/int/hrw/arms/15.

8. Stavrianos, *The Balkans,* p. 590. Of the two million Armenians resident in Anatolia in March 1915, roughly three-quarters of a million were killed or sent into desert exile to die. Turkey has never come to terms with these events. See Moorehead, *Gallipoli,* pp. 84–87, or G. S. Graber, *Caravans to Oblivion: The Armenian Genocide, 1915* (New York: John Wiley & Sons, 1996).

9. See Ted Robert Gurr, *Minorities at Risk: A Global View of Ethnopolitical Conflicts* (Washington, D.C.: U.S. Institute of Peace, 1993), pp. 226–227. For a comprehensive treatment, see David McDowall, *A Modern History of the Kurds* (London: I. B. Tauris, 1996).

10. Gurr, *Minorities at Risk,* p. 227. U.S. Department of State, Bureau of Democracy, Human Rights, and Labor, "Turkey Country Report on Human Rights Practices for 1996," January 30, 1997. Available online at http://www.state.gov/www/global/human_rights/1996_hrp_report/turkey.html.

11. Eric Rouleau, "Turkey: Beyond Atatürk," *Foreign Policy* 103 (Summer 1996): 76.

12. Ibid. The State Department report (note 10, above) gave varying estimates on displacement, from 330,000 to two million, with 560,000 as a "credible estimate."

13. Alan Cowell, "War on Kurds Hurts Turks in U.S. Eyes," *New York Times,* November 17, 1994, p. A3; Cowell, "Muslim Party Threatens Turks' Secular Heritage," *New York Times,* November 30, 1994, p. A5. Stephen Kinzer, "Italy Locked in Stalemate with Turks over Kurd," *New York Times,* December 13, 1998.

14. Rouleau, "Turkey: Beyond Atatürk," p. 74; Stephen Kinzer, "Turkish Military Raises Pressure on Leader of Islamic Government," *New York Times,* June 13, 1997, quoting Gen. Fevzi Turkeri at a rare briefing for prosecutors, professors, journalists, and civic leaders to emphasize the dangers seen in the rise of political Islam.

15. Stephen Kinzer, "Turks' High Court Orders Disbanding of Islamic Party," *New York Times,* January 17, 1998, p. 1.

16. Rouleau, "Turkey: Beyond Atatürk," pp. 80–82.

17. Istanbul has doubled in size every 15 years since the 1950s and by 1994 hovered at 10 million, most of them newly arrived poor. Stephen Kinzer, "In Turkey, A Zealous Pragmatist," *New York Times,* June 30, 1996, p. 6; Cowell, "Muslim Party Threatens Turks' Secular Heritage."

18. Stephen Kinzer, "Mass Grave in Turkey Put Attention on Terror Cell," *New York Times,* January 23, 2000, p. 7; and Kinzer, "Turkey Accused of Arming Terrorist Group," *New York Times,* February 15, 2000, p. A13. The human rights organization Human Rights Watch notes evidence of government links to Hizbullah assembled by the Turkish par-

liament's 1993 Commission on Unsolved Murders. "What is Turkey's Hizbullah? A Human Rights Watch Backgrounder," February 16, 2000. On the Internet: http://www.hrw.org/press/2000/02/tur0216.html. Accessed March 1, 2000.

19. Ibid., pp. 10–12.

20. John Pomfret, "Iraq-Turkey Fuel Smugglers Back in Business," *New York Times,* April 7, 1995, p. A1. "The Kurds: An Ancient Tragedy," *The Economist,* February 20, 1999, p. 52.

21. Stephen Kinzer, "Turkey's Ties to Syria Sink to War in All But the Name," *New York Times,* October 4, 1999, p. 17; Lale Sariibrahimoglu, "Turkey Warns Syria as Exercises Are Set to Go Ahead," *Jane's Defence Weekly,* October 21, 1998, p. 4; Stephen Kinzer, "Accord Set for Syria and Turkey," *New York Times,* October 22, 1999, p. A13.

22. Candice Hughes, "Italy Releases Kurd as Turkey Protests," *Washington Post,* November 22, 1998, p. A37. Tim Weinger, "U.S. Helped Turkey Find and Capture Kurd Rebel," *New York Times,* February 20, 1999, p. 1.

23. Stephen Kinzer, "Kurdish Rebels Tell Turkey They Are Ending Their War," *New York Times,* February 10, 2000, p. A12; Kinzer, "Crackdown on Rebels Renews Fears of War and Terror in Turkey's Kurdish Region," ibid., February 24, 2000, p. A6.

24. Mer Karasapan, "Turkey's Armaments Industries," *Middle East Report* (January-February 1987); Gülay Günlük-Senesen, "An overview of the arms industry modernization programme in Turkey," *SIPRI Yearbook* 1993 (Oxford: Oxford University Press for the Stockholm International Peace Research Institute, 1993), Appendix 10E.

25. Ibid.; Umit Enginsoy, "Turkey Pumps $330 Million into Arms Procurement Fund," *Defense News,* November 28-December 4, 1994, p. 16.

26. Umit Enginsoy, "Turkey Acts to Boost Industry," *Defense News,* March 13–18, 1995, p. 3.

27. Umit Enginsoy, "Defense Industry Boosts Emphasis on Exports," *Defense News,* July 25–31, 1994, p. 10.

28. Sezer, *Turkey's Political and Security Interests and Policies,* pp. 31–34.

29. Umit Enginsoy, "Turkish Welfare Leader Criticizes Arms Deals," *Defense News,* January 8–14, 1996, p. 3; and "Erbakan Pushes Turk Defense Support," *Defense News,* September 2–8, 1996, p. 4.

30. Umit Enginsoy, "Ciller Approves $5 billion for Delayed Arms," *Defense News,* October 21–27, 1996, p. 1.

31. *SIPRI Yearbook 1995* (Oxford: Oxford University Press, 1995), pp. 548–50.

32. Umit Enginsoy, "Turkey Nears Deal to Make Israeli Popeye Missile," *Defense News,* March 24–30, 1997, p. 56. Steve Rodan, "Israel, Turkey to Conclude Cooperation Agreement," *Defense News,* March 9–15, 1998, p. 28.

33. "Turkey Urges U.S. to Lift 'Embargo,'" *The Washington Report on Middle East Affairs,* August-September 1997, p. 39.

34. U.S. Department of Defense, "Memorandum for Correspondents," Memorandum No. 267-M, December 4, 1995 (Turkey ATACMS sale); and ibid., Memorandum No. 159-M, July 12, 1996 (Greece ATACMS sale). Online at http://www.defenselink.mil/news/Dec1995/ml120495_m267–95.html and . . . /news/Jul1996/m071296_ml159–96.html.

35. *Defense News,* November 11, 1996, p. 27. "Excess defense articles" like the Perry-class frigates look cheap but entail substantial support costs. The ships themselves will cost Turkey just $7 million apiece, but technical support, shore spares, crew training, and ammunition costs will boost the total to $150 million per ship. *Defense News,* August 14, 1995, p. 10.

36. Joseph Ward Swain and William H. Armstrong, *Peoples of the Ancient World* (New York: Harper & Row, 1959), pp. 83–87, 204–17, 291–92. Sandra Mackey, *The Iranians: Persia, Islam, and the Soul of a Nation* (New York: Dutton Signet, 1996), p. 57.

37. Mackey, *The Iranians,* pp. 67–69, 71, 79–81.

38. Ibid., pp. 102, 114–15.

39. Ibid., 109–11. Also Jamal S. al-Suwaidi, "The Gulf Security Dilemma," in Jamal S. al-Suwaidi, ed., *Iran and the Gulf: A Search for Stability* (Abu Dhabi: The Emirates Center for Strategic Studies and Research, 1996), p. 331.

40. Mackey, *The Iranians,* pp. 94–95.

41. Ibid., pp. 145–55.

42. Ibid., pp. 167–78.

43. Ibid., pp. 195–207. See also James A. Bill, *The Eagle and the Lion: The Tragedy of American-Iranian Relations* (New Haven: Yale University Press, 1988), pp. 62–71, 90–94.

44. Bill, *The Eagle and the Lion,* pp. 156–61.

45. Ibid., pp. 186–192, 200.

46. Ibid., pp. 219–25, 236, 295.

47. Mackey, *The Iranians,* p. 316, and Behrouz Souresrafil, *The Iran-Iraq War* (Plainview, NY: Guinan Lithographic Co., 1989), p. 23.

48. Mackey, *The Iranians,* pp. 318, 322.

49. Ahmed Salah Hashim, *Iranian National Security Policies under the Islamic Republic: New Defense Thinking and Growing Military Capabilities,* Occasional Paper No. 20 (Washington, D.C.: The Henry L. Stimson Center, July 1994), pp. 7–9. This paper was commissioned in support of this study, with the support of the Ploughshares Fund. Brian D. Smith, "United Nations Iran-Iraq Military Observer Group," in William J. Durch, ed., *The Evolution of UN Peacekeeping: Case Studies and Comparative Analysis* (New York: St. Martin's Press, 1993), pp. 239–41.

50. Mackey, *The Iranians,* pp. 349, 355.

51. Ibid., p. 391.

52. Ibid., p. 360; Mohsen M. Milani, "Iran's Gulf Policy: From Idealism and Confrontation to Pragmatism and Moderation," in al-Suwaidi, *Iran and the Gulf,* p. 92.

53. Clerical foundations, or *bonyads,* originally founded to take over the assets of the Pahlavis and to look after those injured fighting for the revolution or in the Iran-Iraq War, soon became the largest manufacturers, traders, and real estate developers in the country, but still receive large subsidies from the government. Ahmed S. Hashim, *The Crisis of the Iranian State,* Adelphi Paper No. 296 (Oxford: Oxford University Press, July 1995), pp. 5, 7, and Mackey, *The Iranians,* p. 370. Barbara Crossette, "Iran Drops Rushdie Death Threat, and Britain Renews Teheran Ties," *New York Times,* September 25, 1998, p. A1.

54. U.S. Department of State, Office of the Coordinator for Counterterrorism, *Patterns of Global Terrorism,* annual reports for 1997 and 1998, released April 1998 and April 1999 respectively. Internet: http://www.state.gov/www/global/terrorism/1997Report/1997index.html and . . . /1998Report/1998index.html. Also Philip Shenon, "State Dept. Drops Iran as Terrorist Leader," *New York Times,* May 1, 1999, p. A4.

55. Shahram Chubin and Charles Tripp, *Iran-Saudi Arabia Relations and Regional Order,* Adelphi Paper No. 304 (Oxford: Oxford University Press, November 1996), p. 67. Hashim, *The Crisis of the Iranian State,* p. 15. Such troubles did not, however, keep politically connected mullahs from becoming rich at the public's expense.

56. James L. Williams, "Oil Price History and Analysis," *WTRG Economics,* London, Arkansas, 1999. Internet: http://wtrg.com/prices.htm. Accessed June 6, 1999. Crude oil prices in 1998 dropped below the median price level for the entire postwar period, in constant dollars.

57. "The Changing Face of Iranian Politics," Gulf States Newsletter No. 536, Centre for Arab Gulf Studies, University of Exeter; United Kingdom (posted online at http://www.ex.ac.uk/ags/536.htm). Also Richard H. Curtiss, "Khatami's Election May Be a Turning Point," *The Washington Report on Middle East Affairs* (August/September 1997), pp. 10–12.

58. Agence France-Presse, "Iran: Khatami Speaks Out on Reforms," *New York Times,* April 30, 1999, p. A9. George Perkovich, "In Iran, Whispers of Moderation," *Washington Post,* November 30, 1997, p. C1. Douglas Jehl, "Uphill Battle in Iran: Trying to Keep '79 Revolution Alive," *New York Times,* February 5, 1999, p. A1.

59. After Taliban forces killed eight Iranian diplomats and a journalist who were in their custody in August 1998, Iran mobilized 270,000 troops across the border from 10,000 dug-in Taliban militants, demanding freedom for several dozen other Iranians also detained by the Taliban. The latter threatened to shell Iranian border towns with Scud missiles if Iranian troops or aircraft attacked. After exchanges of mortar and artillery fire, and a week of shuttle diplomacy by the UN special envoy to Afghanistan, Lakhdar Brahimi, the Iranians were released. Kathy Gannon, "Iran, Taliban in Dangerous Face-off," *Associated Press Online,* September 19, 1998, 16:26 eastern time. Accessed via Lexis-Nexis Universe October 3, 1998. Douglas Jehl, "Iran's Forces Said to Clash With Afghans Along Border," *New York Times,* October 9, 1998, p. A5; id., "Iran: Taliban Dispute Over, Envoy Says," *New York Times,* October 20, 1998, p. A8.

60. Chubin and Tripp, *Iran-Saudi Arabia Relations,* p. 71.

61. Allessandra Stanley, "Iran's Leader and Pope Seek Better Muslim-Christian Ties," *New York Times,* March 12, 1999, p. A11. Jane Perlez and Steve LeVine, "U.S. Oilmen Chafing at Curbs on Iran," *New York Times,* August 8, 1998, p. 8. U.S. Congress, "Iran-Libya Sanctions Act of 1996," Public Law 104–172 (August 5, 1996), summary Available via THOMAS, the Library of Congress online legislative history service (http://www.thomas.loc.gov/). Accessed June 6, 1999. The U.S. legislation was crafted in the Republican-led Congress to pressure Tehran into ending programs to develop weapons of mass destruction and ballistic missiles and to cease support of international terrorism.

62. "Saudi Arabia, Iran, and Iraq," *Iraq News* (online), May 25, 1999, citing Reuters dispatches. Also Reuters, "Iran's Leader Meets With Syria's, Hoping to Improve Ties With Arab Nations," *New York Times,* May 15, 1999, p. A5. Douglas Jehl, "On Trip to Mend Ties, Iran's President Meets Saudi Prince," *New York Times,* May 17, 1999, p. A3.

63. The UN Special Commission on Iraq (UNSCOM) was established under UN Security Council Resolution 687 in April 1991 to oversee the destruction of Iraq's weapons of mass destruction (WMD) programs and facilities, and all ballistic missiles of greater than battlefield range. Section C of Res. 687, "decides that Iraq shall unconditionally accept, under international supervision, the destruction, removal or rendering harmless of its weapons of mass destruction, ballistic missiles with a range over 150 kilometers, and related production facilities and equipment. It also provides for establishment of a system of ongoing monitoring and verification of Iraq's compliance with the ban on these weapons and missiles." (Internet: http://www.un.org/Depts/unscom/unscom. htm) The International Atomic Energy Agency (IAEA), also an arm of the UN, received a similar mandate to root out Iraqi nuclear weapons programs. (Internet: http://www.iaea.org.) Iraq was to remain under international economic sanctions until these two bodies could certify that their mandates had been accomplished. Iraqi efforts to bar inspections of certain facilities in late 1997 led to confrontation with the United States and Britain in February and again in December 1998. Facing non-cooperation from Iraq and a threat of allied air strikes, UNSCOM inspectors withdrew and U.S. and British forces launched Operation Desert Fox. Subsequent revelations that U.S. intelligence agencies had used UNSCOM as a vehicle for planting surveillance devices in Iraqi facilities virtually assured that the organization would not be able to resume its monitoring tasks. Iraq ceased full cooperation with IAEA in August 1998. While the IAEA could reinspect old facilities, it was prevented from conducting inspections at new or suspected sites. See IAEA's website, Internet: http://www.iaea.org/worldatom/

inforesource/pressrelease/prn2498.html. See also Monterey Institute for International Studies, Center for Nonproliferation Studies, Internet: http://cns.miis.edu/research/ira, and International Institute for Strategic Studies [IISS], *Strategic Survey, 1998–99* (Oxford: Oxford University Press, 1999), pp. 169–75.

64. Hashim, *Iranian National Security Policies,* p. 26.

65. IISS, *The Military Balance, 1998–99* (Oxford: Oxford University Press, 1996), pp. 127, 130, 140.

66. Anthony Cordesman, "Threats and Non-Threats from Iran," in al-Suwaidi, *Iran and the Gulf,* p. 239.

67. Michael Eisenstadt, *Iranian Military Power: Capabilities and Intentions,* Policy Papers No. 42 (Washington, D.C.: The Washington Institute for Near East Policy, 1996), pp. 40–41, 43–44; Hashim, *Iranian National Security Policies,* p. 27; and Kenneth Katzman, "The Politico-Military Threat from Iran," in al-Suwaidi, *Iran and the Gulf,* p. 202.

68. Hashim, *Crisis,* p. 69; Eisenstadt, *Iranian Military Power,* p. 46, 46n; and Cordesman, "Threats and Non-Threats," p. 246.

69. Hashim, *Crisis,* p. 53; Cordesman, "Threats and Non-Threats," p. 225.

70. Eisenstadt, *Iranian Military Power,* pp. 51–54.

71. Dan Caldwell, "Flashpoints in the Gulf: Abu Musa and the Tunb Islands," *Middle East Policy,* 4, no. 3 (March 1996), pp. 50–57.

72. In early 1997, deputy ground forces commander Brigadier General Karim Ebadat, in an interview with the Iranian National News Agency, claimed that "the Iranian army is distinguished for its independence from foreign countries in terms of armaments and equipment." (From an Internet posting by the Neda Rayaneh Institute at http://www.neda.net/iran-wpd/vol1006/i1006–05.htm.)

73. Hashim, *Iranian National Security Policies,* p. 19.

74. Cordesman, "Threats and Non-Threats," p. 231.

75. Steven Erlanger, "Washington Casts Wary Eye at Missile Test," *New York Times,* July 24, 1998, p. A6; "U.S. Says Chinese Will Stop Sending Missiles to Iran," *New York Times,* October 18, 1997, p. A1; and "U.S. Gets Russia's Firm Vow to Halt Missile Aid to Iran," *New York Times,* January 16, 1998, p. A8.

76. Hashim, *Iranian National Security Policies,* p. 32.

77. United States, Director of Central Intelligence, *The Acquisition of Technology Relating to Weapons of Mass Destruction and Advanced Conventional Munitions, July–December 1996,* submitted to the U.S. Congress in response to section 721 of the Fiscal Year 1997 Intelligence Authorization Act, June 1997, p. 4.

78. Judith Miller and William J. Broad, "Bio-Weapons in Mind, Iranians Lure Needy Ex-Soviet Scientists," *New York Times,* December 8, 1998, p. A1.

79. U.S. Department of Defense, *Proliferation, Threat and Response* (Washington, D.C.: November 1997), pp. 3–4. Internet: http://www.defenselink.mil/pubs/prolif97/meafrica.html. Accessed December 23, 1998.

80. Stuart D. Goldman, Kenneth Katzman, and Zachary S. Davis, *Russian Nuclear Reactor and Conventional Arms Transfers to Iran,* CRS Report 95–641 F (Washington, D.C.: Congressional Research Service, May 23, 1995), p. 9.

81. Eisenstadt, *Iranian Military Power,* pp. 10–11, and Hashim, *Crisis,* pp. 60–65.

82. Mackey, *The Iranians,* p. 389.

83. James A. Bill, "The Geometry of Instability in the Gulf: The Rectangle of Tension," in al-Suwaidi, *Iran and the Gulf,* p. 116.

84. Milani, "Iran's Gulf Policy," p. 96.

85. Jamal al-Suwaidi, "The Gulf Security Dilemma," in al-Suwaidi (ed.), *Iran and the Gulf,* pp. 346–47. Also Alon Ben-Meir, "The Dual Containment Strategy is No Longer Viable," *Middle East Politics* 4, no. 3 (March 1996), pp. 58–71, and Joshua Teitelbaum, "The Gulf States and the End of Dual Containment," *Middle East Review of International Af-*

fairs (MERIA) Journal, 2, no. 3 (September 1998). Electronic journal published by the Begin-Sadat Center for Strategic Studies, Bar-Ilan University, Israel. Internet: http://www.biu.ac.il/SOC/besa/meria.html. Accessed September 10, 1998.

86. John F. Burns, "India's Five Decades of Progress and Pain," *New York Times,* August 14, 1997, p. A11.

87. A. G. Noorani, "The tortured and the damned: Human Rights in Kashmir," *Frontline,* January 28, 1994, cited in Noorani, *Easing the Indo-Pakistani Dialogue on Kashmir,* Occasional Paper no. 16 (Washington, D.C.: Henry L. Stimson Center, April 1994), p. 2.

88. One or two border crises have been interpreted as raising the risk of war. For discussion, see Michael Krepon and Mishi Faruqee, eds., *Conflict Prevention and Confidence-Building Measures in South Asia: The 1990 Crisis,* Occasional Paper no. 17 (Washington, D.C.: Henry L. Stimson Center, April 1994), executive summary. Barry Bearak, "Frozen Fury on the Roof of the World," *New York Times,* May 23, 1999, p. A1.

89. Noorani, "The tortured and the damned: Human Rights in Kashmir," p. 4. John Pike, "Border Security Forces," Federation of American Scientists Intelligence Resource Program, May 1998. Internet: http://www.fas.org/irp/world/india/home/bsf.htm. Accessed May 12, 1999. Pamela Constable, "Kashmir Duel Stirs Fears of an Expanded Conflict," *Washington Post,* May 28, 1999, p. A1. Barry Bearak, "2 Indian Warplanes Lost Over Pakistan's Part of Kashmir," *New York Times,* May 28, 1999, p. A3; Bearak, "India Jets Strike Guerrilla Force Now in Kashmir," *New York Times,* May 27, 1998, p. A1. Stephen Kinzer, "India and Pakistan to Discuss Flare-Up in Kashmir," *New York Times,* June 9, 1999, p. A6. Celia W. Dugger, "Kashmir War, Shown on TV, Rallies India's Unity," *New York Times,* July 18, 1999, p. 4. Barry Bearak, "Kashmir a Crushed Jewel Caught in a Vise of Hatred," *New York Times,* August 12, 1999, p. A1.

90. For three years, Berlin-based Transparency International (TI) has been compiling surveys of corruption worldwide, as perceived by the business community. It combines data from six or more surveys undertaken by other polling and risk analysis organizations to create a composite corruption index. Its 1997 index lists 52 countries, rated on a scale from 0 (highly corrupt) to 10 (clean). Denmark and Finland top the list with index scores of 9.5 or better; while Bolivia and Nigeria, at the bottom, score 2.0 or less. Pakistan scores little better than bottom-ranked, at 2.5, while India receives a 2.75. TI stresses the subjectivity of these ratings, but because they reflect the perceptions of business people and the risk analysts who contribute to their investment decisions, the ratings are a barometer of business confidence and, potentially, of profit margin. TI cites a Harvard study's conclusions that "the difference in corruption levels from that of Singapore [8.7] to that of Mexico [2.7] is equivalent to raising the marginal tax rate by over twenty percent. A one percentage point increase in the marginal tax rate [tends to reduce] inward foreign direct investment by about five percent." ("Transparency International publishes 1997 Corruption Perception Index," TI Press Release, Berlin, Germany, July 31, 1997, and "Index 1997," both available on the Internet at http://www.gwdg.de/~uwvw.) In other words, in a competitive world, corruption is to investment what bad breath is to romance.

91. Chris Smith, "Conventional Forces and Regional Stability," in C. Smith, S. Gupta, et al., *Defense and Insecurity in Southern Asia: The Conventional and Nuclear Dimensions,* Occasional Paper no. 21 (Washington, D.C.: Henry L. Stimson Center, May 1995), p. 18. This paper was commissioned in support of this study, with the support of the John Merck Fund.

92. Bharata Janata won a plurality of the seats in the lower house of parliament in the 1996 general elections (195 of 543). The resulting BJP government lost a vote of confidence after two weeks, however, relinquishing power to a coalition government headed by the Janata Dal party, with Congress, the long-time power in Indian politics, as a partner. (Congress lost its majority in parliament in 1996, returning with nearly 100 fewer seats than in

1991, and lost still more seats in 1999.) Alexander Nicoll, "50 Year Diary," *The Financial Times,* August 1, 1997, special section: "India: 50 Years of Independence." Available on the Internet at http://www.ft.com. The second BJP coalition lasted longer and lost another vote of confidence more narrowly. Celia W. Dugger, "Coalition Falls in India, Defeated By Just One Vote," *New York Times,* April 18, 1999, p. 1. Dugger, "Hindu-First Party Wins Solid Victory in India's Election," *New York Times,* October 8, 1999, p. A1.

93. Altogether, the military ruled Pakistan for 24 of its first 50 years, and ruled through closely controlled civilian governments for another four years. John F. Burns, "Pakistan's Bitter Roots, and Modest Hopes," *New York Times,* August 15, 1997, p. A12.

94. Tim Weiner, "Pakistan Military Completes Seizure of All Authority," *New York Times,* October 16, 1999, p. A1.

95. Smith, "Conventional Forces and Regional Stability," p. 25.

96. "Pakistan Takes a Beating," *The Economist,* August 22, 1998, pp. 29–30. Richard W. Stevenson, "IMF Agrees to Resume Pakistan Aid, Cut Off After Atom Tests," *New York Times,* November 26, 1998, p. A17.

97. The result has been gender-selective abortion in India, "deliberate under feeding of both Indian and Pakistani infant girls," and other measures from "killings of Indian brides . . . [for] insufficient dowry" to "fatal stonings of suspected adulterers in Pakistan." Such things combine with pregnancy- and birth-related injury and disease to increase overall female mortality rates. Kenneth J. Cooper, "India and Pakistan, Bound by Pasts," *Washington Post,* August 14, 1997, p A1. See also Siddharth Dube, "Health: Disease, Malnutrition take a heavy toll," *Financial Times,* August 1, 1997; and Mahbub ul Haq, "The Poverty Puzzle," *Financial Times,* August 14, 1997. Both articles are available on the Internet at http://www.ft.com, special section: "India: 50 Years of Independence." Ul Haq singles out illiteracy and failure to reform landholding as the two biggest structural factors holding Pakistan back.

98. Shekhar Gupta, "Nuclear Weapons in the Subcontinent," in C. Smith, S. Gupta, et al., *Defense and Insecurity in Southern Asia,"* pp. 33–34, 41–42.

99. Gupta, "Nuclear Weapons in the Subcontinent," p. 34.

100. See, for example, Ved Marwah, *Uncivil Wars: Pathology of Terrorism in India* (New Delhi: Harper Collins India, 1995).

101. See also Chris Smith, *India's Ad Hoc Arsenal* (Oxford: Oxford University Press for the Stockholm International Peace Research Institute, 1994), chapter 3. Rahul Bedi, "Indian forces to be restructured," *Jane's Defence Weekly,* October 21, 1998, p. 20; Bedi, "Indian service chiefs fall out over supremo," *JDW,* February 3, 1999, p. 15.

102. See U.S. Congress, Office of Technology Assessment, "The Defense Industry of India," in *Global Arms Trade,* OTA-ISC-460 (Washington, D.C.: U.S. Government Printing Office, June 1991), chapter 10.

103. Pravin Sawhney, "Arjun MBT Still in Technical and Fiscal Mire," *Jane's International Defense Review,* 11/1996, p. 15. Rahul Bedi, "Mixed fortunes for India's defense industrial revolution," *Jane's International Defense Review,* 5/1999, pp. 22–30.

104. Smith, "Conventional Forces and Regional Stability," p. 20. Also Smith, *India's Ad Hoc Arsenal,* ch. 7.

105. Associated Press, "U.S. to Pay Pakistan Back for Undelivered Jets," *New York Times,* December 22, 1998, p. A 15.

106. "Pakistan Takes a Beating," *The Economist,* August 22, 1998, p. 30. Celia W. Dugger, "Pakistani Premier Prevails in Clash With General," *New York Times,* October 20, 1998. Ahmed Rashid, "Raise the Crescent," *Far Eastern Economic Review,* December 3, 1998, pp. 20–22.

107. Smith, "Conventional Forces and Regional Stability," pp. 25–26.

108. John Kifner, "Pakistan Sets off Atom Test Again, But Urges 'Peace'," *New York Times,* May 31, 1998, pp. 1, 8; Kifner, "Complex Pressures, Dominated by Islam, Led to Testing," *New York Times,* June 1, 1998, p. A6.

109. Gupta, "Nuclear Weapons in the Subcontinent" pp. 37–38,

110. Rahul Bedi, "India Confirms Nuclear Policy," *Jane's Defence Weekly,* December 23, 1998, p. 7.

111. Ben Sheppard, "Regional rivalries are replayed as India and Pakistan renew ballistic missile tests," *Jane's International Defence Review,* 5/1999, pp. 57–59. Following the Ghauri I missile test in April 1998, Pakistan's president said that new missiles would be named after Muslim rulers and conquerors of the subcontinent. (*Hong Kong Standard,* April 20, 1998. Internet: http://202.86.1.1/clips/news/200498/asia/news003.htm.)

112. For details on this period in Chinese history, see, for example, Henry McAleavy, *The Modern History of China* (New York: Praeger Publishers, 1967), James Pinckney Harrison, *The Long March to Power* (New York: Praeger Publishers, 1972), and Barbara W. Tuchman, *Stilwell and the American Experience in China, 1911 45* (New York: Bantam Books, 1971).

113. IISS, *Strategic Survey, 1998–99,* p. 182.

114. Susan V. Lawrence, "Out of Business: Jiang orders a halt to PLA, Inc.," *Far Eastern Economic Review,* August 6, 1998, p. 68. Also, David Phinney, "Special Report: China's Capitalist Army, Part One: People's Liberation Army, Inc.," ABCNEWS.com (1998). Internet: http://204.202.136.230/sections/world/hongkong97/pla_china1.html. Accessed May 25, 1999.

115. Bruce Gilley, "Stand-down Order," *Far Eastern Economic Review,* September 10, 1998, pp. 18–21. Michael Lanz, "China's Communist Party Orders Itself to Get Out of Business," *Washington Post,* November 30, 1998, p. A26.

116. U.S. Department of Defense [U.S. DoD], Office of the Secretary of Defense, *The Security Situation in the Taiwan Strait,* a report to the Congress pursuant to the FY99 Appropriations Bill. Washington, D.C., February 1, 1999, p. 4. "Active defense," notes China military analyst Kenneth Allen, "involves three phases: strategic defense (blunting an attack), strategic stalemate (stabilizing the battlefield for a shift from defense to offense), and strategic counterattack." K. Allen, "PLAAF Modernization: An Assessment," in *Crisis in the Taiwan Strait,* ed. James R. Lilley and Chuck Downs (Washington, D.C.: National Defense University Press, 1997), p. 222.

117. Australian National University analyst Paul Dibb notes that, in addition to high-technology weapon systems, armed forces looking to benefit from the "revolution in military affairs" (RMA) need very capable intelligence, surveillance and reconnaissance capabilities; computer-augmented command, control, and communications systems; real-time integration of information generated by those systems in a form useful to the war fighter; and joint-force organizations, doctrines, strategies, and tactics. He notes that the last component is "very little practiced in Asia," and that a military hoping to exploit the RMA "depends critically on a new approach to maintenance and the support in combat of weapon systems capable of remaining operational 24 hours a day in all types of weather. Very few Asian countries seem to acknowledge the vital nature of [such] integrated logistic support." Paul Dibb, "The Revolution in Military Affairs and Asian Security," *Survival,* Winter 1997–98, p. 94. Allen, "PLAAF Modernization," notes that air force operational readiness is low due to old, unreliable equipment, a disinclination to preventive maintenance, and a lack of standardized parts ("virtually all the holes in China's aircraft structural components are hand drilled without use of a template, so that the pieces are not interchangeable"). The last point makes it extremely difficult for the air force to operate and sustain aircraft outside their home bases. Training style and intensity also pose problems for implementing an RMA in the PLAAF. Chinese pilots fly only half as many hours per year as their American counterparts, only 15 to 20 percent of those hours can be considered combat training, and much of that is not very challenging (pp. 224–27, 231–33). Unwillingness to risk newer, more complex equipment in realistic training may make it even more difficult to exploit the expanded performance envelope that such equipment represents.

118. Efficient, powerful, and reliable engines seem to be a significant pacing factor. China plans to import Russian Klimov RD-33 aircraft engines, the ones that power the MiG-29, but the same engines have worn out far faster than expected in Indian Air Force service. (See *Jane's Defence Weekly,* February 19, 1994, p. 28; *Flight International,* May 19–25, 1993, p. 15, and July 7–13, 1993, p. 13.)

119. Allen, "PLAAF Modernization," p. 240. IISS, *The Military Balance, 1998–99,* p. 169.

120. IISS, *Strategic Survey, 1997–98* (Oxford: Oxford University Press, 1998), pp. 182–83. "Japan's Constitution: The call to arms," *The Economist,* February 27, 1999, p. 24. Both houses of Japan's legislature ratified the new security cooperation guidelines. Nicholas D. Kristoff, "Japan: Military Pact Approved," *New York Times,* May 25, 1999, p. A10; and Kristoff, "Tokyo Lawmakers Pass Bill to Improve Military Ties with U.S.," *New York Times,* April 28, 1999, p. A5.

121. Two-way trade between China and Taiwan totaled roughly $24 billion in 1996, up from just $1 billion in 1987, with Taiwan running a $17 billion trade surplus. Mark Daly, "Democracy is Taiwan's best shield against China's threat," *Jane's International Defence Review* 4/1999, p. 25.

122. IISS, *Strategic Survey, 1996–97* (Oxford: Oxford University Press, 1997), p. 167. China announced the "closure" of major sea lanes around Taiwan and targeted ballistic missile tests uncomfortably close to the major ports of Taipei and Kaohsiung, at the northern and southern ends of the island. The United States sent two aircraft carrier battle groups to the area in response, sailing one through the Taiwan Strait just to make it clear that Washington did not accept the declared "closures." While the entire exercise may have been largely symbolic in nature—China wasn't about to invade and the United States was unlikely to have to engage Chinese forces—it had a bracing effect on Taiwan and also demonstrated clearly just how much ground China's forces need to gain if they ever expect to be effective against a modern military.

123. Julian Baum, "One-Track Mind, Taipei insists on democracy before unification," *Far Eastern Economic Review,* October 29, 1998.

124. John Pomfret, "Chinese Military Backs Beijing's Latest Warning to Taiwan," *Washington Post,* February 24, 2000, p. A17.

125. The new frigates all have greater full-load displacement than the older destroyers they will be replacing; the change in nomenclature does not mean a decrease in capability. IISS, *The Military Balance, 1998–99,* pp. 171–72, 198).

126. *Aviation Week and Space Technology,* July 4, 1994, p. 20. IISS, *The Military Balance, 1998–99,* p. 172. U.S. DoD, *The Security Situation in the Taiwan Strait,* pp. 8–9.

127. Shannon Selin, *Asia Pacific Arms Buildups, Part One: Scope, Causes, and Problems,* Working Paper No. 6 (Vancouver, Canada: University of British Columbia, Inst. of International Relations, November 1994), p. 32. Barbara Opall-Rome, "U.S. to Deny Taipei Subs, Missiles," *Defense News,* April 20–28, 1998, p. 3.

128. China's capability is theoretical because Chinese nuclear submarines may be noisy enough to be vulnerable to antisubmarine warfare measures. Moreover, all Chinese submarines may lack timely intelligence or good enough communications to allow them to intercept convoys on the open ocean.

129. U.S. DoD, *The Security Situation in the Taiwan Strait,* pp. 5–7.

130. Eric Eckholm and Steven Lee Myers, "Taiwan Asks U.S. to Let It Obtain Top-Flight Arms," *New York Times,* March 1, 2000, p. A1.

131. U.S. DoD, *Report to Congress on Theater Missile Defense Architecture Options for the Asia-Pacific Region,* Washington, D.C., May 1999. Made available on the Internet by New York University's Center for War, Peace, and the News Media and its Global Beat website. Internet: http://www.nyu.edu/globalbeat/asia/DOD0599.html. Accessed May 27, 1999.

132. Ibid., p. 3.

133. South Korea, National Statistical Office, "Growth Rates of GDP (at 1995 prices)." Internet: http://www.nso.go.kr/graph/Month/0–1.htm. Accessed June 2, 1999; and "South Korea: Making a Comeback," *The Economist*, February 20, 1999, pp. 38–39.

134. Robert Karniol, "South Korea Postpones programmes amid crisis," *Jane's Defence Weekly*, January 21, 1998, p. 14.

135. Selin, *Asia Pacific Arms Buildups*, pp. 24, 30. IISS, *The Military Balance, 1998–99*, p. 187.

136. Susan Willett, "East Asia's Changing Defence Industry," *Survival*, Autumn 1997, p. 110.

137. David Reese, *The Prospects for North Korea's Survival*, Adelphi Paper No. 323 (London: IISS, November 1998), pp. 64–65.

138. Richard Halloran, " . . . But Carry a Big Stick," *Far Eastern Economic Review*, December 3, 1998, pp. 26–27.

139. Reese, *Prospects*, pp. 50–51.

140. Donald H. Rumsfeld, et al., *Executive Summary of the Report of the Commission to Assess the Ballistic Missile Threat to the United States*, Washington, D.C., July 15, 1998. Pursuant to Public Law 104–201. Internet: http://www.stimson.org/policy/excbmd.htm.

141. David E. Sanger, "N. Korea Consents to U.S. Inspection of a Suspect Site," *New York Times*, March 17, 1999, p. A1; Philip Shenon, "Suspected North Korean Atom Site Is Empty, U.S. Finds," *New York Times*, May 28, 1999, p. A3.

142. See Barbara Crossette, "Hunger in North Korea: A Relief Aide's Stark Report," *New York Times*, June 11, 1997, p. A5; and Crossette, "Relief Teams Say North Korea Faces Vast Drought Emergency," ibid., August 5, 1997, p. A1. Drought had destroyed 70 percent of the corn harvest and roughly 40 percent of the total harvest.

143. Reese, *Prospects*, pp. 29–32.

144. Ibid., p. 60.

145. For a systematic treatment of the issues involved in the South China Sea disputes, see Mark J. Valencia, *China and the South China Sea Disputes*, Adelphi Paper No. 298 (London: International Institute for Strategic Studies, October 1995).

146. Among the voluminous literature on the Vietnam War, two comprehensive and readable accounts are Stanley Karnow, *Vietnam: A History* (New York: Penguin Books, 1983) and George C. Herring, *America's Longest War: The United States and Vietnam, 1950–1975*, 2nd ed. (New York: Alfred A. Knopf, 1986).

147. James A. Schear, "Riding the Tiger: The United Nations and Cambodia's Struggle for Peace," in *UN Peacekeeping, American Policy, and the Uncivil Wars of the 1990s*, ed. William J. Durch (New York: St. Martin's Press, 1996), pp. 135–92. Janet E. Heininger, *Peacekeeping in Transition: The UN in Cambodia* (New York: Twentieth Century Fund Press, 1994). Michael W. Doyle, Ian Johnstone, and Robert C. Orr, eds., *Keeping the Peace: Multinational UN Operations in Cambodia and El Salvador* (New York: Cambridge University Press, 1997).

148. "Hesitant to Reform, Vietnam Warned It May Be Left Behind," *Global Intelligence Update*, March 9, 1999. Internet: http://www.stratfor.com/servies/giu/030999.asp. Accessed May 27, 1999. Cites a March 4 press conference by the Asian Development Bank's chief economist for Vietnam, Allessandro Pio. Seth Mydans, "Vietnam Awash in Graft Trials, but They Don't Clean Up Graft," *New York Times*, May 25, 1999, p. A5.

149. IISS, *Strategic Survey, 1988–89* (Oxford: Brassey's for the IISS, 1989), p. 137; *Strategic Survey, 1989–90*, p. 129; and *Strategic Survey, 1990–91*, pp. 271–72.

150. P. S. Suryanarayana, "Thailand Plays Host to Myanmar's Junta Chief," *The Hindu*, March 9, 1999, p. 14.

151. U.S. Department of State, Bureau for International Narcotics and Law Enforcement Affairs, *International Narcotics Control Strategy Report, 1998: Burma* (Washington, D.C.: February 1999), pp. 1,2, 4. Internet: http://www.state.gov/www/global/narcotics_law/1998_narc_report/major/Burma.html. Accessed June 3, 1999.

152. Dennis Bernstein and Leslie Kean, "Burma's Dictatorship of Drugs," *The Nation*, December 16, 1996, and William Barnes, "Drugs Keep Burmese Junta Afloat as Traffickers Prosper," *South China Morning Post*, October 22, 1997. Tiffany Danitz, "China Trades Arms to Burma for Heroin," *Insight*, March 24, 1997. All reprinted online at http://www.soros.org/burma/. Accessed November 15, 1997. Rahul Roy-Chaudhury, "Maritime Security in the Indian Ocean Region," *Maritime Studies* (Canberra, Australia), 1996. Internet: http://uniserve.edu.au/law/pub/icl/mStudies/maritimeStudies.html. Accessed June 4, 1999. "China Opening Route Through Myanmar," *Global Intelligence Update*, August 8, 1997. Internet: http://www.stratfor.com/servies/giu/080897.asp. Accessed June 4, 1999.

153. Paul Dibb, David D. Hale, and Peter Prince, "The Strategic Implications of Asia's Economic Crisis," *Survival*, Summer 1998, p.14.

154. For discussion, see Selin, *Asia Pacific Arms Buildups*, pp. 39–48. See also David Harries, "Asia's New Arms Race," *Far Eastern Economic Review*, July 17, 1997, p. 30.

155. Barbara Opall, "Thai F-18s Cost U.S. Navy," *Defense News*, March 16–22, 1998, p. 1. Paul Lewis, "Cash crunch: Arms buying has come to a halt in the troubled South-East Asian Economies," *Flight International*, January 21–27, 1998, pp. 60–62.

156. "The Thai Slump, Deep and Long," *The Economist*, August 8, 1998, pp. 59–60. "1998 Yearend Economic Review," *Bangkok Post* (nd). Internet: http://www.bangkokpost.net/yere/98yere01.html. Accessed June 3, 1999.

157. Steven Radelet and Jeffrey Sachs, "What Have We Learned, So Far, From the Asian Financial Crisis?" Discussion Paper No. 37, Consulting Assistance on Economic Reform II project of the Harvard Institute for International Development, March 1999, p 4. Internet: http://www.hiid.harvard.edu/projects/caer/papers/paper37.html. Accessed June 3, 1999.

158. Martin Feldstein, "Refocusing the IMF," *Foreign Affairs*, March/April 1998, p. 23.

159. Radelet and Sachs, "What Have We Learned," pp. 5–9.

160. Ibid., pp. 9–10. Devesh Kapur, "The IMF: A Cure or a Curse?" *Foreign Policy*, Summer 1998, pp. 123–25. Feldstein, "Refocusing," pp. 30–33.

161. Abdul Razak Abdullah Baginda, "Malaysia's armed forces in the 1990s," *International Defense Review*, 4/1992, pp. 305–08.

162. Barbara Opall, "Malaysia Tempers Defense Spending in New Budget Plan," *Defense News*, November 11–17, 1996, p. 19.

163. Thomas A. Cardamone, ed., "Singapore to Buy Apache Attack Helicopters," *Arms Trade News*, March 1999, p. 4. IISS, *The Military Balance, 1998–99*, p. 174.

164. *Financial Times* (London), February 23, 1995, p. 8.

165. "Philippines Allocates Fund to Start Stalled Military Modernization," *Berdama-DPA*, April 13, 1999. Internet: http://www.skali.com.my/today/gen/199904/13/gen19990413_16.html. Accessed June 4, 1999.

166. John Taylor, "The Economic Challenge for Indonesia," *BBC News Online: Events: Indonesia: Special Report*, June 2, 1999. Internet: http://193.130.149.131/low/english/events/Indonesia/special_report/newsid_358000/ 353422.stm. Accessed June 4, 1999.

167. John McBeth, "Father Knows Best," *Far Eastern Economic Review*, November 25, 1993, p. 25. Anon., "IMF's Latest Aid Plan for Indonesia Calls for Billions More," *New York Times*, June 26, 1998, p. A6. "Will the army defend or defeat Indonesia's new democracy?" *The Economist*, April 10, 1999, p. 39. Catherine Napier, "Politics and the Military," *BBC News Online: Events: Indonesia: Special Report*, June 1, 1999. Internet: http://193.130.149.131/low/english/events/Indonesia/special_report/newsid_358000/35816.stm. Accessed June 4, 1999.

168. Seth Mydans, "New Indonesian Leader Governs with a Light Touch," *New York Times*, December 27, 1999, p. A3; Mydans, "Governing Tortuously: Indonesia's President Outflanks the General," *New York Times*, February 15, 2000, p. A8.

169. See U.S. Department of State, "Indonesia Human Rights Practices, 1995," March 1996 (online at gopher://dosfan.lib.uic.edu:70/1).

170. "East Timor, adrift and violent," *The Economist,* April 10, 1999, p. 40. Seth Mydans, "Thousands in East Timor Fleeing as Militias Attack," *New York Times,* September 7, 1999, p. A8; Mydans, "Indonesia Says No to Timor Peacekeepers," *New York Times,* September 9, 1999, p. A8. Philip Shenon with David E. Sanger, "Indonesia Assisting Militias, U.S. Says," *New York Times,* September 11, 1999, p. A6. Elizabeth Becker, "U.S.-to-Jakarta Messenger: Chairman of the Joint Chiefs," *New York Times,* September 14, 1999, p. A14. General Henry Shelton called Indonesia's chief of staff on Friday, September 10; on Sunday, after a quick fact finding trip to Timor, Wiranto recommended that President Habibie ask for international peacekeepers. Barbara Crossette, "Annan Says U.N. Must Take Over East Timor Rule," *New York Times,* October 6, 1999, p. A1. On October 19, Indonesia's People's Consultative Assembly ratified East Timor's wishes, granting independence. Six days later, the UN Security Council voted unanimously to establish the UN Transitional Administration in East Timor. Some 9,000 UN peacekeepers, 1,600 UN police, and hundreds of administrators finally took responsibility for security and development of East Timor in February 2000. United Nations, Department of Public Information, "East Timor—UNTAET—Background." Internet: http://www.un.org/ peace/etimor/UntaetB.htm. Accessed January 22, 2000. Seth Mydans, "East Timor, Stuck at 'Ground Zero,' Lacks Law, Order, and Much More," *New York Times,* February 16, 2000, p. A8.

171. Associated Press, "Indonesia Cancels Order for U.S. Fighter Jets," *New York Times,* June 7, 1997, p. 3. Thalif Deen, "Indonesia postpones planned arms purchases," *Jane's Defence Weekly,* October 21, 1998, p. 21.

Chapter 4

1. Bernard and Fawn M. Brodie, *From Crossbow to H-Bomb* (Bloomington: University of Indiana Press, 1973), pp. 35–36.

2. The risk of unverified agreements was made clear when President Boris Yeltsin announced in 1992 that the Russian Federation (an early signatory of the BWC) still maintained a vast empire of biological weapons laboratories. Judith Miller and William J. Broad, "Germ Weapons: In Soviet Past or in the New Russia's Future?" *New York Times,* December 28, 1998, p. A1.

3. Maurice A. Mallin, "CTBT and NPT: Options for U.S. Policy," *The Nonproliferation Review,* 2, no. 2 (Winter 1995), pp. 1–11, and Rebecca Johnson, "Viewpoint: The CTBT and the 1997 NPT Prepcom," *The Nonproliferation Review,* 3, no. 3 (Spring–Summer 1996), pp. 55–62.

4. For texts and narratives on major accords to which the United States is a party, see U.S. Arms Control and Disarmament Agency, *Arms Control and Disarmament Agreements: Texts and Histories of Negotiations* (Washington, D.C.: U.S. ACDA, 1990), also on the Internet at: http://www.acda.gov/treaties. For the Ottawa Convention, see the Canadian government's website: http://www.mines.gc.ca. See also Coit D. Blacker and Gloria Duffy, eds., *International Arms Control,* 2nd ed. (Stanford, Calif: Stanford University Press, 1984); Jozef Goldblat, *Arms Control: a Guide to Negotiations and Agreements* (Thousand Oaks, Calif.: Sage Publications, 1994); the gopher site of the Fletcher School of Law and Diplomacy for the text of the UN "Convention on Prohibitions or Restrictions on the Use of Certain Conventional Weapons . . . ," Internet: http://www.tufts.edu/departments/fletcher/multi/texts/BH790.txt. Accessed June 11, 1999; and for the 1995 protocol to that treaty, see "The Vienna Review Conference," *International Review of the Red Cross,* no. 309, November 1, 1995, pp. 672–77.

5. The International Campaign to Ban Landmines reported 10 to 15 million mines destroyed from existing stockpiles between March 1998 and 1999. UNICEF director

Carol Bellamy reported, upon the Ottawa Convention's entry into force, that the number of states producing landmines had dropped from 50 to 15. Among the 15, however, were the United States, Russia, and China, plus Iran, Iraq, Israel, Pakistan, and India, all non-signatories. Sapa-IPS, "UN Seeks Universal Ratification of Landmine Treaty," March 4, 1999. Internet: http://www.undp.org.za/docs/news/ 1999/nz0305.html.

6. U.S. Department of State, Office of the Coordinator for Counterterrorism, *Patterns of Global Terrorism, 1998,* April 1999. Internet: http://www.state.gov/www/global/terrorism/1998Report/1998index.html. See also chapter 3 of this volume, n.62.

7. Kenneth Waltz started the debate about the potential benefits of nuclear proliferation, John Mearsheimer extended it to post–Cold War Europe, and Scott Sagan laid out the opposing view in a debate with Waltz. Kenneth N. Waltz, *The Spread of Nuclear Weapons: More May Be Better,* Adelphi Paper 171 (London: International Institute for Strategic Studies, 1981). John J. Mearsheimer, "Back to the Future: Instability in Europe after the Cold War," in *The Perils of Anarchy,* ed. Michael E. Brown, Sean M. Lynn-Jones, and Steven E. Miller (Cambridge, Mass.: MIT Press, 1995). Kenneth N. Waltz and Scott D. Sagan, *The Spread of Nuclear Weapons: A Debate* (New York: W. W. Norton, 1995).

8. One philosophy is not necessarily more disposed toward arms transfer restraint than the other. A proponent of *realpolitik* might be all for the arms trade as good for business and for building the autonomy of state actors within the international system. A proponent of normative concerns might in fact embrace the libertarian argument that governments should not try to manage one another's affairs any more than they should try to manage the lives of their own citizens. Regional stability derives from states adopting such a perspective of mutual non-interference, while human security derives from the actions and support of civil society rather than from public sector intervention to promote public good. For exposition of and debate about mainstream thinking in the two schools, see David A. Baldwin, ed., *Neorealism and Neoliberalism: The Contemporary Debate* (New York: Columbia University Press, 1993). For creative treatment of the realist perspective, see Barry Buzan, Charles Jones, and Richard Little, *The Logic of Anarchy: Neorealism to Structural Realism* (New York: Columbia University Press, 1993). See also Barry Buzan, *People, States, and Fear,* 2nd ed. (Boulder: Lynne Rienner Publishers, 1991) and, for thoughtful treatment of the need to incorporate moral standards into the theory of international relations, Charles R. Beitz, *Political Theory and International Relations* (Princeton: Princeton University Press, 1979).

9. The terms are from Stephen Walt, *The Origins of Alliances* (Ithaca: Cornell University Press, 1987). "Balancing" means rounding up like-minded states to resist the power of a larger state, as the North Atlantic Treaty Organization (NATO) was created to resist Soviet power in Europe. "Bandwagoning" means, in essence, that what you can't lick, you join. Bandwagoning states go with the flow, hoping to reap some benefits along the way. In a world of one superpower, almost everybody bandwagons because the alternatives are limited, but they can also opt out, because the threats, for most, are limited as well.

10. Most U.S.-provided weapons were turned over for ultimate disposition to the Pakistani Inter-Services Intelligence agency. The American-sponsored arms pipeline to Afghanistan was, according to Chris Smith, the single most adverse development of the last decade for regional stability, hastening Pakistan's fragmentation. See his "Light Weapons and Ethnic Conflict in South Asia," in Jeffrey Boutwell, Michael T. Klare, and Laura W. Reed, eds., *Lethal Commerce: The Global Trade in Small Arms and Light Weapons* (Cambridge, Mass.: American Academy of Arts and Sciences, 1995), pp. 62–64.

11. Of course, there are cases in which one might prefer to take the loss rather than contribute to senseless carnage, but that motivation belongs to the next section, not to *realpolitik.* Raymond Bonner, "Despite Cutoff by U.S., Ethiopia and Eritrea Easily Buy Weapons," *New York Times,* July 23, 1998, p. A8.

12. For discussion of the interwar period, see Edward J. Laurance, *The International Arms Trade* (New York: Lexington Books, 1992), pp. 72–73. For a later take on a similar theme, see Anthony Sampson, *The Arms Bazaar: From Lebanon to Lockheed* (New York: Bantam Books, 1977).

13. Gordon Adams, *The Iron Triangle: The Politics of Defense Contracting* (New Brunswick, NJ: Transaction Books, 1982).

14. Carter's directive was dubbed the "leprosy letter" in defense and foreign policy circles because its no-contact policy treated defense firms' people like lepers, whose disease was long thought to be passed on by touch.

15. Clinton administration policy is contained in Presidential Decision Directive (PDD) 34 and summarized in a White House fact sheet. White House, Office of the Press Secretary, "Fact Sheet on Conventional Arms Transfer Policy," February 17, 1995. Internet: http://www.fas.org/asmp/library/white_house/whfacts.html. Accessed June 12, 1999.

16. U.S. Department of State, *Country Reports on Human Rights Practices,* 1998. Internet: http://www.state.gov/www/services/survey.html. Office of the United Nations High Commissioner for Human Rights, Human Rights Committee, documents (Internet: http://www.unhchr.ch/html/menu2/6/hrc.htm) and UN Human Rights Commission, reports. (Internet: http://www.unhchr.ch/html/menu2/2/chr.htm.) Accessed June 17, 1999.

17. Early 1990s analyses of multilateral intervention include Laura W. Reed and Carl Kaysen, eds., *Emerging Norms of Justified Intervention* (Cambridge, Mass.: American Academy of Arts and Sciences, 1993); Lori Fisler Damrosch and David J. Scheffer, eds., *Law and Force in the New International Order* (Boulder, Colo: Westview Press, 1991); and Lori Fisler Damrosch, ed., *Enforcing Restraint: Collective Intervention in Armed Conflicts* (New York: Council on Foreign Relations Press, 1993).

18. When genocidal violence erupted in Rwanda in April 1994, the United States argued for reduction rather than reinforcement of UN peacekeeping troops there and sent its own troops to help set up refugee camps outside Rwanda after the violence had run its course. In the camps were most of the perpetrators of the violence. On Somalia, see W. J. Durch, "Introduction to Anarchy: Humanitarian intervention and 'State-building' in Somalia," ch. 8, and J. Matthew Vaccaro, "The Politics of Genocide: Peacekeeping and Disaster Relief in Rwanda," ch. 9, in *UN Peacekeeping, American Policy, and the Uncivil Wars of the 1990s,* ed. William J. Durch (New York: St. Martin's Press, 1996).

19. "Elements for Objective Analysis and Advice Concerning Potentially Destabilising Accumulations of Conventional Weapons," paper approved by the plenary meeting of the Wassenaar Arrangement, December 3, 1998. Internet: http://www.wassenar.org/docs/criteria.html. Accessed June 13, 1999.

20. J. Jones, *Stealth Technology: The Art of Black Magic* (np: Tab Aero, nd), cited in "B-2 From Any View," *Air&Space Magazine* (Smithsonian), April/May 1997. Internet: http://www.airspacemag.com/ASM/mag/ supp/am97/stealth.html. Accessed June 13, 1999.

21. Michael J. Gething, "Balancing the pitfalls and potential of aircraft upgrade," *Jane's International Defence Review,* 12/1998, pp. 30–37. Ann Markusen, "The Rise of World Weapons," *Foreign Policy* (Spring 1999), pp. 40–51. See also the discussion in chapter 1 of this volume.

22. For a concise review of hegemonic stability theory and its critics, see Joseph M. Grieco, *Realist International Theory and the Study of World Politics,* in *New Thinking in International Relations Theory,* ed. Michael W. Doyle and G. John Ikenberry (Boulder: Westview Press, 1997), pp. 176–77.

23. Most models are, in any case, quite sensitive to analytical assumptions about key combat factors such as adequacy of training and command and control, force-to-space ratios (how much force can be concentrated at any one point, or how little force is

needed to defend a particular position), and combat logistics, as well as measures of fire-power, self-protection, and mobility of combat forces themselves. Altered assumptions can alter outcomes substantially. For a critical view of combat modeling, see T. N. Dupuy, "Can We Rely Upon Computer Combat Simulations?" *Armed Forces Journal International* (August 1987), pp. 58–63. For a good example of a two-party attrition model (used to evaluate the impact of different reductions from Cold War force levels in Europe), see Joshua Epstein, *Conventional Forces: A Dynamic Assessment* (Washington, D.C.: The Brookings Institution, 1990). For an earlier assessment of the same theater with a different approach, see Barry R. Posen, "Measuring the European Conventional Balance: Coping with Complexity in Threat Assessment," *International Security* (Winter 1984–85), pp. 47–88. Simulating combat with more than two sides or with fluid areas of fighting (for example, the Bosnian civil war) becomes very complex very fast.

24. James A. Bill, *The Eagle and the Lion: The Tragedy of Iranian-American Relations* (New Haven: Yale University Press, 1988), p. 220.

25. Ian Anthony, arms trade specialist at the Stockholm International Peace Research Institute (SIPRI), calls the third category "capabilities-based" restraint, which, in his formulation, aims at limiting the distribution of certain classes of technologies (such as those with dual civilian and military uses). Since technology supplier states tend to worry about particular states acquiring such capabilities, "in practice, making a distinction between recipient-based and capability-based controls is a problem." (Ian Anthony, "Conventional Arms Transfer Control," *Pacific Research,* May 1995, p. 43.)

26. U.S. sales totaled $1.5 billion in fiscal year 1971, and $13.1 billion in fiscal 1976. Henry T. Simmons, "U.S. Arms Sales and the Carter White House," *International Defense Review,* March 1979, p. 340.

27. Mark S. Sternman, "Mechanics of an Arms Sale," fact sheet (Washington, D.C.: Peace Action, April 1993); William D. Hartung, *And Weapons for All* (New York: Harper Collins Publishers, 1994), p. 61. According to Hartung, the Congress has never actually voted down a proposed arms sale, although the threat of such a vote has altered the size or composition of some packages. Although the AECA originally contemplated simple majority votes in House and Senate that were not subject in turn to presidential veto, subsequent federal court decisions overturned that provision. Thus, effective congressional opposition to arms deals under the AECA now requires a two-thirds majority in both houses.

28. "Conventional Arms Transfer Policy, Statement by the President, May 19, 1977," *Department of State Bulletin,* 13, no. 21, 1977.

29. Possible candidates for restraint under this approach included weapons sold by just a handful of suppliers (such as shoulder-fired surface-to-air missiles); weapons whose effects are horrific (for example, napalm); or weapons whose capabilities were closely associated with the delivery of weapons of mass destruction (for example, long-range ballistic missiles).

30. For good accounts of the CAT talks and the bureaucratic politics of U.S. policy formulation, see Barry M. Blechman, Janne E. Nolan, and Alan Platt, "Pushing Arms," *Foreign Policy* 46 (Spring 1982), esp. pp. 144–45; Janne E. Nolan, "The U.S.-Soviet Conventional Arms Transfer Talks," in *U.S. Soviet Security Cooperation,* edited by Alexander George, et al. (Oxford University Press, 1988); and Hartung, *And Weapons for All,* ch. 4.

31. Opposition to the arms transfer restraint policy was spearheaded by National Security Advisor Zbigniew Brzezinski and helped fuel his ongoing struggle with Secretary of State Cyrus Vance over control of American foreign policy.

32. Adherents to the MTCR include the original seven members (Canada, [West] Germany, France, Italy, Japan, UK, and U.S.), plus Argentina, Australia, Austria, Belgium, Brazil, Denmark, Finland, Greece, Hungary, Iceland, Ireland, Luxembourg, Netherlands, New Zealand, Norway, Portugal, Russia, South Africa, Spain, Sweden, Switzerland, and

Turkey. (See U.S. Arms Control and Disarmament Agency, "The Missile Technology Control Regime," fact sheet, September 15, 1997, available on the Internet at: http://acda.gov/factshee/exptcon/mtcr96.htm.) In addition, China, Israel, Romania, South Korea, and Ukraine say they adhere to the principles of the regime.

33. K. Scott McMahon and Dennis M. Gormley, *Controlling the Spread of Land-Attack Cruise Missiles* (Marina del Rey, Calif.: American Institute for Strategic Cooperation, January 1995), pp. 84–85. The MTCR establishes the presumption of export denial for rockets and missiles capable of delivering at least a 500 kilogram payload to a range of 300 kilometers. The regime is also intended to restrain the export of supporting technologies, from engines and navigation systems to test and production facilities.

34. John R. Harvey and Uzi Rubin, "Controlling Ballistic Missiles: How Important? How To Do It?" *Arms Control Today* (March 1992), pp. 13–18. For a detailed technical assessment, see John R. Harvey, "Regional Ballistic Missiles and Advanced Strike Aircraft: Comparing Military Effectiveness," *International Security* (Fall 1992), pp. 41–83. See also a background paper by the Arms Control Association, "The Missile Technology Control Regime (MTCR)," Washington, D.C., September 1993.

35. For data on anti-ship cruise missiles, see David Foxwell, "Anti-ship Cruise Missiles Home-in on Littoral Requirements," *Jane's International Defence Review*, 8/1996, 63.

36. Janne E. Nolan (chair), Edward Randolph Jayne, II, Ronald E. Lehman, Dave E. McGiffert, and Paul C. Warnke, *Report of the Presidential Advisory Board on Arms Proliferation Policy*, Washington, D.C., 1996.

37. Ibid., p. 6.

38. Ibid.

39. Ibid., pp. 7–8.

40. Marcy Agmon, et al., *Arms Proliferation Policy: Support to the Presidential Advisory Board*, MR-771-OSD (Santa Monica, Calif.: RAND Corporation, 1996), pp. 66–67.

41. Ibid., pp. 14, 19.

42. The argument is repeated and amplified in William W. Keller and Janne E. Nolan, "The Arms Trade: Business As Usual?" *Foreign Policy* (Winter 1997–98), pp. 113–125.

43. Andrew J. Pierre, *The Global Politics of Arms Sales* (Princeton: Princeton University Press, 1982), p. 203; Hartung, *And Weapons for All*, pp. 199–200.

44. "Bush Unveils Long-Awaited Middle Eastern Arms Control Plan," *Arms Control Today* (June 1991), pp. 27–28.

45. Natalie Goldring, "President Bush's Middle East Arms Control Initiative: One Year Later," Arms Control Association press conference, *Arms Control Today* (June 1992), p. 12. In the same issue, see also "Third Round of Arms Sales Talks Fails to Resolved Notification Issue," p. 21.

46. For the texts of all public documents of the Wassenaar Arrangement, see its website. Internet: http://wassenaar.org/doc/press_3.html. Accessed June 11, 1999.

47. Bruce Odessey, "Officials Defend U.S. Participation in Wassenaar Arrangement," posted online by the United States Mission to Italy, at http://www.usis.it/wireless/wf970514/97051417.htm, accessed April 8, 1998.

48. Ibid.

49. Theresa Hitchens, "Experts Tout Post-COCOM Regime, Despite Shortfalls," *Defense News*, January 29–February 4, 1996. U.S. Department of State, Office of the Spokesman, "The Wassenaar Arrangement," address by Under Secretary of State for Arms Control and International Security Affairs Lynn E. Davis, at the Carnegie Endowment for International Peace, Washington, D.C., January 23, 1996. Sarah Walkling, "Wassenaar Members End Plenary: First Data Exchange Falls Short," *Arms Control Today* (Jan/Feb 1997) and Wade Boese, "Divisions Still Impede Wassenaar Export Control Regime at Plenary," *Arms Control Today* (Nov/Dec 1997). Available online at http://www.armscontrol.org/ACT/JAN_FEB/wass.html and . . . /ACT/novdec97/wassnov.htm.

50. Although intergovernmental exchanges under Wassenaar are confidential, the U.S. Arms Control and Disarmament Agency declassified the "List of Dual-Use Goods and Technologies and Munitions List" submitted to Wassenaar's July 1996 plenary meeting. See U.S. Arms Control and Disarmament Agency, *List of Dual Use Goods and Technologies and Munitions List, submitted to the Plenary Meeting in Vienna,* July 11–12, 1996, ACDA document No. 97023, declassified April 25, 1997. Annexes 1 and 2. Available on the Internet at: http://jya.com/wa/watoc.htm. (Accessed April 8, 1998.)

51. By mid-1999, U.S. computer makers were chafing at the definitions of computer and chip capabilities found in Wassenaar's original technologies list, as high-end desktop computers approached the capacity of early supercomputers.

52. "Administration Encryption Efforts with Wassenaar Arrangement Come up Short," *PR Newswire,* December 9, 1998. Robert MacMillan, "Don't Bet on Any More Encryption Changes This Year," *Newsbytes,* April 8, 1999. Both accessed via Lexis/Nexis Universe, June 11, 1999.

53. Keith Perine, "Feds Release Revised Crypto Export Rules," *The Standard,* January 12, 2000. Internet: http://www.thestandard.net/article/display/0,1151,8780,00.html. Accessed March 2, 2000. U.S. Department of Commerce, Bureau of Export Administration, "Fact Sheet: Administration Implements Updated Encryption Export Policy," January 12, 2000, accessed via the Center for Democracy and Technology. Internet: http://www.cdt.org/crypto/admin/000112commercefactsheet.shtml. Accessed March 2, 2000.

54. The United Nations Register on Conventional Arms was created in 1992. It is discussed at length in chapter 5.

55. Walkling, "Wassenaar Members End Plenary."

56. See the Saferworld report, "Developing the Wassenaar Arrangement: A new arms export control regime" (London, September 1996), downloadable from the Internet at http://www.gn.apc.org/sworld/wassen.html.

57. Boese, "Divisions Still Impede Wassenaar."

58. Wassenaar Arrangement, "Public Statement," Vienna, Austria, December 3, 1998. Internet: http://www.wassennaar.org/docs/press_4.html. Accessed June 14, 1999.

59. "Arms Sales to African countries raise concern," *Star Tribune* (Minneapolis), December 6, 1998, p. 13A. (Accessed via Lexis/Nexis Universe.) Associated Press [AP], "Ethiopia-Eritrea War Rages On," *World Africa Network,* April 1, 1999. Internet: http://www.wanonline.com/news/news6503.html. Accessed June 14, 1999. AP, "Fighting Erupts in Ethiopia-Eritrea Border War," *Washington Post,* June 11, 1999.

60. Saferworld, "Developing the Wassenaar Arrangement."

61. Walkling, "Wassenaar Members End Plenary."

62. John Stremlau, *Sharpening International Sanctions: Toward a Stronger Role for the United Nations,* a report to the Carnegie Commission on Preventing Deadly Conflict (Washington, D.C., November 1996), pp. 21–35. For a comprehensive treatment of sanctions regimes up to 1990, see G. C. Hufbauer, et al., *Economic Sanctions Reconsidered: History and Current Policy* (Washington, D.C.: Institute for International Economics, 1990).

63. Applied to a situation like Rwanda's, such end-conflict-at-all-costs logic would have opposed the rebel Rwandan Patriotic Front's surge back into the country to drive out the perpetrators of genocide. In the event, moral equivalencing and a fixation on tactical action led both the international relief community and the U.S. military to feed, clothe, and shelter the perpetrators of genocide in refugee camps along the Rwanda-Zaire border.

64. Michael R. Gordon, "President Orders End to Enforcing Bosnian Embargo," *New York Times,* November 11, 1994, p. A1. As the Associated Press pointed out at the time, "Significant amounts of weapons and ammunition have been reaching Bosnian government forces since spring, so the U.S. decision not to enforce the arms embargo will have lit-

tle immediate impact." Maud S. Beelman, "Bosnia-Embargo Effect," *Associated Press* newswire, November 11, 1994, 1:48 pm.

65. John Lancaster, "Annan, Iraq Agree on Inspections," *Washington Post,* February 23, 1998, p. A1. The Annan accord also very likely saved the U.S. administration a good deal of political grief. Five days beforehand, the U.S. secretaries of state and defense, and the president's national security advisor, attended a "town hall meeting" in Columbus, Ohio, on Iraq and U.S. policy that drew a large and raucous crowd asking pointed questions about U.S. policy and objectives. For a transcript, see the following Internet web-page of the U.S. Department of State, http://secretary.state.gov/www/statements/1998/980218.html. (Accessed April 8, 1998.) For a review of events in 1998, see chapter 3, n. 63. As of June 1999, Iraq remained under UN arms embargo but was forging black market and gray market links to rebuild its military and probably its WMD programs. Iraq's channels for breaking the UN embargoes reportedly included Syria (for military spare parts and illicit petroleum exports), Serbia (for mutual support in air defense), and Russia (for technical assistance, spare parts, and upgrades for MiG fighter aircraft and air defense systems). The Russian connection represents an especially egregious end-run of UN sanctions, as Russia is one of five permanent members of the UN Security Council. An accord to provide assistance and spares to the Iraqi military reportedly was reached in quiet retaliation for U.S. and British bombing of Iraq in December 1998 (Operation Desert Fox). Michael Evans, "Damascus to rearm Saddam," *London Times,* March 8, 1999. Philip Shenon, "Serbs Seek Iraqi Help for Defense, Britain Says," *New York Times,* April 1, 1999, p. A16. Con Coughlin, "More on Serb Iraq Arms Pact: Milosevic and Saddam plot joint revenge," *Sunday Telegraph* (London), March 28, 1999, via *Iraq News,* March 28, 1999. Jamie Dettmer, "Russians said to sell missiles to Saddam," *Washington Times,* February 22, 1999. Con Coughlin, "Russian Experts Confirm Baghdad Connection," *Sunday Telegraph* (London), February 21, 1999, via *Iraq News,* February 21, 1999.

66. Nuclear Age Peace Foundation, "Nobel Laureates' International Code of Conduct on Arms Transfers," available online at http://www.wagingpeace.org/codeofconduct.html or at http:www.igc.apc.org/basic/code_itl.htm. (Accessed April 8, 1998.)

67. Friends Committee on National Legislation, "Legislative History of the Code of Conduct on Arms Transfers," available on the Internet at: http://www.fcnl.org/pub/fcnl/codehist.htm. (Accessed April 9, 1998.)

68. Similarly, the Foreign Assistance Act of 1961 requires the president to take into account U.S. foreign aid recipients' military spending and "the amount of foreign exchange used to procure military equipment." However, "no country has ever been denied funding" on the basis of such reviews. See Nicole Ball, *Pressing for Peace: Can Aid Induce Reform?* Policy Essay No. 6 (Washington, D.C.: Overseas Development Council, 1992), pp. 22–23.

69. Federation of American Scientists, "Code of Conduct—Legislative History," available on the Internet at: http://www.fas.org/asmp/campaigns/code/codehist.html. Accessed March 2, 2000. The Code of Conduct on Arms Transfers Act of 1999 (HR 2269) was included in the *Foreign Relations Authorization Act, Fiscal Year 2000 and 2001,* which was in turn folded into the Consolidated Appropriations Act for FY 2000.

70. British American Security Information Council, "EU Code of Conduct," available on the Internet at: http://basicint.org/eucode.htm. (Accessed April 8, 1998.) See also Saferworld, "The EU Code of Conduct on the arms trade: Final Analysis," Internet: http://ww.gn.apc.org/SWORLD/ARMSTRADE/code.html. Accessed June 14, 1999.

71. Saferworld, "The EU Code."

72. The annual Freedom House evaluations of civil and political liberty around the world have been published for a quarter-century and are highly regarded, based on evaluations

of 21 different variables. The ratings and methodology on which they are based are published in the February issue of *Freedom Review* and are now available on the organization's Internet website: http://www.freedomhouse.org/.

73. For discussion of monitoring requirements, see Wolfgang H. Reinecke, "From Denial to Disclosure: The Political Economy of Export Controls and Technology Transfer," in *Bridging the Nonproliferation Divide: The United States and India,* edited by Francine Frankel (Lanham, MD: University Press of America, 1995).

74. Joan M. Nelson and Stephanie J. Eglinton, *Global Goals, Contentious Means: Issues of Multiple Aid Conditionality,* Policy Essay No. 10 (Washington, D.C.: The Overseas Development Council, 1993), p. 32. Other forms include "policy-based lending," in which a donor releases funds "contingent on the recipient making prearranged policy reforms" (ibid.), or "hybrid/project conditionality" that may address the procedural details of a specific development project. (p. 33)

75. Ibid., executive summary.

76. Janne E. Nolan, "The Global Arms Market After the Gulf War: Prospects for Control," *The Washington Quarterly* (Summer 1991), p. 132.

77. Geoffrey Kemp, "Regional Security, Arms Control, and the End of the Cold War," *The Washington Quarterly* (Autumn 1990), p. 46.

78. Brad Roberts, "From Nonproliferation to Antiproliferation," *International Security* (Summer 1993), pp. 165–66.

Chapter 5

1. Ted Greenwood, "Experience from European and US-Soviet Agreements," in Sumit Ganguly and Ted Greenwood, eds., *Mending Fences: Confidence- and Security-building Measures in South Asia* (Boulder, Colo.: Westview Press, 1996), pp. 92–110.

2. Jack Child, "Confidence-building Measures and Their Application in Latin America," in Augusto Varas, James A. Schear, and Lisa Owens, eds., *Confidence-building Measures in Latin America: Central America and the Southern Cone,* Report No. 16 (Washington, D.C.: The Henry L. Stimson Center, February 1995), p. 8.

3. Michael Krepon, "Conflict Avoidance, Confidence-building, and Peacemaking," in M. Krepon, et al., eds., *A Handbook of Confidence-building Measures for Regional Security,* 3rd ed. (Washington, D.C.: The Henry L. Stimson Center, March 1998), pp. 4–11.

4. Marie-France Desjardins, *Rethinking Confidence-building Measures,* Adelphi Paper No. 307 (New York: Oxford University Press for the International Institute for Strategic Studies [IISS], December 1996), pp. 5, 8–11, 19–20, 61–62.

5. Ibid., p. 18.

6. William J. Durch, "Introduction," in *The Evolution of UN Peacekeeping: Case Studies and Comparative Analysis,* ed., W. J. Durch (New York: St. Martin's Press, 1993), p. 4.

7. William J. Durch, "Keeping the Peace: Politics and Lessons of the 1990s," in *UN Peacekeeping, American Politics, and the Uncivil Wars of the 1990s,* ed. W. J. Durch (New York: St. Martin's Press, 1996), pp. 4–6, 7. The UN Preventive Deployment (UNPREDEP) mission in the former Yugoslav republic of Macedonia (1992–1999) is an example of an operation deployed against a one-way threat. A light force, it was never challenged militarily, so the issue of fighting power or reinforcement was never tested.

8. Cathleen S. Fisher, "The Preconditions of Confidence-Building: Lessons from the European Experience," in Krepon, et al., *A Handbook of Confidence-Building Measures,* pp. 268–70.

9. B. Blechman, W. Durch, W. Eaton, and J. Werbel, *Effective Transitions from Peace Operations to Sustainable Peace: Final Report,* prepared for the U.S. Department of Defense, Office of Peacekeeping and Humanitarian Affairs, OSD/OASD/S&R/PK&HA, contract no. DASW01–96-C-0075, September 1997.

10. Malcolm Chalmers, Owen Greene, Edward J. Laurance, and Herbert Wulf, eds., *Developing the UN Register of Conventional Arms,* Bradford Arms Register Studies No. 4 (Trowbridge, UK: Redwood Books for the University of Bradford, 1994), pp. 3–5.

11. Edward J. Laurance and Herbert Wulf, "Lessons from the First Year," in Chalmers, et al., *Developing the UN Register,* pp. 37–38.

12. Ibid. pp. 39–42.

13. Ibid., p. 43.

14. Malcolm Chalmers and Owen Greene, *Taking Stock: The UN Register After Two Years,* Bradford Arms Register Studies No. 5 (Trowbridge, UK: Redwood Books for Westview Press, 1995), pp. 36–38, 44, 46–47, 151–53.

15. Ibid., pp. 70–75.

16. The analysis of the 1994 data is based on numbers reported by Edward J. Laurance and Tracy Keith, "An Evaluation of the Third Year of Reporting to the United Nations Register of Conventional Arms," *Nonproliferation Review* (Spring 1996), and cited earlier in Susannah L. Dyer and Natalie J. Goldring, "U.S. and Germany Dominate World Weapons Exports," *BASIC Reports* No. 48 (November 20, 1995) and No. 49 (December 1, 1995).

17. Other countries with above-average levels of trade with non-participants included Slovakia, Italy, Ukraine, France, Russia, the UK, and China.

18. Chalmers and Greene, *Taking Stock,* pp. 96–115.

19. United Nations, General Assembly, *General and Complete Disarmament: Transparency in Armaments, United Nations Register of Conventional Arms, Report of the Secretary-General,* A/51/300 (August 20, 1996) and addenda 1 through 5 (July 14, 1997).

20. In 1997, the register moved onto the Internet, and every participant's data submissions can be viewed by accessing. UN Register of Conventional Arms, country reports, Internet: http://domino.un.org/REGISTER.NSF. Accessed June 21, 1999.

21. United Nations, General Assembly, *General and Complete Disarmament: Transparency in Armaments, United Nations Register of Conventional Arms, Report of the Secretary-General,* A/52/312 (August 15, 1997) and Add.1 (October 17, 1997), annex.

22. Assume that an exporter delivered a shipload of tanks to country X in January 1995. The report to the register including that delivery was not due until April 1996, and there are no sanctions for late presentation of reports, which could be delivered in June, August, October, or even later.

23. Cathleen S. Fisher, "Preconditions of Confidence-building," p. 28.

24. Coit D. Blacker and Gloria Duffy, *International Arms Control: Issues and Agreements,* 2nd ed. (Stanford: Stanford University Press, 1984), p. 289.

25. Raymond L. Garthoff, *Détente and Confrontation: American-Soviet Relations from Nixon to Reagan* (Washington, D.C.: The Brookings Institution, 1985), p. 474; and U.S. Arms Control and Disarmament Agency, "Confidence and Security Building Measures in Europe," updated to May 20, 1996. Internet: http://www/acda.gov/factshee/secbldg/csbms394.html.

26. Garthoff, *Détente and Confrontation,* p. 475.

27. Ibid., pp. 478–79.

28. "The Paris Agreements: CFE and CSCE," *Strategic Survey 1990–91* (London: Brassey's for the IISS, 1991), p. 29.

29. In postconflict situations, outside efforts to implement peace accords have done better when the implementers have been able to exercise full supervision over the electoral process, the better to guard against irregularities. In Bosnia, the antagonistic local parties controlled the voter registration lists, and in the municipal elections, displaced voters cast ballots not for their current places of residence but for the towns and villages in which they lived before the war. In six municipalities, displaced persons won a majority of council seats, and in 42 others they won between 20 and 49 percent of the

seats. For a report on the election results, see the International Crisis Group, "ICG Analysis of the 1997 Municipal Election Results," ICG Bosnia Project Press Release, October 14, 1997. Internet: http://www.intl-crisis-group.org/projects/bosnia/reportb-hxxpr10.htm. Accessed April 10, 1998.

30. One sign that the two superpowers were not anxious to reach an MBFR agreement: no private, "back channel" bilateral talks. Back channel negotiations were common in the two states' strategic arms limitation efforts and were considered a hallmark of seriousness. See Garthoff, *Détente and Confrontation,* p. 483.

31. Of the newly independent states, the three Baltic republics (Estonia, Latvia, and Lithuania) chose not to remain in the Treaty when the Soviet Union broke up. Most Western states had never recognized their 1940 incorporation into the USSR, in any case.

32. Paul F. Pineo and Lora Lumpe, "Recycled Weapons: American Exports of Surplus Arms, 1990–1995," a study by the Arms Sales Monitoring Project of the Federation of American Scientists (Washington, D.C.: FAS, May 1996). Internet: http:www.fas.org/asmp/library/publications/recycle.html. See also Joseph P. Hanrahan and John C. Kuhn, III, *On-Site Inspections Under the CFE Treaty* (Washington, D.C.: U.S. Onsite Inspection Agency, 1997), ch. 7, "The Reduction Years." Hanrahan and Kuhn also note that the USSR removed 57,000 pieces of otherwise treaty-limited equipment from the zone of reductions to locations east of the Ural Mountains, outside the ATTU, while talks were underway and before the required data exchanges to establish baseline obligations for TLE cuts. This move generated heated negotiations in the Joint Consultative Group, as it was a blatant effort to undermine the force-balancing objectives of CFE. In the end, Moscow agreed in June 1991 to destroy "at least 14,000 pieces of that equipment." Equipment destruction outside the ATTU could not be verified by on-site inspection, however, only by satellite reconnaissance.

33. Stockholm International Peace Research Institute, *SIPRI Yearbook 1994: Armaments, Disarmament, and International Security* (New York: Oxford University Press, 1994), pp. 522, 538, 544.

34. Hanrahan and Kuhn, *On-Site Inspections Under the CFE Treaty,* ch. 7.

35. Colonel Jeffrey D. McCausland, "The Conventional Armed Forces in Europe (CFE) Treaty: Threats from the Flank," nd. Internet: http://www.cdsar.af.mil/apj/mccaus.html.

36. Jorgen Dragsdahl, "CFE Talks Advance Amid Controversy," *Basic Reports* No. 69 (March 1999), pp. 3–6. Internet: http://www.basicint.org/br69.htm. Roland Eggleston, "Europe/Russia: Basic Agreement Reached on Conventional Arms Accord," *RFE/RL Online,* April 12, 1999. Internet: http://euro.rferl.org/nca/features/1999/04/F.RU.990412125005.html. Wade Boese, "CFE Parties Outline Adapted Treaty: Limits to Allow NATO Growth," *Arms Control Today* (March 1999). Internet: http://www.armscontrol.org/ACT/march99/cfemr99.html. Accessed June 21, 1999.

37. Defense Threat Reduction Agency [DTRA], "Treaty on Open Skies," fact sheet, as of March 1999. Internet: http://www.dtra.mil/news/facts/os99.html. Accessed June 21, 1999.

38. Ibid. See also U.S. Arms Control and Disarmament Agency, "Open Skies Consultative Commission," fact sheet, as of May 20, 1996. Internet: http://www.acda.gov/factshee/secbldg/opnskicc.html.

39. For details and pictures of the trial flights, see the onsite inspection pages of the DTRA website. Internet: http://www.dtra.mil/.

40. Gilbert M. Khadiagala, "Confidence-building Measures in Sub-Saharan Africa," in Krepon, ed., *A Handbook on Confidence-building Measures,* pp. 103–5.

41. For details, see Internet: http://www.sadc-online.com/sadc/about/about.htm, or http://www.sadc- usa.net/overview/history.html.

42. Khadiagala, "Confidence-building Measures in Sub-Saharan Africa," pp. 106–7.

43. "SADC: Mandela, Mugabe feud over crucial security organ," *The Financial Gazette* (Harare), September 25, 1997. Internet: http://www.fellesraadet.africainfo.no/africain-dex/Update/archive/SADC/mandela.htm.

44. Nelson Mandela, "New Directions for SADC," speech at the concluding session of the Southern Africa International Dialogue, Kasane, Botswana, May 6, 1997. Internet: http://www.sadc- online.com/sadc/said/spemand.htm.

45. Howard Barrell, "Angola broke as Unita tightens noose," *Daily Mail and Guardian* (Johannesburg), May 28, 1999. Internet edition: http://www.mg.co.za/mg/news/99may2/28may-angola.html. Chris Gordon, "Diamonds are a guerrilla's best friend," *Daily Mail and Guardian,* June 15, 1999. Internet edition: http://www.mg.co.za/mg/news/99jun2/15jun-diamonds.html. UN Office for the Coordination of Humanitarian Affairs, Integrated Regional Information Network [IRIN], "Angola: DeBeers says complying with diamond embargo," June 17, 1999. Internet: http://www.reliefweb.int/IRIN/sa/countrystories/angola/19990617a.htm. "Chiluba Appeals to UN and OAU for International Verifiers," Johannesburg SAPA, February 12, 1999, 14:43 GMT. Transcribed by U.S. Foreign Broadcast Information Service [FBIS], Daily Report FBIS-AFR-1999-0212, February 12, 1999.

46. J. Matthew Vaccaro, "The Politics of Genocide: Peacekeeping and Disaster Relief in Rwanda," in Durch, ed., *The Evolution of UN Peacekeeping,* pp. 367–408. See also Bruce D. Jones, "Military Intervention in Rwanda's Two Wars: Partisanship and Indifference," in Barbara F. Walter and Josh Snyder, eds., *Civil Wars, Insecurity, and Intervention* (New York: Columbia University Press, 1999).

47. Howard W. Adelman, "Early Warning and Humanitarian Intervention: Zaire -March to December 1996," paper written for the Forum on Early Warning and Early Response, revised September 1998. Mimeo, pp. 6–7, 14–18. On drug trafficking, see the "Final report of the UN International Commission of Inquiry (Rwanda)," UN Document S/1998/1096, November 18, 1998. Annex, p. 6.

48. Lynne Duke, "Revolt in Congo Had Multiethnic Genesis," *Washington Post,* October 27, 1998, p. A20.

49. International Commission of Inquiry (Rwanda), Annex, pp. 4–6. When pushed from the Rwandan border zones by the RPA in 1996–97, ex-FAR and *interhamwe* fighters spread into the Central African Republic, Congo-Brazzaville, Sudan, UNITA-held parts of Angola, northwestern Zambia, and the southwestern DRC. Many reportedly regrouped in the DRC after the outbreak of war to fight on behalf of the Kabila government. Also, Brittain, Victoria. "The Congo war redraws African alliances," *Mail & Guardian* (Johannesburg), March 26, 1999. Internet: http://www.mg.co.za/mg/news/99mar2/26mar-congo.html. Accessed May 20, 1999. IISS, *Strategic Survey, 1998–99* (London: Oxford University Press for the IISS), pp. 235–37.

50. "Zimbabwe Asks SADC to Commit Troops to Angola," *The Financial Gazette* (Harare), Internet edition, March 18, 1999. Transcribed in FBIS-AFR-1999-0318. Iden Wethrell, "Mugabe crafts new Africa defence pact," *Daily Mail and Guardian* (Johannesburg), April 16, 1999. R. W. Johnson, "Around Africa: Mugabe's intolerance must not taint region," *The Sunday Independent* (Johannesburg), February 14, 1999, p. 16. Trans. in FBIS-AFR-1999-0214. "RSA Editorial Urges SADC to Act Against Zimbabwe," *Johannesburg Sowetan,* January 26, 1999, p. 12. Trans. FBIS- AFR-1999-0126.

51. South Africa contributed about 2,800 of the 4,000 soldiers and police participating in the exercise. The other contributors sent company-sized units of 140 to 200 troops each. The DRC was not participating, while Angola and Zimbabwe sent only staff officers to observe the exercise. Stephen Laufer, "Blue Crane will be Africa's largest operation yet," *Oneworld.com,* March 19, 1999. Internet: http://oneworld.org/saep/forDB/forDB9903/ SADCWARbluecrane990319.htm. Accessed June 18, 1999. IRIN, "Peacekeeping exercise opens in South Africa," April 15, 1999. Internet:

http://www.reliefweb.int/IRIN/sa/countrystories/other/19990415.htm. Accessed June 21, 1999. Hans Pienaar, "R20m wargames end, but . . . Peace-keepers not ready," *Cape Town Times* (Internet edition), April 28, 1999. Trans. FGIS-AFR-1999–0428.

52. "Chissano to Mediate in Lesotho," *Panafrican News Agency,* September 15, 1998. Internet: http://www.africanews.org/PANA/news/19980115/feat7.htm. "Additional South African Troops for Lesotho," *Africa Online* (a Prodigy, Inc., Company), September 29, 1998. Internet: http://africaonline.com/AfricaOnline/newsstand/PANA/pana19980 929.html. "South Africans botch military intervention in Lesotho," *Freedom Press International,* September 1998. Internet: http://freedom.tao.ca/FIN/en121.html. Accessed June 22, 1999. Mariette le Roux, "South Africa: Lesotho 'On Track'; SADC to 'Scale Down' Involvement," *Johannesburg SAPA,* December 11, 1998, 12:11 GMT. Trans. FBIS-AFR-98–345. "Lesotho: Parties Report 'Major Advances' in Peace Talks," *Johannesburg Sowetan,* January 29, 1999, p. 4. Trans. FBIS-AFR-99–029. Chantelle Benjamin, "Peacekeeping Strategy Needed in Light of Lesotho," *Johannesburg SAPA,* March 25, 1999, 1859 GMT. Remarks of Rocklyn Williams, director, operations policy and planning division, South African Defence Secretariat. Trans. FBIS-AFR- 1999–0325. "Lesotho Bids Farewell to 'Last Troops' of SADC Mission," *Maseru Radio Lesotho,* May 15, 1999, 16:00 GMT. Trans. FBIS-AFR-1999–0515. "Training of Lesotho Army to Start Next Week: SADC," *Johannesburg SAPA,* June 4, 1999, 13:37 GMT. Trans. FBIS-AFR-1999–0604.

53. For a keen assessment of the DRC's problems that stresses both internal and external factors, see Mahmood Mamdani, "Why foreign invaders can't help Congo," *Electronic Mail and Guardian* (Johannesburg), November 2, 1998. Internet: http://www.mg.co.za/mg/news/98nov1/2nov-congo.html. Accessed November 15, 1998. Mamdani directs the African Studies Centre at the University of Cape Town. See also testimony by U.S. assistant secretary of state Susan Rice, "Central African Conflict and Its Implications for the Future of U.S. Policy Goals and Strategies," before the U.S. Senate Foreign Relations Committee, Subcommittee on African Affairs, June 8, 1999. Posted on the Internet at: http://members.tripod.com/~heritiersjus/rice.htm. Accessed June 21, 1999.

54. "Special Report on the problems of policing a future peace agreement in the DRC," IRIN, June 10, 1999. Internet: http://www.reliefweb.int/IRIN/cea/countrystories/drc/19990610a.htm. Accessed June 18, 1999.

55. Khadiagala, "Confidence-building Measures in Sub-Saharan Africa," pp. 122–25.

56. Unless otherwise noted, this section summarizes the analysis presented in Jill R. Junnolla, "Confidence-building Measures in the Middle East," in Krepon, ed., *A Handbook of Confidence-building,* pp. 47–76.

57. The Middle East is the birthplace of traditional armed peacekeeping operations. Following the 1956 Suez Crisis, UN peacekeepers were deployed to the Sinai and remained in place between Israel and Egypt until 1967, when asked to depart on the eve of the Six-Day War. They returned after the October War in 1973, deploying both in the Sinai and along the Golan Heights, between Israeli and Syrian forces. In 1978, following Israel's invasion of Lebanon to fight the Palestine Liberation Organization, the UN deployed its "Interim Force in Lebanon" (UNIFIL) to oversee Israeli withdrawal, which never quite happened. That force remained deployed in southern Lebanon, functioning largely as a humanitarian safety zone during Lebanon's long civil war (1975–1989) but otherwise ignored by all sides. With the election of Ehud Barak as prime minister of Israel in 1999, partly on a promise to withdraw Israeli forces within one year, there arose the prospect that UNIFIL might actually come to fulfill its original mandate. For further information on these operations, see Mona Ghali, "United Nations Emergency Force I: 1956–1967," "United Nations Emergency Force II: 1973–1979," "United Nations Disengagement Observer Force," and "United Nations Interim Force in Lebanon: 1978–Present," chs. 7–9, 11, in Durch, ed., *The Evolution of*

UN Peacekeeping, and Ghali, *The Multinational Forces: Non-UN Peacekeeping in the Middle East,* Occasional Paper No. 12 (Washington, D.C.: The Henry L. Stimson Center, May 1993).

58. Junnolla, "Confidence-Building Measures in the Middle East," pp. 60–64.

59. For the text of Simla, see Ganguly and Greenwood, *Mending Fences,* Appendix 1.

60. Following a rise of unrest in Kashmir in late 1989, and subsequent Pakistani support for Kashmiri militants, political tensions between Pakistan and India rose, along with rumors of military movements. The White House dispatched Deputy National Security Adviser Robert Gates to Moscow, Islamabad, and New Delhi in April 1990 with a message that the United States would not support any party that started a conflict and to suggest a range of confidence-building measures that the two sides might institute. Michael Krepon and Mishi Faruqee, eds., *Conflict Prevention and Confidence-Building Measures in South Asia: The 1990 Crisis,* Occasional Paper No. 17 (Washington, D.C.: The Henry L. Stimson Center, April 1994), esp. pp. vii, 8, 9. See also Kanti P. Bajpai, "Conflict, Cooperation, and CSBMs with Pakistan and China. A View from New Delhi," p. 2, and "Statement of Government of Pakistan on CSBMs between Pakistan and India," appendix 4, p. 236, in Ganguly and Greenwood, *Mending Fences.*

61. John Sandrock, "Prerequisites for Success," in Ganguly and Greenwood, *Mending Fences,* pp. 209–12.

62. Ibid., and Matthew C. J. Rudolph, "Confidence-building Measures between Pakistan and India," in Michael Krepon, ed., *A Handbook of Confidence-building Measures for Regional Security,* 2nd ed. (Washington, D.C.: The Henry L. Stimson Center, January 1995), pp. 54–55.

63. See Desjardins, *Rethinking Confidence-Building Measures,* esp. pp. 51–53. The author focuses most of her fire on India and Pakistan.

64. See, for example, Neil Joeck, *Maintaining Nuclear Stability in South Asia,* Adelphi Paper No. 312 (New York: Oxford University Press for the IISS, September 1997), esp. ch. 4, and Sandrock, "Prerequisites for Success," pp. 212–16. Sandrock doubts CSBMs would generate any major political breakthroughs, however. In his view, they require a leader of the caliber of an Anwar Sadat who is willing to take political risks for peace. He attributes much of the current governments' travails precisely to their leaders' aversion to risk.

65. Sheen Rajmaira, "Indo-Pakistani Relations: Reciprocity in Long-Term Perspective," *International Studies Quarterly* 41, no. 3 (September 1997), pp. 557–59.

66. Kenneth J. Cooper, "Pakistan, India Discuss Kashmir," *Washington Post,* October 18, 1998, p. A29. Barry Bearak, "India Leader Pays Visit to Pakistan," *New York Times,* February 21, 1999, p. 1. Celia W. Dugger, "Pakistan Tests a New Missile, Matching India in Arms Race," *New York Times,* April 15, 1999, p. A9.

67. Rajmaira, "Indo-Pakistani Relations," p. 559.

68. See Bajpai, "Conflict, Cooperation, and CSBMs," p. 8. For the charter and history of the South Asian Association for Regional Cooperation (SAARC) see the website maintained by the government of Sri Lanka at: http://www.lanka.net/saarc. Accessed June 25, 1999.

69. Editorial, "SAARC: Only a Talking Shop," *Economic Times* (New Delhi), May 25, 1997. Internet: http://www.economictimes.com/250597/opin2.htm.

70. John F. Burns, "India and Pakistan to Hold Talks About Kashmir," *New York Times,* June 24, 1997.

71. Michael Leifer, *The ASEAN Regional Forum: Extending ASEAN's model of regional security,* Adelphi Paper No. 302 (London: Oxford University Press for the IISS, July 1996), p. 13.

72. Leifer, *The ASEAN Regional Forum,* pp. 9–10.

73. Ibid., p. 11.

74. Ibid., p. 41. The equivalent of "comfortable" in both Chinese and Indonesian languages is a sense of being at ease.

75. Robert A. Manning, "Building Community or Building Conflict? A Typology of Asia Pacific Security Challenges," in Ralph A. Cossa, ed., *Asia Pacific Confidence and Security Building Measures* (Washington, D.C.: Center for Strategic and International Studies, 1995), pp. 29–30.

76. Jeannie Henderson, *Reassessing ASEAN*, Adelphi Paper 328 (London: IISS, May 1999).

77. "The Limits of Politeness," *The Economist*, February 28, 1998, p. 43. Keith B. Richburg, "Regional Crises Leave ASEAN in Disarray," *Washington Post*, December 17, 1998, p. A54. Jusuf Wanandi, "ASEAN's Challenges for its Future," *Pacific Forum CSIS PacNet* No. 3 (January 22, 1999). Internet: http://www.nyu/globalbeat/asia/Wanandi012399.html. Accessed May 27, 1999. Seth Mydans, "Cambodia Gains Full Status in Southeast Asian Group," *New York Times*, April 30, 1999, p. A10.

78. Henderson, *Reassessing ASEAN*, p. 49.

79. Richburg, "Regional Crises."

80. Ralph A. Cossa, "Asia Pacific Confidence and Security Building Measures," in Cossa, ed., *Asia Pacific Confidence and Security Building Measures*, pp. 1, 21, and Col. Larry M. Wortzel, *The ASEAN Regional Forum: Asian Security without an American Umbrella* (Carlisle Barracks, PA: U.S. Army War College, Strategic Studies Institute, December 1996), pt. 3. Internet: http://carlisle-www.army.mil/usassi/ssipubs/pubs96/asean/aseanai.html.

81. Leifer, *The ASEAN Regional Forum*, pp. 21–22.

82. Ibid., pp. 32–33.

83. "The difference in name between two identical entities was again an attempt to accommodate Chinese objections, in this case to any impression of continuous institutionalized activities." Leifer, *The ASEAN Regional Forum*, p. 42.

84. Australia, Department of Foreign Affairs and Trade [AUS/DFAT], "Agreement between the Government of Australia and the Government of the Republic of Indonesia on Maintaining Security," Australian Treaty Series 1996 No. 13. Internet: http://www.austlii.edu.au/au/other/dfat/treaties/19960013.html. Accessed December 3, 1997.

85. Wortzel, *The ASEAN Regional Forum: Asian Security Without an American Umbrella*, p. 4. In the 1976 incident, American military police trimming a tree in the DMZ were attacked with axes by North Korean troops. The United States deployed naval forces and strategic bombers to the region in response, flying B-52s at low level along the DMZ. Most people who experience a B-52 hurtling past at close range remember that experience, even if they aren't targets. IISS, *Strategic Survey 1976* (London: IISS, 1977), pp. 103–04.

86. The 22 participants as of July 1998 are Australia, Brunei, Burma (Myanmar), Cambodia, Canada, China, European Union, India, Indonesia, Japan, Laos, Malaysia, Mongolia, New Zealand, Papua New Guinea, Philippines, Republic of Korea, Russia, Singapore, Thailand, United States, and Vietnam.

87. AUS/DFAT, "Chairman's Statement, Third ASEAN Regional Forum," Jakarta, 23 July 1996. Internet: http://www.dfat.gov.au//arf/arf3.html. Accessed December 3, 1997.

88. AUS/DFAT, "Intersessional Support Group on Confidence Building Measures," March 6–8, 1997, Beijing, China, Internet: http://www.dfat.gov.au/arf/arf_1_confid.html. Accessed December 3, 1997.

89. "The fourth meeting of the ASEAN regional forum," *CSCAP Newsletter*, October 1997. Internet: http://www.dfat.gov.au/arf/regional_97.html. Also AUS/DFAT, "ARF Meetings and Activities, 1994–99," Internet: http://www.dfat.gov.au/arf/arf_meet.html. The listing has links to each available online report for ARF ministerials and intersessional meetings. For synopses of agreed CSBMs and their status, see

AUS/DFAT, "Implementation of ARF Agreed CBMs," Internet: http://www.dfat. gov.au/arf/arfhomeMatrix.html. Accessed June 23, 1999.

90. Leifer, *The ASEAN Regional Forum*, p. 59. A summer 1997 piece in *Beijing Review* emphasized that Asian regional security arrangements should involve no arms control elements, no "collective interference" in "international disputes" (i.e., the Spratlys), and no discussion of "unresolved territorial disputes left from history" (i.e., the status of Taiwan or Tibet). A. Ying, "New Security Mechanism Needed for Asia-Pacific Region," *Beijing Review,* August 1997. Internet: http://www.china.or.cn/bjreview/97Aug/ 97-33-6.html.

91. Kenneth W. Allen, "Confidence-Building Measures and the People's Liberation Army," paper presented at the Conference on The PRC's Reform: A Reappraisal after Twenty Years, National Chengchi University, Taipei, Taiwan, April 8-9, 1999. Mimeo. (Copies available from the author at the Henry L. Stimson Center, Washington, D.C.)

92. Institute of Contemporary International Problems, Russian Diplomatic Academy, Moscow. *DPRK Report No. 7,* May-June 1997. Distributed online at http://www.nautilus.org/pub/ftp/napsnet/RussiaDPRK/DPRK_7.txt.

93. Keith B. Richburg, "Beyond a Wall of Secrecy, Devastation," *Washington Post,* October 19, 1997, p. A1. Richburg, the first American journalist to travel extensively in the North, reported the food crisis as "just one more sign of total systemic collapse" that had left North Korea's economy in "free fall." Extensive interviews of North Korean refugees in China conducted by South Korean NGOs indicated that, by late 1997, child mortality in North Korea (ages 0-6) may have reached 45 percent, and mortality among the elderly, who give food rations to their grandchildren, may have reached 74 percent. (Korean Buddhist Sharing Movement, "Famine Witnessed by 472 Korean Refugees Interviewed in China," background and executive summary, March 4, 1998, available on the Internet at: ftp://ftp.nautilus.org/napsnet/special_reports/Survey_on_ North_Korean_Refugees.txt.)

94. *Associated Press,* April 11, 1998, 11:42 EDT. Internet: http://www.nytimes.com/aponline/i/ AP-Koreas-Dialogue.html.

95. Seth Faison, "One Korean Certainty: No More Business as Usual," *New York Times,* January 4, 1998, IV, p. 1.

96. International Institute for Strategic Studies, *Strategic Survey, 1994-95* (London: Oxford University Press for the IISS, 1995), p. 180. U.S. General Accounting Office, *Nuclear Nonproliferation: Implementation of the U.S./North Korean Agreed Framework on Nuclear Issues,* GAO/RCED/NSIAD-97-165, June 1997, pp. 50-51.

97. GAO, *Nuclear Nonproliferation,* pp. 33-34, and Ralph Cossa, "Monitoring the Agreed Framework," PacNet Report 43 (Honolulu: Pacific Forum CSIS, October 24, 1997). Reprinted by the Nautilus Policy Forum Online, at http://www.nautilus.org/napsnet/fora/11a_Cossa.html.

98. GAO, *Nuclear Nonproliferation,* pp. 33-34, 37. Korean Peninsula Energy Development Organization [KEDO], "KEDO Executive Board Agrees on Cost Sharing for Light Water Reactor Project," KEDO Press Release, November 10, 1998. Internet: http://www.nyu.edu/globalbeat/asia/KEDO111098.html. Accessed May 27, 1999.

99. David E. Sanger, "North Korea: Missile Talks Open," *New York Times,* March 30, 1999, p. A6. Elizabeth Becker, "U.S. Says Photos Show North Korea Preparing for Missile," *New York Times,* June 18, 1999, p. A3. Jane Perlez, "US-North Korea Talks on Missile Program Set for March," *New York Times,* January 31, 2000, p. A3.

100. "Background Briefing by Senior American Official on Four-Party Preparatory Talks," at the Columbia University Kellogg Center for Continuing Education, School of Public and International Affairs, New York, November 21, 1997. Transcript distributed by the Northeast Asia Peace and Security Network (NAPSNET) on the Internet at: http://www.nautilus.org/pub/ftp/napsnet/special_reports/US_Official_on_Four-

Party_Prepatory_Talks.txt. See also "United States, Four-Party Peace Talks," NAP-SNET Daily Report for March 23, 1998, citing *Reuters* newswire. Available at: http://www.nautilus.org/napsnet/recent_daily_reports/03_98_reports/MAR23.html.

101. Elizabeth Olsen, "Koreas Take a First Step to Work Out Formal Peace," *New York Times,* October 25, 1998, p. 9. U.S. Information Agency, "Background Briefing on Korea Four-Party Talks," April 27, 1999. Internet: http://www.nyu.edu/globalbeat/asia/USIA050399.html. Accessed May 27, 1999.

102. Elaine Kurtenbach, "Koreas Conclude 1st Round of Talks," *Associated Press,* April 11, 1998, 10:31 EDT. Internet: http://www.washingtonpost.com/wp-srv/inatl/longterm/korea/korea.htm. Michael Laris and Kevin Sullivan, "After One-Day Delay, North and South Korea Hold Talks in Beijing," *Washington Post,* June 22, 1999, p. A10. The sinking of the North Korean vessel off the peninsula's western coast came after a week-long standoff over North Korean navy attempts to escort fishing vessels into waters south of the "northern limit line" claimed by South Korea, a demarcation not recognized by the North. Twenty North Korean sailors died in the 30-minute fire fight. Sheryl WuDunn, "Korean Fight at Sea Evolves into Exchange of Charges," *New York Times,* June 16, 1999, p. A10.

103. David E. Sanger, "Korean Clash May Ruin U.S. Reconciliation Bid," *New York Times,* June 17, 1999, p. A14. Kevin Sullivan and Mary Jordan, "S. Korea's Kim Defends Détente with North," *Washington Post,* June 24, 1999, p. A1. (Interview with South Korean President Kim Dae Jung.)

104. University of California, Institute on Global Conflict and Cooperation [UC-IGCC], "About Northeast Asia Cooperation Dialogue." Internet: http://www-igcc.ucsd.edu/igcc2/neacd/aboutneacd.html. Accessed June 24, 1999. Each participating country sends five delegates to a meeting: one policy-level official each from the foreign and defense ministries, a uniformed military officer, and two participants from "private research facilities, think-tanks, or universities." The 1997 principles of cooperation included: respect sovereignty and territorial integrity; accept that other countries have different political, economic, social, and cultural systems; refrain from threat or use of force; commitment to the protection and promotion of human rights in accordance with the purposes and principles of the UN Charter; promote dialogue and transparency; respect freedom of navigation; promote economic cooperation; cooperate on transnational issues; and cooperate in the provision of humanitarian assistance. "NEACD VII Reaches Watershed Agreements," *NEACD Newsletter,* April 29, 1998. Internet: http://neacd.llnl.gov/news/neacd.html. Accessed June 24, 1999.

105. For a report on the first Georgia Tech meeting, see ftp://ftp.nautilus.org/napsnet/special_reports, Military-to- Military_Security_Dialogue.txt.

106. Brian Job, "Report on the Fourth Meeting of the NPWG, November 1998," CSCAP, March 31, 1999. Internet: http://www.cscap.org/pacific.htm. Accessed June 23, 1999.

107. Nautilus is also host for a number of peace-and-security-related networks. Internet: http://www.nautilus.org.

108. UC-IGCC, "Virtual Dialogue in Northeast Asia: A Virtual Research Center Supporting Regional Multilateral Peacebuilding," and "Virtual Track-two Diplomacy, A Proposal to the U.S. Institute of Peace." Internet: http://www-igcc.ucsd.edu/igcc2/NEACD/USIPNEAC.html. The initial phase of IGCC's project relied heavily on donated time of researchers and technologists. Recognizing the uneven distribution of networking technology in Northeast Asia (as in, virtually none in North Korea), IGCC proposed to provide the necessary hardware and software. Whether the current regime in North Korea will buy into even the focused and limited-access information system that IGCC has in mind is not clear.

109. Moon Chung-in, "Arms Control and Peace on the Korean Peninsula," paper no. 5, *NAPSNet Policy Forum Online,* June 26, 1997. Available at http://www.nautilus.org/napsnet/fora/5a_Moon.html.

110. Ralph Cossa, "South Korea's Package Deal," *PacNet* no. 10, March 12, 1999. Internet: http://www.nyu.edu/ globalbeat/asia/Cossa031299.html. Accessed May 27, 1999.

111. Center for Contemporary International Problems (Moscow), "Prospects for North Korean-US Relations," *DPRIK Report* No. 18, May–June 1999. Published by the Center for Nonproliferation Studies, Monterey, Calif. Available at Internet: http://www.nyu.edu/globalbeat/asia/CNS0599.html. Accessed May 27, 1999.

112. Ibid.

113. Domingo E. Acevedo and Claudio Grossman, "The Organization of American States and the Protection of Democracy," in *Beyond Sovereignty: Collectively Defending Democracy in Latin America,* ed. Tom Farer (Baltimore: Johns Hopkins University Press, 1996), p. 135.

114. Tom Farer, "Collectively Defending Democracy in the Western Hemisphere," in Farer, *Beyond Sovereignty,* p. 1.

115. For analysis of Latin America's transition to democracy, see Jorge I. Dominguez, *Democratic Politics in Latin America and the Caribbean* (Baltimore: Johns Hopkins University Press, 1998), and J. Samuel Fitch, *The Armed Forces and Democracy in Latin America* (Baltimore: Johns Hopkins University Press, 1998).

116. For an overview of the democratic peace thesis, see James Lee Ray, *Democracy and International Conflict* (Columbia: University of South Carolina Press, 1995).

117. "Eight Latin American Governments Sign Declaration Aimed at Limiting Armaments," *U.N. Monthly Chronicle* (March 1975), pp. 54–57, cited in Pierre, *Global Politics of Arms Sales,* p. 283.

118. Pierre, *Global Politics of Arms Sales,* 284.

119. Ibid., and Coit D. Blacker and Gloria Duffy, eds., *International Arms Control: Issues and Agreements,* 2nd ed. (Stanford: Stanford University Press, 1984), p. 333.

120. Child, "Confidence-building Measures and Their Application in Latin America," pp. 16–18.

121. Acevedo and Grossman, "The Organization of American States," p. 136.

122. Ibid.

123. OAS AG/Res. 1080 (XXI-0/91), June 1991. Cited in Acevedo and Grossman, p. 137.

124. Ibid., pp. 142–45.

125. Ibid., pp. 140–42. Fitch, *The Armed Forces and Democracy in Latin America,* p. 59.

126. OAS, *Annual Report of the Secretary General, 1997–98,* Washington, D.C., 1998. Introduction. Internet: http://www.cidi.oas.org/SGannualchap-e.html. Accessed June 18, 1999.

127. Child, "Confidence-building Measures and Their Application in Latin America," p. 19.

128. *Keesing's Record of World Events,* December 1995, p. 40862.

129. U.S. Arms Control and Disarmament Agency, "The Central American Democratic Security Treaty," fact sheet, May 20, 1996. Internet: http://www.acda.gov/factshee/secbldg/cademosb.html.

130. U.S. Congress, House International Relations Committee, "Testimony of Deputy Assistant Secretary of State John Hamilton before the Subcommittee on the Western Hemisphere," 105th cong., 1st sess., June 25, 1997. Internet: http://www.house.gov/international_relations/wh/wswh6252.html.

131. Joaquín Tacsan, Director, Center for Peace and Reconciliation, Arias Foundation, "The Failure in San Pedro Sula," *Inforpress Central America* no. 3 (January 1996). Internet: http://www.us.net/cip/dialogue/9601in01.html. The Arias Foundation supports the Central American Dialogue for Security and Demilitarization, which promotes the efforts of regional NGOs toward "a greater civilian role in questions of security."

132. U.S. Arms Control and Disarmament Agency, "Organization of American States Conference on Confidence- and Security Building Measures," fact sheet, May 20, 1996. Internet: http://www.acda.gov/factshee/secbldg/orgsbp.htm

133. U.S. Department of State, Bureau of Inter-American Affairs, "Major Accomplishments of the OAS General Assembly, Lima, Peru, June 1–5, 1997," June 27, 1997. Internet:

http:/www.state.gov/www/regions/ara/fs_oasga_accomp062797.html. U.S. Arms Control and Disarmament Agency, "Organization of American States (OAS) General Assembly Resolution on Conventional Arms Transparency and Confidence-Building in the Americas," July 30, 1997. Internet: http://www.acda.gov/factshee/secbldg/unoas.htm. Also, OAS, *Annual Report . . . 1997–98.* The Carter Center, in Atlanta, Georgia, hosts a track II effort called the Council of Freely Elected Heads of Government, which counts among its members 10 current and 22 former heads of governments in the Western Hemisphere from 19 countries as part of its Latin America and Caribbean Program. The program focuses on election monitoring and mediating, building democratic institutions and transparency in government, and promoting hemispheric cooperation in "stemming the flow of illegal drugs, resolving disputes over borders, limiting arms sales, and reducing social inequities and corruption." Internet: http://www.cartercenter.org/latinamerica.html. Accessed June 26, 1999.

134. John G. Ruggie, *Winning the Peace,* A Twentieth Century Fund Book (New York: Columbia University Press, 1996), pp. 104–05.

135. Ibid., pp. 80–88, citing the earlier work of Karl Deutsch, et al., *Political Community and the North Atlantic Area* (Princeton: Princeton University Press, 1957).

136. Kalevi J. Holsti has extensive discussion of South America's transformation over the past century from a "zone of war" to a "zone of peace" edging toward a security community. See his *The State, War, and the State of War* (New York: Cambridge University Press, 1995), ch. 8.

137. North Korea briefly detained a South Korean tourist who had "preached defection" to her tour guide. Associated Press, "S. Korean President Warns N. Korea Against Missile Test," *Washington Post* online, June 25, 1999. Internet: http://www.washingtonpost.com/wp-sr/inatl/daily/june99/warn25.htm. Accessed June 29, 1999.

Chapter 6

1. For an innovative analysis of the relative strengths and weaknesses of different forms of organization, see David F. Ronfeldt, "Tribes, Institutions, Markets, and Networks: A Framework About Societal Evolution," RAND Paper P-7967 (Santa Monica, Calif.: RAND Corporation, 1996).

2. Andrew J. Pierre, "Toward an International Regime for Conventional Arms Sales," in *Cascade of Arms: Managing Conventional Weapons Proliferation,* A. J. Pierre, ed. (Washington, D.C.: The Brookings Institution Press, 1997), pp. 374–75.

3. "G7 Statement," June 18, 1999, and "Communique, Köln 1999," June 20, 1999. G8 Summit, Köln, Germany, June 18–20, 1999. Reprinted by the University of Toronto G8 Information Centre. Internet: http://www.library.utoronto.ca/g7/99koln. Accessed July 5, 1999.

4. Thomas L. Friedman, "The Reverse Domino," *New York Times,* March 19, 1999, p. A21 (on Thailand).

5. Stephen D. Biddle, "Recent Trends in Armor, Infantry, and Artillery Technologies: Developments and Implications," in W. Thomas Wander, Eric H. Arnett, and Paul Bracken, eds., *The Diffusion of Advanced Weaponry: Technologies, Regional Implications, and Responses* (Washington, D.C.: American Association for the Advancement of Science, 1994), p. 122.

6. Ibid., pp. 123–25.

7. Laurie Nathan, *The Changing of the Guard: Armed Forces and Defence Policy in a Democratic South Africa* (Pretoria: HSRC Publishers, 1994), pp. 2, 85–86. Cited in Joao Honwana, "Implementing Peace Agreements in Civil Wars: The Case of Mozambique," paper presented to the authors' conference of the project on Peace Implementation, Stanford University Center for International Security and Cooperation, October 24–26, 1998, Stanford, California, p. 36.

8. Honwana, "Implementing Peace Agreements in Civil Wars," p. 36.
9. Council of Freely Elected Heads of Government, "Statement," in *Final Report: Agenda for the Americas for the 21st Century* (Atlanta: The Carter Center, April 1997), pp. 7–15.
10. Steven Lee Myers, "The Latin Arms Explosion That Fizzled," *New York Times,* December 3, 1998, p. A3. International Institute for Strategic Studies, *The Military Balance, 1998–1999* (London: IISS, 1998), pp. 207–09.
11. Frederic S. Pearson, *The Global Spread of Arms: Political Economy of International Security* (Boulder, Colo.: Westview Press, 1994), p. 92, suggests outside support for peacekeeping measures plus security guarantees to states participating in restraint regimes.
12. U.S. government and business—covered by a comparable American statute for 20 years—assiduously sought the OECD action. Organization for Economic Cooperation and Development, *Convention on Combating Bribery of Foreign Public Officials in International Business Transactions,* signed in Paris, December 17, 1997. For the full text online, see: http://www.unog.ch/frames/disarm/review/conweap.htm. For OECD interpretation of the convention, adopted by the OECD's negotiating conference on November 21, 1997, see: http://www.oecd.org/daf/cmis/bribery/20nov2e.htm.
13. Negotiations that aim to end some competitive activity, whether fighting or arms buying, can function as a temporary accelerant to that activity unless steps are taken in advance to anticipate and reduce the accelerant effect. Generally, the greater the prospect of accord, and the deeper the anticipated impact of that accord, the greater the accelerant effect, as the would-be signatories jockey for the very best position they can manage before an agreed freeze or constraint goes into effect. One way to control this phenomenon is to put the activity under discussion into temporary stasis *before* talks begin—by means of an immediate cease-fire in place, for example, where a fight is the focus of discussion—and to monitor compliance closely, as United Nations peacekeepers have done in many cease-fire situations. The downside of a stasis agreement is that talks conducted under its umbrella can continue for years, as the UN has found, to its chagrin, in places like Cyprus.
14. Jozef Goldblat, *Agreements for Arms Control: A Critical Survey* (Cambridge, Mass.: Oelgeschlager, Gunn & Hain, Inc., 1982), treaty text pp. 162–70 and commentary pp. 63–68.
15. The United States was sufficiently leery of the land mines convention that it refrained from signing the 1997 treaty, citing the continuing need for such devices in Korea as long as the military standoff there continues. By opposing all prohibitions on antipersonnel mines, the United States missed a potential opportunity to write into the agreement some sort of exception for self-disabling or "smart" mines that become nonfunctional after a specified time interval. Land mines were singled out for special attention by the NGO community precisely because the vast majority of the mines used in conflicts in the developing world outlive their wars and were, moreover, sown in unmarked and uncatalogued areas, often with the objective of terrorizing or displacing the civilian population. The debate on land mines is thus polarized between military officers and analysts who start from the perspective of force protection, for which this class of weapon can be very useful in certain wartime situations, and those who start from the perspective of women, children, and old people with missing eyes and limbs who stumbled into an unmarked and/or left-behind minefield. To date, these perspectives have proven very difficult to reconcile.
16. A number of peace settlements since the early 1990s have called for the collection of such arms as part of the process of demobilizing fighting forces and returning their members to civilian life. But since those responsible for demobilization cannot know precisely how many small arms a given force has and are unlikely to be told the truth, it is next to impossible to collect all of the small arms related to a conflict even if a peace accord calls for it and all parties have agreed to the provision. Since jobs for ex-fighters are often scarce in postconflict situations but arms are plentiful, such situations are frequently plagued by

high rates of violent crime. Because local security forces may themselves have been torn apart by a country's internal strife or may be closely associated with one or another political faction, the local populace also has every incentive to arm itself and may resist official efforts to seize personal arms as vigorously as it supports official efforts to remove land mines. Both impulses have their roots in a desire for personal security in a semi-anarchic environment. For a good compendium of views on the issue, see E. A. Zawels, et al., *Managing Arms in Peace Processes: The Issues* (Geneva: United Nations Institute for Disarmament Research, 1996).

17. *Executive Summary of the Report of the Commission to Assess the Ballistic Missile Threat to the United States* ["The Rumsfeld Report"], Pursuant to Public Law 201, 104th Congress, July 15, 1998, esp. pp. 12–13. Barbara Crossette, "Iraq Has Network of Outside Help on Arms, Experts Say," *New York Times,* November 20, 1998, p. A3.

18. J. Jerome Holton, Lora Lumpe, and Jeremy J. Stone, "Proposal for a Zero Ballistic Missile Regime," *1993 Science and International Security Anthology* (Washington, D.C.: American Association for the Advancement of Science, 1993), pp. 379–96. Lora Lumpe, "Zero Ballistic Missiles and the Third World," *Arms Control,* 14, no. 1, April 1994, and Lumpe, "Arms Control Options for Delivery Vehicles," *Physics and Society,* 24, January 2, 1995. All articles are available online from the Federation of American Scientists at http://www.fas.org/asmp/library/articles/. See also Janne E. Nolan, *Trappings of Power: Ballistic Missiles in the Third World* (Washington, D.C.: The Brookings Institution, 1991), pp. 155–67, and Steve Fetter, "Ballistic Missiles and Weapons of Mass Destruction," *International Security,* 16, no. 1 (Summer 1991), pp. 5–42.

19. International Institute for Strategic Studies, *The Military Balance, 1997–1998* (Oxford: Oxford University Press, 1997), pp. 129, 141.

20. "MGM-140 ATACMS," in *Forecast International/DMS Market Intelligence Report: Missile Forecast,* January 1996, p. 1.

21. John R. Harvey, "Regional Ballistic Missiles and Advanced Strike Aircraft: Comparing Military Effectiveness," *International Security,* 17, no. 2 (Fall 1992), esp. pp. 79–83.

22. Alton Frye, "Banning Ballistic Missiles," *Foreign Affairs,* Nov/Dec 1996, pp. 99–112. U.S. Senate, Committee on Armed Services, *Hearing Before the Strategic Subcommittee on Strategic Nuclear Forces and Policy in Review of the Defense Authorization Request for Fiscal Year 2000 and the Future Years Defense Program,* testimony of Hon. Edward L. Warner, III, Assistant Secretary of Defense for Strategy and Threat Reduction, April 14, 1999. Mimeo. Internet: http://www.clw.org/coalition/warner0599.htm.

23. Holton, Lumpe, and Stone, "Proposal for a Zero Ballistic Missile Regime."

24. See, for example, Brian P. Curran, William J. Durch, and Jolie M. F. Wood, "Alternating Currents: Technology and the New Context for U.S. Foreign Policy," Foreign Policy Project Occasional Paper No. 5 (Washington, D.C.: The Henry L. Stimson Center, October 1997), p. 2.1.

25. Wolfgang Reinecke, "Cooperative Security and the Political Economy of Nonproliferation," in *Global Engagement,* ed. Janne E. Nolan (Washington, D.C.: Brookings Institution, 1994), pp. 187–88.

Dual-use technology with applications to WMD should be subject to more stringent constraints, consistent with the Nuclear Non-Proliferation Treaty and the Chemical Weapons Convention. Biotechnology presents perhaps the greatest challenge to the future of technology transfer control regimes. The Biological and Toxin Weapons Convention (BWC) of 1975 prohibits development, production, stockpiling or other acquisition of types or quantities of "microbial or other biological agents or toxins" that "have no justification for prophylactic, protective, or other peaceful purposes." The Geneva Protocol of 1925 had already prohibited use but like the BWC, the Geneva Protocol lacks enforcement clauses. Many of the tools applicable to legitimate genetic and other biological research are equally applicable to research and development of biological weaponry and the tools and techniques for maintaining a defensive ("prophylactic, protective") capability against biological warfare may differ only in scale from those associated with offensive capability, making compliance verification very difficult and commercial sensitivities to intrusive monitoring especially acute.(For texts, see U.S. Arms Control and Disarmament Agency [USACDA], *Arms Control and Disarmament Agreements: Texts and Histories of Negotiations,* 5th ed. ([Washington, D.C.: USACDA, 1982] For an updated list of agreements, see Internet: http://www.acda/gov.)

26. Reinecke, "Cooperative Security and the Political Economy of Nonproliferation," pp. 211–14.
27. Gansler, *Defense Conversion,* pp. 96–97.
28. James Lee Ray, *Democracy and International Conflict: An Evolution of the Democratic Peace Proposition* (Columbia: University of South Carolina Press, 1995).
29. Kenneth E. Boulding, *Three Faces of Power* (Newbury Park, Calif.: SAGE Publications, 1990), p. 10.
30. I would have preferred to get beyond rank-ordering and create quasi-interval scales for both ease and importance but do not as yet have good enough definitions of the underlying dimensions of either scale.
31. The United States was one of seven states that voted against the establishment of an International Criminal Court at a UN conference in Rome, July 1998. For details on the evolution of the ICC, see the United Nations website. Internet: http://www.un.org/icc/pressrel/lrom22.htm. July 17, 1998. Accessed November 2, 1998.
32. David C. Gompert, "Sharing the Burdens of Global Security," FPP Occasional Paper No. 1 (Washington, D.C.: Henry L. Stimson Center and Overseas Development Council, August 1996). A broader and more institutionally structured "Intercontinental Community of Democracies" is the subject of James R. Huntley in *Pax Democratica: A Strategy for the 21st Century,* foreword by Lawrence Eagleburger (New York: St. Martin's Press, 1998). Huntley is former president of the Atlantic Council.
33. Ibid., p. 28.
34. For criticism, see Marcy Agmon, et al., *Arms Proliferation Policy: Support to the Presidential Advisory Board,* MR-771 (Santa Monica, Calif.: RAND Corporation, 1996), pp. 61–63.
35. For discussion of the implications of institution-building for continuity of the end-century structure and distribution of power in the international system, see G. John Ikenberry, "Institutions, Strategic Restraint, and the Persistence of American Postwar Order," *International Security,* 23, no. 3 (Winter 1998–99).

Appendix A

1. Barry Buzan, *People, States, and Fear: An Agenda for International Security Studies in the Post–Cold War Era,* 2nd ed. (Boulder, Colo.: Lynne Rienner Publishers, 1991), pp. 187–192.
2. Ibid., pp. 193, 196.
3. Mexico does not consider itself a Central American country, and its membership in the North American Free Trade Agreement reinforces the distinction. Mexico can be considered a transitional state like Burma, but like Canada, it spends little to guard against military threats from the United States and can be more accurately said to belong to a North American security community in which military threats are no longer part of interstate relations.
4. For an interesting comparative analysis of British and French colonial and postcolonial policies in Africa, see A. F. Mullins, *Born Arming: Development and Military Power in New States* (Stanford: Stanford University Press, 1987), esp. ch. 2.

Appendix B

1. Michael Brzoska and Thomas Ohlson, *Arms Transfers to the Third World, 1971–85* (New York: Oxford University Press, 1987).

2. For discussion, see *SIPRI Yearbook 1996, Armaments, Disarmament and International Security* (Oxford: Oxford University Press, 1996), Appendix 11C: Sources and methods. Every edition has a similar section.

3. U.S. Arms Control and Disarmament Agency, *World Military Expenditures and Arms Transfers 1997* (Washington, D.C.: ACDA, 1998). Internet:

4. Richard F. Grimmett, *Conventional Arms Transfers to the Third World, 1986–93,* CRS report 94–612F (Washington, D.C.: Congressional Reference Service, 1994); and Grimmett, *Conventional Arms Transfers to Developing Nations, 1989–96,* CRS Report 97–778F (Washington, D.C.: Congressional Reference Service, 1997).

5. U.S. Department of Defense, Office of the Undersecretary of Defense (Comptroller), *National Defense Budget Estimates for FY 1999,* March 1998. Of several price deflators presented, the U.S. Gross Domestic Product Implicit Price Deflator is "generally regarded as the best single measure of broad price movements in the [US] economy." The conversions from current dollars using that deflator can be considered at best approximations of the actual value of the arms trade for any given year, because the exchange rate value of the U.S. dollar against different countries' currencies has varied a great deal over time, sometimes stronger, sometimes weaker. But since the annual tallies graphed in figure B.1 are at best educated guesses as to the value of the arms trade, we're not looking at two-decimal-point precision in any case, and the conversion to constant dollars certainly gives us a more realistically shaped historical curve of arms transfer values.

6. Conventional arms transfer data for 1950–1985 and 1998 are drawn from the project "Arms Transfers and the Onset of War," copyright © 1993 by William H. Baugh and Michael L. Squires. Used by permission of the authors.

7. Lewis W. Snider, *Arabesque: Untangling the Patterns of Supply of Conventional Arms to Israel and the Arab States,* Denver Monograph Series in World Affairs, vol. 15, book one (Denver: University of Denver, 1977). Ronald G. Sherwin and Edward J. Laurance, "Arms Transfers and Military Capability: Measuring and Evaluating Conventional Arms Transfers," *International Studies Quarterly,* 23, no. 3 (September 1979), pp. 360–389. For a comparative assessment of factor analysis, multi-attribute utility analysis, and canonical correlation as measurement devices for determining military capabilities, see Jeffrey T. Richelson, Lewis W. Snider, and Abraham R. Wagner, *Arms Transfer Control Criteria: Quantitative Measures and Analytical Approach,* report TR-6001/78 for the U.S. Arms Control and Disarmament Agency (Marina del Ray, Calif.: Analytical Assessments Corporation, July 1978). For the army's use of MAU, see U.S. Army Concepts Analysis Agency, War Gaming Directorate, *Weapon Effectiveness Indices/Weighted Unit Values III (WEI/WUV III),* CAA-SR-79–12 (Bethesda, Md.: CAA, November 1979). Declassified December 31, 1986.

8. U.S. Congressional Budget Office, *A Look at Tomorrow's Tactical Air Forces,* A CBO Study (Washington, D.C.: U.S. GPO, January 1997), pp. 14–15; U.S. Navy, Office of Naval Intelligence, *Worldwide Challenges to Naval Strike Warfare* (Washington, D.C.: ONI, February 1997), and Robert P. Berman, *Soviet Air Power in Transition* (Washington, D.C.: The Brookings Institution, 1978), p. 52.

9. For an elaborate but still static multi-attribute utility analysis of armor capabilities, see U.S. Army, WEI/WUV III, Appendix E, Tank WEI Calculation.

10. In the point-counterpoint of military technology, some antitank guided missile (ATGM) designers have adapted to reactive armor by giving their missiles two charges, the smaller one mounted in a nose probe designed to trigger reactive armor plates a few milliseconds prematurely, the larger charge to blast through the main armor be-

neath. The newest generation of ATGMs overfly their target and fire shaped charges di
rectly downward onto the parts of the target that are traditionally least heavily protected.

Appendix C

1. James F. Dunnigan and William Martel, *How to Stop a War: The Lessons of Two Hundred Years of War and Peace* (New York: Doubleday, 1987), p. 181.
2. Joshua Goldstein, "A Conflict-Cooperation Scale for WEIS Events Data," *Journal of Conflict Resolution* 36, no. 2 (June 1992), pp. 369–85.
3. Ted Robert Gurr, *Why Minorities Rebel: A Global View of Ethnopolitical Conflicts* (Washington, D.C.: U.S. Institute of Peace, 1993), p. 95.
4. Frank Whelon Wayman and J. David Singer, "Evolution and Directions for Improvement in the Correlates of War Project Methodologies," in Singer and Diehl, eds., *Measuring the Correlates of War*, pp. 263–65.
5. The MID data set is drawn from a wide variety of sources and has been in existence long enough to have gone through several iterations. Version 2.10 is currently available on the Internet and catalogues threats and displays of force, uses of force, and war from 1816 to 1992. A description of the data set's structure may be found in Charles S. Gochman and Zeev Maoz, "Militarized Interstate Disputes, 1816–76," *Journal of Conflict Resolution* 28, no 4 (1984), pp. 585–615. The online.zip file contains files identifying the variables used in the data set and identifying the disputes coded. It is presented by the Peace Science Society International on the website of the political science department of Pennsylvania State University, at http://www.pss.la.psu.edu/mid_data.htm. (Internet address verified March 17, 1998.)
6. Ted Robert Gurr uses the same coding practice. See his *Minorities at Risk,* p. 94.
7. Sources used include Neil Grant, *Chronicle of 20th Century Conflict* (New York: Smithmark Publishers, Inc., 1993); James F. Dunnigan and William Martel, *How to Stop a War;* Stockholm Peace Research Institute, *Yearbook of Armaments and Disarmament* (annual), and its listings of major conflicts; and John W. Wright, ed., *The Universal Almanac,* 1991 and 1994 eds. (Kansas City: Andrews and McMeel, 1990, 1993). Supplementary data from Paul Beaver, ed., "Global Update: Flashpoints and Conflicts," special report no. 2, *Jane's Defence Weekly,* 1994.
8. Perhaps the best-known quantitative model of arms races is that devised by the late Louis Fry Richardson, who posited that the rate at which country A acquires arms is a linear function of "the perceived external threat, the economic burden of military competition, and the magnitude of grievances against the other party" (Joshua M. Epstein, "Security Regimes and Arms Race Dynamics," presentation to the Defense and Arms Control Studies Seminar, Massachusetts Institute of Technology, Cambridge, Mass., December 2, 1992). For an exhaustive bibliography of the extensive literature on Richardson models, see Craig Etcheson, *Arms Race Theory: Strategy and Structure of Behavior* (New York: Greenwood Press, 1989). Analysts have also applied game theory to arms race behavior (George W. Downs, David M. Rocke, and Randolph M. Siverson, "Arms Races and Cooperation," *World Politics* 38 [1985], pp. 118–46), as well as non-linear models, which warn that competitive political-military interactions may be subject to chaotic instabilities (Alvin M. Saperstein, "Alliance Building versus Independent Action: A Non-linear Modeling Approach to Comparative International Stability," *Journal of Conflict Resolution,* 36, no. 3 [September 1992], pp. 518–45, and Murray Wolfson, Anil Puri, and Mario Martelli, "The Non-linear Dynamics of International Conflict," *Journal of Conflict Resolution,* 36, no. 1 [March 1992], pp. 119–49).
9. See, for example, the debates among scholars over the relative fighting capabilities of NATO and the Warsaw Pact in late-1980s Europe, stimulated by concerns over the

relationship of the conventional force balance in Europe to the potential use of nuclear weapons in a NATO-Warsaw Pact conflict. The debate was given added impetus by pending conventional force reduction talks and the need to get the final force ratio "right." One escalating debate was touched off by "Policy Focus: The European Conventional Balance," *International Security*, 12, no. 4 (Spring 1988), pp. 152–202, with contributions by Joshua M. Epstein, Kim R. Holmes, John J. Mearsheimer, and Barry R. Posen. It was followed by Eliot Cohen, "Toward Better Net Assessment: Rethinking the European Conventional Balance," *International Security*, 13, no. 1 (Summer 1988); John J. Mearsheimer, "Assessing the Conventional Balance: The 3:1 Rule and Its Critics," and Joshua M. Epstein, "The 3:1 Rule, the Adaptive Dynamic Model, and the Future of Security Studies," *International Security*, 13, no. 4 (Spring 1989), pp. 54–89, 90–127, and correspondence relating to the Cohen piece by Mearsheimer, Posen, and Cohen, pp. 128–79.

10. For a discussion of the strictures on using path analysis for purposes of causal influence, see Thomas D. Cook and Donald T. Campbell, *Quasi-Experimentation: Design and Analysis Issues for Field Settings* (Boston: Houghton Mifflin Co., 1979), pp. 296–97.

11. Gordon Hilton, *Intermediate Politometrics* (New York: Columbia University Press, 1976), pp. 82–84, 93ff.

12. For a good discussion, see Hilton, *Intermediate Politometrics*, pp. 103–105.

13. The test parameter is a ratio of residual mean square error in the model to "pure error," or "the portion of the sample error that cannot be explained or predicted no matter what form the model uses for the X variables." The higher the ratio, the lower the probability that "lack of fit" can be *ruled out* as a serious problem. A relatively conservative probability (0.05) was selected as the cutoff point below which "lack of fit" is deemed a statistically significant problem. For discussion, see SAS Institute, *JMP Statistics and Graphics Guide,* vol. 3.1 (Cary, N.C.: SAS Institute, Inc., 1995), pp. 147–49.

14. For discussion, see Yuhang Shi, "What to Gain from Technical Sophistication?" in *PS: Political Science and Politics* (September 1995), pp. 505–506.

15. Hubert M. Blalock, Jr., *Social Statistics,* 2nd ed. (New York: McGraw-Hill Book Co., 1972) pp. 499–500. The omitted dummy variable is a perfect linear combination of the others. If you have just four regional categories, and country X has a score of zero in regions one, two, and three, it is always going to have a score of one in category four. If ordinary least squares is to function, either that redundant fourth variable category or the regression intercept, alpha, must be suppressed.

16. Fox recommends that dummy variable sets be preserved when pruning a regression model. John Fox, *Applied Regression Analysis, Linear Models, and Related Methods* (Thousand Oaks, Calif.: SAGE Publications, 1997), p. 359.

17. Ibid., pp. 357–58.

18. Variables for 1970–1974 were treated as exogenous, that is, as predictors only. For this application of ordinary least squares regression to "breaking a [path] diagram down into a set of sub-diagrams," see David R. Heise, *Causal Analysis* (New York: John Wiley & Sons, 1975), p. 158.

19. Hilton, *Intermediate Politometrics,* pp. 220–24.

Bibliography

Chapter 1

Ball, Nicole. *Security and Economy in the Third World*. Princeton: Princeton University Press, 1988.

Beal, Clifford, and Barbara Starr. "Fighter Upgrades: The Only Game in Town." *International Defense Review* 6 (1993): 445–50.

Berman, Robert P. *Soviet Air Power in Transition*. Washington, D.C.: The Brookings Institution, 1978.

Blacker, Coit D., and Gloria Duffy, eds. *International Arms Control, Issues and Agreements*, 2nd ed. Stanford: Stanford University Press, 1984.

Brandt, Craig M. "American Weapon Sales and the Defense Industrial Base: Can Arms Transfers Save Defense Production?" A paper presented at the annual meeting of the International Studies Association. San Diego, Calif., April 16–20, 1996.

de Briganti, Giovanni. "French Move Belies Policy." *Defense News* (April 7, 1997): 1.

———. "Policy Disputes Stymie Europe Consolidation." *Defense News* (May 12, 1997): 6.

de Briganti, Giovanni, and Stephen C. LeSueur. "As Procurement Sags, Upgrade markets Boom." *Defense News* (July 4–10, 1994): 8.

Dismukes, Bradford, and James M. McConnell, eds. *Soviet Naval Diplomacy*. New York: Pergamon Press, 1979.

Donnelly, John, and Colin Clark. "Merger Mania Hits $53 billion This Year, So Far." *Defense Week* (July 7, 1997): 1.

Felgengauer, Pavel. "An Uneasy Partnership: Sino-Russian Defense Cooperation and Arms Sales." In *Russia in the World Arms Trade*. Edited by Andrew Pierre and Dmitri V. Trenin. Washington, D.C.: Brookings Institution Press, 1997.

Forsberg, Randall, ed. *The Arms Production Dilemma: Contraction and Restraint in the World Combat Aircraft Industry*. CSIA Studies in International Security 7. Cambridge, Mass.: The MIT Press, 1994.

Foye, Stephen. "Russian Arms Exports after the Cold War." *RFE/RL Research Report* 2:13 (1993): 62.

Gaddy, Clifford G., and Melanie L. Allen. "Russian Arms Sales Abroad: Policy, Practice, and Prospects." Washington, D.C.: Brookings Institution, Foreign Policy Studies Program, September 1993.

Gamba, Virginia, ed. *Society Under Siege: Licit Responses to Illicit Arms*. Toward Collaborative Peace Series: Vol. II. Halfway House, South Africa: Institute for Security Studies, 1998.

Garthoff, Raymond. *Détente and Confrontation: American-Soviet Relations from Nixon to Reagan*. Washington, D.C.: The Brookings Institution, 1985.

Gerasev, Mikhail I., and Viktor M. Surikov. "The Crisis in the Russian Defense Industry: Implications for Arms Exports." In *Russia in the World Arms Trade*. Edited by Andrew J. Pierre and Dmitri V. Trenin. Washington, D.C.: Carnegie Endowment for International Peace, 1997.

Graham, Norman A., ed. *Seeking Security and Development: The Impact of Military Spending and Arms Transfers*. Boulder, Colo.: Lynne Rienner Publishers, 1994.

334 • Constructing Regional Security

Hammond, Grant T. "The Role of Offsets in Arms Collaboration." In *Global Arms Production.* Edited by Ethan B. Kapstein. Lanham, Md.: University Press of America, 1992.

Hastings, Max, and Simon Jenkins. *The Battle for the Falklands.* New York: W.W. Norton, 1983.

International Institute for Strategic Studies [IISS]. *The Military Balance, 1997–98.* Oxford: Oxford University Press, 1997.

Karp, Aaron. "The Rise of Black and Gray Markets." *The Annals,* 535 (September 1994): 187–89.

Keller, Bill. *Arm in Arm: The Political Economy of the Global Arms Trade.* New York: Basic Books, 1995.

Klare, Michael T. "The Subterranean Arms Market: Black-Market Sales, Covert Operations, and Ethnic Warfare." In *Cascade of Arms: Managing Conventional Weapons Proliferation.* Edited by Andrew J. Pierre. Washington, D.C.: Brookings Institution Press, 1997.

Laurance, Edward J. *The International Arms Trade.* New York: Lexington Books, 1992.

———. *Light Weapons and Intrastate Conflict.* A report to the Carnegie Commission on Preventing Deadly Conflict. Washington, D.C.: July 1998.

Lok, Joris J. "Scandinavians Study Joint Procurement." *Jane's Defence Weekly* (February 11, 1995): 2.

Louscher David J., and James Sperling. "Arms Transfers and the Structure of International Power." In *Seeking Security and Development.* Edited by Norman A. Graham. Boulder, Colo.: Lynne Rienner Publishers, 1994.

Martin, David C., and John Walcott. *Best Laid Plans: The Inside Story of America's War Against Terrorism.* New York: Simon and Schuster, 1988.

Michell, Simon. "Remake, Remodel, Retrofit, Remarket." *Jane's Defence Weekly* (December 17, 1994): 25.

Morris, Michael A. "Confidence-Building Measures in the Maritime Domain." In *Confidence-Building Measures in Latin America: Central America and the Southern Cone.* Edited by Augusto Varas et al. Report No. 16, Stimson Center/FLACSO-Chile. Washington, D.C.: The Henry L. Stimson Center, February 1995.

Opall, Barbara. "East Asia Allies Distrust Sino-Russian Strategic Pact." *Defense News* (May 19, 1997): 8.

Pierre, Andrew J. *Cascade of Arms: Managing Conventional Weapons Proliferation.* Washington, D.C.: Brookings Institution Press, 1997.

Prézelin, Bernard, ed. *Combat Fleets of the World 1993.* Trans. A. D. Baker III. Annapolis, Md.: Naval Institute Press, 1993.

Ross, Andrew L. "Developing Countries." In *Russia in the World Arms Trade.* Edited by Andrew J. Pierre and Dmitri V. Trenin. Washington, D.C.: Carnegie Endowment for International Peace, 1997.

Smith, Chris. "Light Weapons and Ethnic Conflict in South Asia." In *Lethal Commerce: The Global Trade in Small Arms and Light Weapons.* Edited by Jeffrey Boutwell, Michael T. Klare, and Laura W. Reed. Cambridge, Mass.: American Academy of Arts and Sciences, 1995.

Stefanick, Tom. *Strategic Antisubmarine Warfare and Naval Strategy.* Lexington, Mass.: Lexington Books, 1987.

Stockholm International Peace Research Institute [SIPRI]. *SIPRI Yearbook 1995: Armaments, Disarmament, and International Security.* New York: Oxford University Press, 1995.

Stremlau, John. *Sharpening International Sanctions: Toward a Stronger Role for the United Nations.* Washington D.C.: Carnegie Endowment for International Peace, November 1996.

Taylor, Michael J. H. *Jane's Encyclopedia of Aviation.* Rev. ed. New York: Crescent Books, 1993.

Tower, John, Edmund Muskie, and Brent Scowcroft. *The Tower Commission Report.* New York: Bantam Books, 1987.

Trice, Robert H. "Transnational Industrial Cooperation in Defense Programs" In *Global Arms Production.* Edited by Ethan B. Kapstein. Lanham, Md.: University Press of America, 1992.

Tusa, Francis. "'Any Old Iron' in Tank Market." *Armed Forces Journal International* (May 1993): 36.

U.S. Arms Control and Disarmament Agency. *World Expenditures and Arms Transfers, 1991–92* and *1996*. Washington, D.C.: ACDA, 1992, 1997.

U.S. Congress. Office of Technology Assessment. *Global Arms Trade.* Washington, D.C.: US GPO, June 1991.

U.S. Congressional Budget Office. *A Look at Tomorrow's Tactical Air Forces.* CBO Study. Washington, D.C.: GPO, January 1997.

U.S. Defense Security Assistance Agency. *Foreign Military Sales, Foreign Military Construction Sales and Military Assistance Facts.* Annual (1978 -). Washington, D.C.: DSAA.

Wagner, Mark. "Fighting Over the Scraps." *Flight International* (June 1–7, 1994): 23–28.

Working Group on the Future of the Defence Industry. "Conclusions of the First Meeting, June 19, 1998." Centre for European Reform, London. Internet: http://www.cer.org.uk/wp_edi_1.htm. Accessed March 7, 1999.

Yudin, Pyotr. "Russia Offers Upgrade Deals." *Defense News* (December 8–14, 1997).

———. "Russian Exports Fall Short of Expectations." *Defense Week* (December 8, 1997): 16.

Chapter 2

Allison, Graham T. "Questions About the Arms Race: Who's Racing Whom? A Bureaucratic Perspective." In *Contrasting Approaches to Strategic Arms Control.* Edited by Robert L. Pfaltzgraff, Jr. Lexington, Mass.: Lexington Books, 1974.

Ayoob, Mohammed. *The Third World Security Predicament: State Making, Regional Conflict, and the International System.* Boulder, Colo.: Lynne Rienner Publishers, 1995.

Azar, Edward, and Moon Chung-In. "Legitimacy, Integration, and Policy Capacity." In *National Security in the Third World.* Edited by Edward Azar and Moon Chung-In. Cambridge, UK: Cambridge University Press, 1988.

Biddle, Stephen D. *The Determinants of Offensiveness and Defensiveness in Conventional Land Warfare.* Ph.D dissertation, Harvard University, May 1992.

———. "Recent Trends in Armor, Infantry, and Artillery Technology: Developments and Implications." In *The Diffusion of Advanced Weaponry: Technologies, Regional Implications, and Responses.* Edited by W. Thomas Wander, Eric H. Arnett, and Paul Bracken. Washington, D.C.: American Association for the Advancement of Science, 1994.

Blechman, Barry M., and Tamara Wittes. "Defining Moment: The Threat and Use of Force in American Foreign Policy Since 1989." Foreign Policy Project Occasional Paper 6. Washington, D.C.: The Henry L. Stimson Center and the Overseas Development Council, May 1998.

Brzoska, Michael, and Frederic Pearson. *Arms and Warfare: Escalation, Deescalation, and Negotiation.* Columbia: University of South Carolina Press, 1994.

Buzan, Barry. *People, States, and Fear,* 2nd ed. Boulder, Colo.: Lynne Rienner Publishers, 1991.

Center for Defence and International Security Studies. "Missile Threats and Responses." Lancaster University, United Kingdom. Internet: http://www.cdiss.org/tempor1.htm. Accessed September 10, 1999.

Cohen, Lenard J. *Broken Bonds: The Disintegration of Yugoslavia.* Boulder, Colo.: Westview Press, 1993.

Colton, Joel, and R. R. Palmer. *A History of the Modern World,* 8th ed. New York: Knopf, 1995.

Cook, Thomas D., and Donald T. Campbell. *Quasi-Experimentation: Design and Analysis Issues for Field Settings.* Boston: Houghton Mifflin Co., 1979.

Davis, James W., Jr., et al. "Correspondence: Taking Offense at Offense-Defense Theory." *International Security* 23:3 (Winter 1998–1999): 179–206.

Diehl, Paul F. "Armaments Without War: An Analysis of Some Underlying Effects." *Journal of Peace Research* 22 (1985): 249–59.

———, and Jean Kingston. "Messenger or Message? Military Buildups and the Initiation of Conflict." *Journal of Politics* 49 (1987): 801–13.

Downs, George W. "Arms Races and War" In *Behavior, Society, and Nuclear War*. Vol. 2. Edited by Philip E. Tetlock, et al. New York: Oxford University Press, 1991.

———, David M. Rocke, and Randolph M. Siverson. "Arms Races and Cooperation." *World Politics* 38 (1985):118–46.

Draper, N. R., and H. Smith. *Applied Regression Analysis*. New York: John Wiley and Sons, 1966.

Durch, William J., and James A. Schear. "Faultlines: UN Operations in the Former Yugoslavia." In *UN Peacekeeping, American Policy, and the Uncivil Wars of the 1990s*. Edited by William J. Durch. New York: St. Martin's Press, 1996.

Etcheson, Craig. *Arms Race Theory, Strategy and Structure of Behavior*. New York: Greenwood Press, 1989.

Flanagan, Stephen J. "Nonprovocative and Civilian-Based Defenses." In *Fateful Visions: Avoiding Nuclear Catastrophe*. Edited by Joseph S. Nye, Jr., Graham T. Allison, and Albert Carnesale. Cambridge, Mass.: Ballinger Publ. Co., 1988.

Freedman, Lawrence. *The Evolution of Nuclear Strategy*. New York: St. Martin's Press, 1983.

Gaddis, John Lewis. "The Long Peace: Elements of Stability in the Post-War International System." *International Security* (Spring 1986): 99–142.

Glaser, Charles L. *Analyzing Strategic Nuclear Policy*. Princeton: Princeton University Press, 1990.

———, and Chaim Kaufmann. "What is the Offense-Defense Balance and Can We Measure It?" *International Security* 22:4 (Spring 1998): 44–82.

Goodpaster, Gen. Andrew, et al. *An American Legacy: Building a Nuclear-Free World*. Final report of the steering committee, Project on Eliminating Weapons of Mass Destruction. Washington, D.C.: The Henry L. Stimson Center, 1997.

Graham, Norman A. "The Quest for Security and Development." In *Seeking Security and Development* Edited by Norman A. Graham. Boulder, Colo.: Lynne Rienner Publishers, 1994.

Gray, Colin S. "The Arms Race Phenomenon." *World Politics* (October 1971): 39–59.

———. "The Urge to Compete: Rationales for Arms Racing." *World Politics* (January 1974): 207–33.

Green, Philip. *Deadly Logic: The Theory of Nuclear Deterrence*. New York: Schocken Books, 1966.

Greenfeld, Liah. *Nationalism: Five Roads to Modernity*. Cambridge, Mass.: Harvard University Press, 1992.

Hall, Brian. *The Impossible Country: A Journey through the Last Days of Yugoslavia*. Boston: David R. Godine Publisher, 1994.

Hardin, Russell. *One for All: The Logic of Group Conflict*. Princeton: Princeton University Press, 1995.

Haseman, John. "Indonesia cuts back on spending as crisis bites." *Jane's Defence Weekly* (January 21, 1998): 6.

Hiro, Dilip. *The Longest War: The Iran-Iraq Military Conflict*. New York: Routledge, 1991.

Holsti, K. J. "International Theory and Domestic War in the Third World: The Limits of Relevance." Paper presented at the annual meeting of the International Studies Association, Toronto, Canada, February 1997.

Huntington, Samuel P. "Arms Races, Prerequisites and Results." *Public Policy* 8 (1958).

International Crisis Group. *Kosovo Spring*. March 20, 1998. Internet: http://www.intl-crisis-group.org. Click Publications, South Balkans, scroll to the title. Accessed April 1, 1999.

Jervis, Robert. *Perception and Misperception in World Politics*. Princeton: Princeton University Press, 1976.

Job, Brian. "The Insecurity Dilemma: National, Regime, and State Securities in the Third World." In *The Insecurity Dilemma, National Security of Third World States*. Edited by Brian Job. Boulder, Colo.: Lynne Rienner Publishers, 1992.

Johnson, John J. "The Latin American Military as a Politically Competing Group in a Transitional Society." In *The Role of the Military in Underdeveloped Countries*. Edited by John J. Johnson. Princeton: Princeton University Press, 1962.

Karniol, Robert. "South Korea postpones programmes amid crisis." *Jane's Defence Weekly* (January 21, 1998): 14.

Kinsella, David. "Arms Transfers, Dependence, and Regional Stability: Isolated Effects or General Patterns?" Paper presented at the annual meeting of the International Studies Association, Washington, D.C., February 16–21, 1999.

Levy, Jack S. "The Offensive/Defensive Balance of Military Technology: A Theoretical and Historical Analysis." *International Studies Quarterly* 28 (1984).

Lewis, Paul. "Cash Crunch: Arms buying has come to a halt in the troubled Southeast Asian economies." *Flight International* (January 21–27, 1998): 60–62.

Lin, Herbert. *New Technologies and the ABM Treaty*. Washington: Pergamon-Brassey's, 1988.

Mazaar, Michael J. *Nuclear Weapons in a Transformed World: The Challenge of Virtual Nuclear Arsenals*. New York: St. Martin's Press, 1997.

Mearsheimer, John J. *Conventional Deterrence*. Ithaca, NY: Cornell University Press, 1983.

Mullins, A. F. *Born Arming: Development and Military Power in New States*. Stanford: Stanford University Press, 1987.

Navias, Martin S., and E. R. Hooton. *Tanker Wars: The Assault on Merchant Shipping during the Iran-Iraq Conflict, 1980–1988*. New York: I. B. Tauris, 1996.

Nutter, John Jacob. "An Analysis of Threat." In *Seeking Security and Development: The Impact of Military Spending and Arms Transfers*. Edited by Norman A. Graham. Boulder, Colo.: Lynne Rienner Publishers, 1994.

Opall, Barbara. "Thai F-18s Cost U.S. Navy." *Defense News*, March 16–22, 1998, 1.

Pace, Scott, et al. *The Global Positioning System: Assessing National Policies*. MR-614-OSTP. Santa Monica, Calif.: RAND, 1995.

——— , et al. *A Policy Direction for the Global Positioning System: Balancing National Security and Commercial Interests*. RAND Research Brief 1501. Santa Monica, Calif.: RAND, 1995.

Payne, James L. *Why Nations Arm*. New York: Basil Blackwell, 1989.

Posen, Barry R. "The Security Dilemma and Ethnic Conflict." In *Ethnic Conflict and International Security*. Edited by Michael E. Brown. Princeton: Princeton University Press, 1993.

Riggs, Robert E., and Jack C. Plano. *The United Nations: International Organization and World Politics*, 2nd ed. Belmont, Calif.: Wadsworth Publishing Co., 1994.

Roberts, Brad, ed. *Biological Weapons: Weapons of the Future?* Washington, D.C.: Center for Strategic and International Studies, 1993.

Roche, James. "Proliferation of Tactical Aircraft and Ballistic and Cruise Missiles in the Developing World." In *The Diffusion of Advanced Weaponry: Technologies, Regional Implications, and Responses*. Edited by W. Thomas Wander, Eric H. Arnett, and Paul Bracken. Washington, D.C.: American Association for the Advancement of Science, 1994.

Sagan, Scott D., and Kenneth N. Waltz. *The Spread of Nuclear Weapons: A Debate*. New York: W. W. Norton and Company, 1995.

Schelling, Thomas C. *The Strategy of Conflict*. New York: Oxford University Press, 1960.

Segal, David. "The Iran-Iraq War: A Military Analysis." *Foreign Affairs* 66 (Summer 1988).

Smithson, Amy E. *Chemical Weapons Disarmament in Russia: Problems and Prospects*. Report no. 17. Washington, D.C.: The Henry L. Stimson Center, 1995.

———. *Separating Fact from Fiction: The Australia Group and the Chemical Weapons Convention*. Occasional paper no. 34. Washington, D.C.: The Henry L. Stimson Center, 1997.

———. *Biological Weapons Proliferation: Reasons for Concern, Courses of Action*. Report no. 24. Washington, D.C.: The Henry L. Stimson Center, 1998.

Spector, Leonard S., and Mark G. McDonough with Even S. Medeiros. *Tracking Nuclear Proliferation: A Guide in Maps and Charts, 1995*. Washington, D.C.: The Carnegie Endowment for International Peace, 1995.

Tilly, Charles. "Western State-making and Theories of Political Transformation." In *Formation of National States in Western Europe*. Edited by Charles Tilly. Princeton: Princeton University Press, 1975.

United Nations Development Program. *Human Development Report, 1996*. New York: Oxford University Press, 1996.

U.S. Congress. Office of Technology Assessment. *Global Arms Trade*. Washington, D.C.: GPO, June 1991.

Van Evera, Stephen. "Offense, Defense, and the Causes of War." *International Security* 22 (Spring 1998): 5–43.

von Bulow, Andreas. "Defensive Entanglement: An Alternative Strategy for NATO." In *The Conventional Defense of Europe*. Edited by Andrew J. Pierre. New York: Council on Foreign Relations, 1988.

Walt, Stephen M. *The Origins of Alliances*. Ithaca, NY: Cornell University Press, 1987.

Woodward, Susan. *Balkan Tragedy: Chaos and Dissolution After the Cold War*. Washington, D.C.: The Brookings Institution, 1994.

Zartman, I. William, ed. *Collapsed States: The Disintegration and Restoration of Legitimate Authority*. Boulder, Colo.: Lynne Rienner Publishers, 1995.

Chapter 3

al-Suwaidi, Jamal S. "The Gulf Security Dilemma." In *Iran and the Gulf: A Search for Stability*. Edited by Jamal S. al-Suwaidi. Abu Dhabi: The Emirates Center for Strategic Studies and Research, 1996.

Allen, Kenneth. "PLAAF Modernization: An Assessment." In *Crisis in the Taiwan Strait*. Edited by James R. Lilley and Chuck Downs. Washington, D.C.: National Defense University Press, 1997.

Baginda, Abdul Razak Abdullah. "Malaysia's Armed Forces in the 1990s." *International Defense Review* 4(1992): 305–308.

Baum, Julian. "One-Track Mind, Taipei Insists on Democracy Before Unification." *Far Eastern Economic Review* (October 29, 1998).

Bedi, Rahul. "India Confirms Nuclear Policy." *Jane's Defence Weekly* (December 23, 1998): 7.

————. "Indian Service Chiefs Fall Out Over Supremo." *Jane's Defence Weekly* (February 3, 1999): 15.

————. "Mixed Fortunes for India's Defense Industrial Revolution." *Jane's International Defence Review* 5 (1999): 22–30.

Ben-Meir, Alon. "The Dual Containment Strategy is No Longer Viable." *Middle East Policy* 4 (March 1996): 58–71.

Bernstein, Dennis, and Leslie Kean. "Burma's Dictatorship of Drugs." *The Nation* (December 16, 1996). Internet at: http://www.soros.org/burma/. Accessed November 15, 1997.

Bill, James A. "The Geometry of Instability in the Gulf: The Rectangle of Tension." In *Iran and the Gulf: A Search for Stability*. Edited by Jamal S. al-Suwaidi. Abu Dhabi: The Emirates Center for Strategic Studies and Research, 1996.

————. *The Eagle and the Lion: The Tragedy of American-Iranian Relations*. New Haven: Yale University Press, 1988.

Caldwell, Dan. "Flashpoints in the Gulf: Abu Musa and the Tunb Islands." *Middle East Policy* 4 (March 1996): 50–57.

Cardamone, Thomas A., ed. "Singapore to Buy Apache Attack Helicopters." *Arms Trade News* (March 1999): 4.

Centre for Arab Gulf Studies. "The Changing Face of Iranian Politics." *Gulf States Newsletter No. 536*. University of Exeter, United Kingdom. Internet: http://www.ex.ac.uk/ags/536.htm.

Chubin, Shahram, and Charles Tripp. *Iran-Saudi Arabia Relations and Regional Order*. Adelphi Paper No. 304. Oxford: Oxford University Press, November 1996.

Cordesman, Anthony. "Threats and Non-Threats from Iran." In *Iran and the Gulf: A Search for Stability*. Edited by Jamal S. al-Suwaidi. Abu Dhabi: The Emirates Center for Strategic Studies and Research, 1996.

Curtiss, Richard H. "Khatami's Election May Be a Turning Point." *The Washington Report on Middle East Affairs* (August/September 1997).

Daly, Mark. "Democracy is Taiwan's Best Shield Against China's Threat." *Jane's International Defence Review* 4 (1999): 25.

Danitz, Tiffany. "China Trades Arms to Burma for Heroin." *Insight* (March 24, 1997). Internet at: http://www.soros.org/burma/. Accessed November 15, 1997.

Deen, Thalif. "Indonesia Postpones Planned Arms Purchases." *Jane's Defence Weekly* (October 21, 1998): 21.

Dibb, Paul. "The Revolution in Military Affairs and Asian Security." *Survival* (Winter 1997–98): 93–116.

———, David D. Hale, and Peter Prince. "The Strategic Implications of Asia's Economic Crisis." *Survival* (Summer 1998): 5–26.

Doyle, Michael W., Ian Johnstone, and Robert C. Orr, eds. *Keeping the Peace: Multinational UN Operations in Cambodia and El Salvador.* New York: Cambridge University Press, 1997.

Duran, Khalid. "Islamists on the March." *Freedom Review* 28 (January–February 1997): 146.

Eisenstadt, Michael. *Iranian Military Power: Capabilities and Intentions.* Policy Papers No. 42. Washington, D.C.: The Washington Institute for Near East Policy, 1996.

Enginsoy, Umit. "Ciller Approves $5 billion for Delayed Arms." *Defense News* (October 21–27, 1996): 1.

———. "Defense Industry Boosts Emphasis on Exports." *Defense News* (July 25–31, 1994): 10.

———. "Erbakan Pushes Turk Defense Support." *Defense News* (September 2–8, 1996): 4.

———. "Turkey Acts to Boost Industry." *Defense News* (March 13–18, 1995): 3.

———. "Turkey Nears Deal to Make Israeli Popeye Missile." *Defense News* (March 24–30, 1997): 56.

———. "Turkey Pumps $330 Million into Arms Procurement Fund." *Defense News* (November 28–December 4, 1994): 16.

———. "Turkish Welfare Leader Criticizes Arms Deals." *Defense News* (January 8–14, 1996): 3.

Feldstein, Martin. "Refocusing the IMF." *Foreign Affairs* (March/April 1998): 23.

Gilley, Bruce. "Stand-down Order." *Far Eastern Economic Review* (September 10, 1998): 18–21.

Goldman, Stuart D., Kenneth Katzman, and Zachary S. Davis. *Russian Nuclear Reactor and Conventional Arms Transfers to Iran.* CRS Report 95–641 F. Washington, D.C.: Congressional Research Service, May 23, 1995.

Graber, G. S. *Caravans to Oblivion: The Armenian Genocide, 1915.* New York: John Wiley & Sons, 1996.

Günlük-Senesen, Gülay. "An Overview of the Arms Industry Modernization Programme in Turkey." *SIPRI Yearbook 1993.* Oxford: Oxford University Press for the Stockholm International Peace Research Institute, 1993, appendix 10E.

Gupta, Shekhar. "Nuclear Weapons in the Subcontinent." In *Defense and Insecurity in Southern Asia: The Conventional and Nuclear Dimensions.* Occasional Paper No. 21. Edited by C. Smith, S. Gupta, et al. Washington, D.C.: The Henry L. Stimson Center, May 1995.

Halloran, Richard. " . . . But Carry a Big Stick." *Far Eastern Economic Review* (December 3, 1998): 26–27.

Harries, David. "Asia's New Arms Race." *Far Eastern Economic Review* (July 17, 1997): 30.

Harrison, James Pinckney. *The Long March to Power.* New York: Praeger Publishers, 1972.

Hashim, Ahmed S. *The Crisis of the Iranian State.* Adelphi Paper No. 296. Oxford: Oxford University Press, July 1995.

———. *Iranian National Security Policies under the Islamic Republic: New Defense Thinking and Growing Military Capabilities.* Occasional Paper No. 20. Washington, D.C.: The Henry L. Stimson Center, July 1994.

Heininger, Janet E. *Peacekeeping in Transition: The UN in Cambodia.* New York: Twentieth Century Fund Press, 1994.

Heper, Metin. "The State, the Military, and Democracy in Turkey." *The Jerusalem Journal of International Relations* 9, no. 3 (1987): 52–64.

Herring, George C. *America's Longest War: The United States and Vietnam, 1950–1975,* 2nd ed. New York: Alfred A. Knopf, 1986.

Human Rights Watch. "Turkey—Translator and Publisher of HRW Report Charged in Court." Press release, October 16, 1996. Available online at gopher://gopher.igc.apc.org:5000/00/int/hrw/arms/15.

———. "What is Turkey's Hizbullah?" Backgrounder, February 16, 2000. Internet: http://www.hrw.org/press/2000/02/tur0216.html. Accessed March 1, 2000.

International Institute for Strategic Studies. *Strategic Survey, 1989–90, 1990–91*. New York: Brassey's, 1989, 1990.

International Institute for Strategic Studies [IISS]. *Strategic Survey, 1996–97, 1997–98, and 1998–99*. Oxford: Oxford University Press, 1997, 1998, 1999.

Kaplan, Robert D. "After the Ottoman Empire." Review of *Turkey Unveiled: A History of Modern Turkey*. *New York Times Book Review* (January 17, 1999): 8.

Kapur, Devesh. "The IMF: A Cure or a Curse?" *Foreign Policy* (Summer 1998): 123–25.

Karasapan, Mer. "Turkey's Armaments Industries." *Middle East Report* (January–February 1987).

Karniol, Robert. "South Korea Postpones Programmes amid Crisis." *Jane's Defence Weekly* (January 21, 1998): 14.

Karnow, Stanley. *Vietnam: A History*. New York: Penguin Books, 1983.

Katzman, Kenneth. "The Politico-Military Threat from Iran." In *Iran and the Gulf: A Search for Stability*. Edited by Jamal S. al-Suwaidi. Abu Dhabi: The Emirates Center for Strategic Studies and Research, 1996.

Korea, Republic of. National Statistical Office. "Growth Rates of GDP (at 1995 prices)." Internet: http://www.nso.go.kr/graph/Month/0–1.htm. Accessed June 2, 1999.

Krepon, Michael, and Mishi Faruqee, eds. *Conflict Prevention and Confidence-Building Measures in South Asia: The 1990 Crisis*. Occasional Paper No. 17. Washington, D.C.: The Henry L. Stimson Center, April 1994.

Lawrence, Susan V. "Out of Business: Jiang Orders a Halt to PLA, Inc." *Far Eastern Economic Review* (August 6, 1998): 68.

Lewis, Paul. "Cash Crunch: Arms Buying Has Come to a Halt in the Troubled South-East Asian Economies." *Flight International* (January 21–27, 1998): 60–62.

Mackey, Sandra. *The Iranians: Persia, Islam, and the Soul of a Nation*. New York: Dutton Signet, 1996.

Marwah, Ved. *Uncivil Wars: Pathology of Terrorism in India*. New Delhi: Harper Collins India, 1995.

McAleavy, Henry. *The Modern History of China*. New York: Praeger Publishers, 1967.

McBeth, John. "Father Knows Best." *Far Eastern Economic Review* (November 25, 1993): 25.

McDowall, David. *A Modern History of the Kurds*. London: I. B. Tauris, 1996.

Milani, Mohsen M. "Iran's Gulf Policy: From Idealism and Confrontation to Pragmatism and Moderation." In *Iran and the Gulf: A Search for Stability*. Edited by Jamal S. al-Suwaidi. Abu Dhabi: The Emirates Center for Strategic Studies and Research, 1996.

Moorehead, Alan. *Gallipoli*. New York: Ballantine Books, 1958.

Napier, Catherine. "Politics and the Military." *BBC News Online: Events: Indonesia: Special Report*, June 1, 1999. Internet: http://193.130.149.131/low/english/events/Indonesia/special_report/ newsid_358000/35816.stm. Accessed June 4, 1999.

Noorani, A. G. "The tortured and the damned: Human Rights in Kashmir." *Frontline* (January 28, 1994). Cited in A. G. Noorani, *Easing the Indo-Pakistani Dialogue on Kashmir*. Occasional Paper No. 16. Washington, D.C.: The Henry L. Stimson Center, April 1994.

Opall, Barbara. "Malaysia Tempers Defense Spending in New Budget Plan." *Defense News* (November 11–17, 1996): 19.

———. "Thai F-18s Cost U.S. Navy." *Defense News* (March 16–22, 1998): 1.

Opall-Rome, Barbara. "U.S. to Deny Taipei Subs, Missiles." *Defense News* (April 20–28, 1998): 3.

Pike, John. "Border Security Forces." Federation of American Scientists Intelligence Resource Program, May 1998. Internet: http://www.fas.org/irp/world/india/home/bsf.htm. Accessed May 12, 1999.

Radelet, Steven, and Jeffrey Sachs. "What Have We Learned, So Far, From the Asian Financial Crisis?" Discussion Paper No. 37. Consulting Assistance on Economic Reform II Project of the Harvard Institute for International Development, March 1999. Internet: http://www.hiid.harvard.edu/projects/caer/papers/paper37.html. Accessed June 3, 1999.

Rashid, Ahmed. "Raise the Crescent." *Far Eastern Economic Review* (December 3, 1998): 20–22.

Reese, David. *The Prospects for North Korea's Survival.* Adelphi Paper No. 323. London: IISS, November 1998.

Rodan, Steve. "Israel, Turkey to Conclude Cooperation Agreement." *Defense News* (March 9–15, 1998): 28.

Rouleau, Eric. "Turkey: Beyond Atatürk." *Foreign Policy* 103 (Summer 1996): 76.

Roy-Chaudhury, Rahul. "Maritime Security in the Indian Ocean Region." *Maritime Studies.* Canberra, Australia, 1996. Internet: http://uniserve.edu.au/law/pub/icl/mStudies/maritimeStudies.html. Accessed June 4, 1999.

Rumsfeld, Donald H., et al. *Executive Summary of the Report of the Commission to Assess the Ballistic Missile Threat to the United States.* Pursuant to Public Law 104 201. Washington, D.C., July 15, 1998. Internet: http://www.stimson.org/policy/excbmd.htm.

Sariibrahimoglu, Lale. "Turkey Warns Syria as Exercises Are Set to Go Ahead." *Jane's Defence Weekly* (October 21, 1998): 4.

Sawhney, Pravin. "Arjun MBT Still in Technical and Fiscal Mire." *Jane's International Defense Review* 11 (1996): 15.

Schear, James A. "Riding the Tiger: The United Nations and Cambodia's Struggle for Peace." In *UN Peacekeeping, American Policy, and the Uncivil Wars of the 1990s.* Edited by William J. Durch. New York: St. Martin's Press, 1996.

Selin, Shannon. *Asia Pacific Arms Buildups, Part One: Scope, Causes, and Problems.* Working Paper No. 6. Vancouver, Canada: University of British Columbia, Inst. of International Relations, November 1994.

Sezer, Duygu B. *Turkey's Political and Security Interests and Policies in the New Geostrategic Environment of the Expanded Middle East.* Occasional Paper No. 19. Washington, D.C.: The Henry L. Stimson Center, July 1994.

Sheppard, Ben. "Regional rivalries are replayed as India and Pakistan renew ballistic missile tests." *Jane's International Defence Review* 5 (1999): 57–59.

Smith, Brian D. "United Nations Iran-Iraq Military Observer Group." In *The Evolution of UN Peacekeeping: Case Studies and Comparative Analysis.* Edited by William J. Durch. New York: St. Martin's Press, 1993.

Smith, Chris. "Conventional Forces and Regional Stability." In *Defense and Insecurity in Southern Asia: The Conventional and Nuclear Dimensions.* Occasional Paper No. 21. Washington, D.C.: The Henry L. Stimson Center, May 1995.

———. *India's Ad Hoc Arsenal.* Oxford: Oxford University Press for the Stockholm International Peace Research Institute, 1994.

Souresrafil, Behrouz. *The Iran-Iraq War.* Plainview, NY: Guinan Lithographic Co., 1989.

Stavrianos, L. S. *The Balkans Since 1453.* New York: Holt, Rinehart, and Winston, 1958.

Stockholm International Peace Research Institute. *SIPRI Yearbook 1995: Armaments, Disarmament, and International Security.* New York: Oxford University Press, 1995.

Swain, Joseph Ward, and William H. Armstrong. *Peoples of the Ancient World.* New York: Harper & Row, 1959.

Taylor, John. "The Economic Challenge for Indonesia." *BBC News Online: Events: Indonesia: Special Report* (June 2, 1999). Internet: http://193.130.149.131/low/english/events/Indonesia/special_report/ newsid_358000/ 353422.stm. Accessed June 4, 1999.

Teitelbaum, Joshua. "The Gulf States and the End of Dual Containment." *Middle East Review of International Affairs (MERIA) Journal* 2:3 (September 1998). Internet: http://www.biu.ac.il/SOC/besa/meria.html. Accessed September 10, 1998.

Tuchman, Barbara W. *Stilwell and the American Experience in China, 1911–45.* New York: Bantam Books, 1971.

U.S. Congress. "Iran-Libya Sanctions Act of 1996." Public Law 104–172 (August 5, 1996). Internet: http://www.thomas.loc.gov/. Accessed June 6, 1999.

———. Office of Technology Assessment. *Global Arms Trade.* Washington, D.C.: GPO, June 1991.

U.S. Department of Defense. Office of the Secretary of Defense. *The Security Situation in the Taiwan Strait.* A report to the Congress pursuant to the FY99 Appropriations Bill. Washington, D.C.: February 1, 1999.

———. "Memorandum for Correspondents." Memorandum No. 267-M (December 4, 1995). Internet: http://www.defenselink.mil/news/Dec1995/ml120495_m267–95.html.

———. "Memorandum for Correspondents." Memorandum No. 159-M (July 12, 1996). Internet: http://www.defenselink.mil/news/Jul1996/m071296_ml159–96.html.

———. *Proliferation, Threat and Response.* Washington, D.C.: November 1997. Internet: http://www.defenselink.mil/pubs/prolif97/meafrica.html.

———. *Report to Congress on Theater Missile Defense Architecture Options for the Asia-Pacific Region,* Washington, D.C., May 1999. Internet: http://www.nyu.edu/globalbeat/asia/DOD0599.html. Accessed May 27, 1999.

U.S. Department of State. Bureau of Arms Control. *World Military Expenditures and Arms Transfers, 1998.* Washington, D.C.: 1999.

U.S. Department of State. Bureau of Democracy, Human Rights, and Labor. "Turkey Country Report on Human Rights Practices for 1996." Washington, D.C.: January 1997. Internet: http://www.state.gov/www/global/human_rights/1996_hrp_report/turkey.html.

———. "Indonesia Human Rights Practices, 1995." Washington, D.C.: March 1996. Internet: gopher://dosfan.lib.uic.edu:70/1.

U.S. Department of State. Bureau for International Narcotics and Law Enforcement Affairs. *International Narcotics Control Strategy Report, 1998: Burma.* Washington, D.C.: February 1999. Internet: http://www.state.gov/www/global/narcotics_law/1998_narc_report/major/Burma.html.

U.S. Department of State. Office of the Coordinator for Counterterrorism. *Patterns of Global Terrorism, 1997.* Washington, D.C.: April 1998. Internet: http://www.state.gov/www/global/terrorism/1997Report/1997index.html.

———. *Patterns of Global Terrorism, 1998.* Washington, D.C.: April 1999. Internet: http://www.state.gov/www/global/terrorism/1998Report/ 1998index.html.

U.S. Director of Central Intelligence. *The Acquisition of Technology Relating to Weapons of Mass Destruction and Advanced Conventional Munitions, July–December 1996.* Submitted to the U.S. Congress in response to section 721 of the Fiscal Year 1997 Intelligence Authorization Act. Washington, D.C.: June 1997.

Valencia, Mark J. *China and the South China Sea Disputes.* Adelphi Paper No. 298. London: International Institute for Strategic Studies, October 1995.

Willett, Susan. "East Asia's Changing Defence Industry." *Survival* (Autumn 1997).

Williams, James L. "Oil Price History and Analysis." *WTRG Economics.* London, Arkansas, 1999. Internet: http://wtrg.com/prices.htm.

Chapter 4

Adams, Gordon. *The Iron Triangle: The Politics of Defense Contracting.* New Brunswick: Transaction Books, 1982.

Agmon, Marcy, et al. *Arms Proliferation Policy: Support to the Presidential Advisory Board.* MR-771-OSD. Santa Monica, Calif.: RAND Corporation, 1996.

Anthony, Ian. "Conventional Arms Transfer Control." *Pacific Research* (May 1995): 43.

Arms Control Association. "The Missile Technology Control Regime (MTCR)." Background paper. Washington, D.C.: September 1993.

Baldwin, David A. ed. *Neorealism and Neoliberalism: The Contemporary Debate*. New York: Columbia University Press, 1993.

Ball, Nicole. *Pressing for Peace: Can Aid Induce Reform?* Policy Essay No. 6. Washington, D.C.: Overseas Development Council, 1992.

Beitz, Charles R. *Political Theory and International Relations*. Princeton: Princeton University Press, 1979.

Blechman, Barry M., Janne E. Nolan, and Alan Platt. "Pushing Arms." *Foreign Policy* 46 (Spring 1982).

Bocse, Wade. "Divisions Still Impede Wassenaar Export Control Regime at Plenary." *Arms Control Today* (Nov./Dec. 1997). Internet: http://www.armscontrol.org/ACT/JAN_FEB/wass.html.

British American Security Information Council. "EU Code of Conduct." Internet: http://basicint.org/ eucode.htm.

Brodie, Bernard and Fawn M. *From Crossbow to H-Bomb*. Bloomington: University of Indiana Press, 1973.

Buzan, Barry. *People, States, and Fear.* 2nd ed. Boulder, Colo.: Lynne Rienner Publishers, 1991.

———, Charles Jones, and Richard Little. *The Logic of Anarchy: Neorealism to Structural Realism*. New York: Columbia University Press, 1993.

Damrosch, Lori Fisler, ed. *Enforcing Restraint: Collective Intervention in Armed Conflicts.* New York: Council on Foreign Relations Press, 1993.

———, and David J. Scheffer, eds. *Law and Force in the New International Order*. Boulder, Colo.: Westview Press, 1991.

Dupuy, T. N. "Can We Rely Upon Computer Combat Simulations?" *Armed Forces Journal International* (August 1987): 58–63.

Durch, W. J. "Introduction to Anarchy: Humanitarian intervention and 'State-building' in Somalia." In *UN Peacekeeping, American Policy, and the Uncivil Wars of the 1990s*. Edited by William J. Durch. New York: St. Martin's Press, 1996.

Epstein, Joshua. *Conventional Forces: A Dynamic Assessment*. Washington, D.C.: The Brookings Institution, 1990.

Federation of American Scientists. "Code of Conduct—Legislative History." Internet: http://www.fas.org/asmp/campaigns/code/codehist.html. Accessed March 2, 2000.

Fletcher School of Law and Diplomacy. "Convention on Prohibitions or Restrictions on the Use of Certain Conventional Weapons . . ." Internet: http://www.tufts.edu/departments/fletcher/multi/texts/BH790.txt. Accessed June 11, 1999.

Foxwell, David. "Anti-ship cruise missiles Home-in on Littoral Requirements." *Jane's International Defense Review* (8/1996): 63.

Freedom House. *Freedom Review*. February issue. Annual. Internet: http://www.freedomhouse.org/.

Friends Committee on National Legislation. "Legislative History of the Code of Conduct on Arms Transfers." Internet: http://www.fcnl.org/pub/fcnl/codehist.htm. Accessed April 9, 1998.

Gething, Michael J. "Balancing the pitfalls and potential of aircraft upgrade." *Jane's International Defence Review* (December 1998): 30–37.

Goldblat, Jozef. *Arms Control: A Guide to Negotiations and Agreements.* Thousand Oaks, Calif.: Sage Pub., 1994.

Goldring, Natalie. "President Bush's Middle East Arms Control Initiative: One Year Later." *Arms Control Today* (June 1992): 12.

Government of Canada. "Ottawa Convention." Internet: http://www.mines.gc.ca. Accessed November 1999.

Grieco, Joseph M. "Realist International Theory and the Study of World Politics." In *New Thinking in International Relations Theory*. Edited by Michael W. Doyle and G. John Ikenberry. Boulder, Colo.: Westview Press, 1997.

Hartung, William D. *And Weapons for All*. New York: Harper Collins Publishers, 1994.

Harvey, John R. "Regional Ballistic Missiles and Advanced Strike Aircraft: Comparing Military Effectiveness." *International Security* (Fall 1992): 41–83.

———, and Uzi Rubin. "Controlling Ballistic Missiles: How Important? How To Do It?" *Arms Control Today* (March 1992): 13–18.

Hitchens, Theresa. "Experts Tout Post-COCOM Regime, Despite Shortfalls." *Defense News* (January 29–February 4, 1996).

Hufbauer, G. C., et al. *Economic Sanctions Reconsidered: History and Current Policy.* Washington, D.C.: Institute for International Economics, 1990.

International Committee of the Red Cross. "The Vienna Review Conference." *International Review of the Red Cross* 309 (November 1, 1995): 672–77.

Johnson, Rebecca. "Viewpoint: The CTBT and the 1997 NPT Prepcom." *The Nonproliferation Review* 3:3 (Spring-Summer 1996): 55–62.

Jones, J. *Stealth Technology: The Art of Black Magic* (np: Tab Aero, nd.). Cited in "B-2 From Any View." *Air&Space Magazine* (Smithsonian), April/May 1997. Internet: http://www.air-spacemag.com/ASM/mag/ supp/am97/stealth.html. Accessed June 13, 1999.

Keller, William W., and Janne E. Nolan. "The Arms Trade: Business As Usual?" *Foreign Policy* (Winter 1997–98): 113–25.

Kemp, Geoffrey. "Regional Security, Arms Control, and the End of the Cold War." *The Washington Quarterly* (Autumn 1990).

Laurance, Edward J. *The International Arms Trade.* New York: Lexington Books, 1992.

Mallin, Maurice A. "CTBT and NPT: Options for U.S. Policy." *The Nonproliferation Review* 2:2 (Winter 1995): 1–11.

Markusen, Ann. "The Rise of World Weapons." *Foreign Policy* (Spring 1999): 40–51.

McMahon, K. Scott, and Dennis M. Gormley. *Controlling the Spread of Land-Attack Cruise Missiles.* Marina del Rey, Calif.: American Institute for Strategic Cooperation, January 1995.

Mearsheimer, John J. "Back to the Future: instability in Europe after the Cold War." In *The Perils of Anarchy.* Edited by Michael E. Brown, Sean M. Lynn-Jones, and Steven E. Miller. Cambridge, Mass.: MIT Press, 1995.

Nelson, Joan M., and Stephanie J. Eglinton. *Global Goals, Contentious Means: Issues of Multiple Aid Conditionality.* Policy Essay No. 10. Washington, D.C.: The Overseas Development Council, 1993.

Nolan, Janne E. "The Global Arms Market after the Gulf War: Prospects for Control." *The Washington Quarterly* (Summer 1991): 132.

———. "The U.S.-Soviet Conventional Arms Transfer Talks." In *U.S. Soviet Security Cooperation.* Edited by Alexander George, et al. Oxford University Press, 1988.

Nolan, Janne E. (chair), Edward Randolph Jayne, II, Ronald E. Lehman, Dave E. McGiffert, and Paul C. Warnke. *Report of the Presidential Advisory Board on Arms Proliferation Policy.* Washington, D.C.: 1996.

Nuclear Age Peace Foundation. "Nobel Laureates' International Code of Conduct on Arms Transfers." Internet: http://www.wagingpeace.org/codeofconduct.html or at http://www.igc.apc.org/basic/code_itl.htm. Accessed April 8, 1998.

Odessey, Bruce. "Officials Defend U.S. Participation in Wassenaar Arrangement." Internet: http://www.unhchr.ch/html/menu2/6/hrc.htm. Accessed June 17, 1999.

Perine, Keith. "Feds Release Revised Crypto Export Rules." *The Standard,* January 12, 2000. Internet: http://www.thestandard.net/article/display/0,1151,8780,00.html. Accessed March 2, 2000.

Pierre, Andrew J. *The Global Politics of Arms Sales.* Princeton: Princeton University Press, 1982.

Posen, Barry R. "Measuring the European Conventional Balance: Coping with Complexity in Threat Assessment." *International Security* (Winter 1984–85): 47–88.

Reed, Laura W., and Carl Kaysen, eds. *Emerging Norms of Justified Intervention.* Cambridge, Mass · American Academy of Arts and Sciences, 1993.

Reinecke, Wolfgang H. "From Denial to Disclosure: The Political Economy of Export Controls and Technology Transfer." In *Bridging the Nonproliferation Divide: The United States and India.* Edited by Francine Frankel. Lanham, Md.: University Press of America, 1995.

Roberts, Brad. "From Nonproliferation to Antiproliferation." *International Security* (Summer 1993): 165–66.

Saferworld. "Developing the Wassenaar Arrangement: A new arms export control regime." London, September 1996. Internet: http://www.gn.apc.org/sworld/wassen.html.

———. "The EU Code of Conduct on the arms trade: Final Analysis." Internet: http://ww.gn.apc.org/SWORLD/ARMSTRADE/code.html. Accessed June 14, 1999.

Sampson, Anthony. *The Arms Bazaar: From Lebanon to Lockheed.* New York: Bantam Books, 1977.

Simmons, Henry T. "U.S. Arms Sales and the Carter White House." *International Defense Review* (March 1979).

Smith, Chris. "Light Weapons and Ethnic Conflict in South Asia." In *Lethal Commerce: The Global Trade in Small Arms and Light Weapons.* Edited by Jeffrey Boutwell, Michael T. Klare, and Laura W. Reed. Cambridge, Mass.: American Academy of Arts and Sciences, 1995.

Sternman, Mark S. "Mechanics of an Arms Sale." Fact sheet. Washington, D.C.: Peace Action, April 1993.

Stremlau, John. *Sharpening International Sanctions: Toward a Stronger Role for the United Nations.* Report to the Carnegie Commission on Preventing Deadly Conflict. Washington, D.C., November 1996.

United Nations. Department of Public Information. "East Timor—NTAET—Background." Internet: http://www.un.org/ peace/etimor/UntaetB.htm. Accessed January 22, 2000.

UN Human Rights Commission. Reports. Internet: http://www.unhchr.ch/html/menu2/2/chr.htm. Accessed June 17, 1999.

U.S. Arms Control and Disarmament Agency. *Arms Control and Disarmament Agreements: Texts and Histories of Negotiations.* Washington, D.C. Internet at: http://www.acda.gov/treaties.

———. *List of Dual Use Goods and Technologies and Munitions List, submitted to the Plenary Meeting in Vienna, 11–12 July 1996.* ACDA document No. 97023, declassified April 25, 1997. Annexes 1 and 2. Internet: http://jya.com/wa/watoc.htm. Accessed April 8, 1998.

———. "The Missile Technology Control Regime." Fact sheet. September 15, 1997. Internet: http://acda.gov/factshee/exptcon/mtcr96.htm.

U.S. Department of Commerce. Bureau of Export Administration. "Fact Sheet: Administration Implements Updated Encryption Export Policy." January 12, 2000. Accessed March 2, 2000 via the Center for Democracy and Technology. Internet: http://www.cdt.org/crypto/admin/000112commercefactsheet.shtml.

U.S. Department of State. "Conventional Arms Transfer Policy, Statement by the President, May 19, 1977." *Department of State Bulletin* 13, no. 21 (1977).

———. *Country Reports on Human Rights Practices,* 1998. Internet: http://www.state.gov/www/services/survey.html.

U.S. Department of State. Office of the Spokesman. "The Wassenaar Arrangement." Address by Under Secretary of State for Arms Control and International Security Affairs Lynn E. Davis, at the Carnegie Endowment for International Peace, Washington, D.C., January 23, 1996.

U.S. Department of State. Office of the Coordinator for Counterterrorism. *Patterns of Global Terrorism, 1998,* April 1999. Internet: http://www.state.gov/www/global/terrorism/1998Report/1998index.html.

Vaccaro, J. Matthew. "The Politics of Genocide: Peacekeeping and Disaster Relief in Rwanda." In *UN Peacekeeping, American Policy, and the Uncivil Wars of the 1990s.* Edited by William J. Durch. New York: St. Martin's Press, 1996.

Walkling, Sarah. "Wassenaar Members End Plenary: First Data Exchange Falls Short." *Arms Control Today* (Jan./Feb. 1997). Internet: http://www.armscontrol.org/ACT/novdec97/wassnov.htm.

Waltz, Kenneth N. *The Spread of Nuclear Weapons: More May Be Better.* Adelphi Paper 171. London: International Institute for Strategic Studies, 1981.

———, and Scott D. Sagan. *The Spread of Nuclear Weapons: A Debate.* New York: W.W. Norton, 1995.

Wassenaar Arrangement. "Elements for Objective Analysis and Advice Concerning Potentially Destabilising Accumulations of Conventional Weapons." Paper approved by the plenary meeting of the Wassenaar Arrangement, December 3, 1998. Internet: http://www.wassenar.org/docs/criteria.html. Accessed June 13, 1999.

———. "Public Statement," Vienna, Austria, December 3, 1998. Internet: http://www.wassennaar.org/docs/press_4.html. Accessed June 14, 1999.

———. Texts of all public documents. Internet: http://wassenaar.org/doc/press_3.html. Accessed June 11, 1999.

White House. Office of the Press Secretary. "Fact Sheet on Conventional Arms Transfer Policy." February 17, 1995. Internet: http://www.fas.org/asmp/library/white_house/whfacts.html. Accessed June 12, 1999.

Chapter 5

Acevedo, Domingo E., and Claudio Grossman. "The Organization of American States and the Protection of Democracy." In *Beyond Sovereignty: Collectively Defending Democracy in Latin America.* Edited by Tom Farer. Baltimore: Johns Hopkins University Press, 1996.

Adelman, Howard W. "Early Warning and Humanitarian Intervention: Zaire—March to December 1996." Paper written for the Forum on Early Warning and Early Response, revised September 1998. Photocopy.

Allen, Kenneth W. "Confidence-Building Measures and the People's Liberation Army." Paper presented at the Conference on The PRC's Reform: A Reappraisal after Twenty Years, National Chengchi University, Taipei, Taiwan, April 8–9, 1999. Photocopy.

Australia, Department of Foreign Affairs and Trade. "Agreement between the Government of Australia and the Government of the Republic of Indonesia on Maintaining Security." Australian Treaty Series 1996 No. 13. Internet: http://www.austlii.edu.au/au/other/dfat/treaties/19960013.html. Accessed December 3, 1997.

———. "ARF Meetings and Activities, 1994–99." Internet: http://www.dfat.gov.au/arf/arf_meet.html.

———. "Chairman's Statement, Third ASEAN Regional Forum." Jakarta, 23 July 1996. Internet: http://www.dfat.gov.au//arf/arf3.html. Accessed December 3, 1997.

———. "Implementation of ARF Agreed CBMs." Internet: http://www.dfat.gov.au/arf/arfhomeMatrix.html. Accessed June 23, 1999.

———. "Intersessional Support Group on Confidence Building Measures, 6–8 March 1997, Beijing, China." Internet: http://www.dfat.gov.au/arf/arf_1_confid.html. Accessed December 3, 1997.

Bajpai, Kanti P. "Conflict, Cooperation, and CSBMs with Pakistan and China: A View from New Delhi." In *Mending Fences: Confidence- and Security-Building Measures in South Asia.* Edited by Šumit Ganguly and Ted Greenwood. Boulder, Colo.: Westview Press, 1996.

———. "Statement of Government of Pakistan on CSBMs between Pakistan and India." In *Mending Fences: Confidence- and Security-Building Measures in South Asia.* Edited by Šumit Ganguly and Ted Greenwood. Boulder, Colo.: Westview Press, 1996.

Blechman, Barry M., William J. Durch, Wendy Eaton, and Julie Werbel. *Effective Transitions from Peace Operations to Sustainable Peace: Final Report.* Prepared for the U.S. Department of Defense, Office of Peacekeeping and Humanitarian Affairs, OSD/OASD/S&R/PK&HA, contract No. DASW01–96-C-0075, September 1997.

Boese, Wade. "CFE Parties Outline Adapted Treaty: Limits to Allow NATO Growth." *Arms Control Today* (March 1999). Internet: http://www.armscontrol.org/ACT/march99/cfemr99.html. Accessed June 21, 1999.

The Carter Center. Latin America and Caribbean Program. "Council of Freely Elected Heads of Government." Internet: http://www.cartercenter.org/latinamerica.html. Accessed June 26, 1999.

Center for Contemporary International Problems (Moscow). "Prospects for North Korean-U.S. Relations." *DPRIK Report* No. 18, May–June 1999. Published by the Center for Nonproliferation Studies, Monterey, Calif. Available at Internet: http://www.nyu.edu/globalbeat/asia/CNS0599.html. Accessed May 27, 1999.

Chalmers, Malcolm, and Owen Greene. *Taking Stock: The UN Register After Two Years.* Bradford Arms Register Studies No. 5. Trowbridge, UK: Redwood Books for Westview Press, 1995.

Chalmers, Malcolm, Owen Greene, Edward J. Laurance, and Herbert Wulf, eds. *Developing the UN Register of Conventional Arms.* Bradford Arms Register Studies No. 4. Trowbridge, UK: Redwood Books for the University of Bradford, 1994.

Child, Jack. "Confidence-building Measures and Their Application in Latin America." In *Confidence-building Measures in Latin America: Central America and the Southern Cone.* Edited by Augusto Varas, James A. Schear, and Lisa Owens. Report No. 16. Washington, D.C.: The Henry L. Stimson Center, February 1995.

Cossa, Ralph A. "Asia Pacific Confidence and Security Building Measures." In *Asia Pacific Confidence and Security Building Measures.* Edited by Ralph A. Cossa. Washington, D.C.: Center for Strategic and International Studies, 1995.

———. "Monitoring the Agreed Framework." PacNet Report 43. Honolulu: Pacific Forum CSIS, October 24, 1997.

———. "South Korea's Package Deal." *PacNet* No. 10, March 12, 1999. Internet: http://www.nyu.edu/ globalbeat/asia/Cossa031299.html. Accessed May 27, 1999.

Defense Threat Reduction Agency [DTRA]. "Treaty on Open Skies." Fact Sheet, as of March 1999. Internet: http://www.dtra.mil/news/facts/os99.html. Accessed June 21, 1999.

Desjardins, Marie-France. *Rethinking Confidence-Building Measures.* Adelphi Paper No. 307. New York: Oxford University Press for the International Institute for Strategic Studies [IISS], December 1996.

Deutch, Karl, et al. *Political Community and the North Atlantic Area.* Princeton: Princeton University Press, 1957.

Dominguez, Jorge I. *Democratic Politics in Latin America and the Caribbean.* Baltimore: Johns Hopkins University Press, 1998.

Dragsdahl, Jorgen. "CFE Talks Advance Amid Controversy." *Basic Reports* No. 69. March 1999. Internet: http://www.basicint.org/br69.htm. Accessed June 21, 1999.

Durch, William J. "Keeping the Peace: Politics and Lessons of the 1990s." In *UN Peacekeeping, American Politics, and the Uncivil Wars of the 1990s.* Edited by William J. Durch. New York: St. Martin's Press, 1996.

———., ed. *The Evolution of UN Peacekeeping: Case Studies and Comparative Analysis.* New York: St. Martin's Press, 1993.

Dyer, Susannah L., and Natalie J. Goldring. "U.S. and Germany Dominate World Weapons Exports." *BASIC Reports* No. 48 (November 20, 1995).

Eggleston, Roland. "Europe/Russia: Basic Agreement Reached on Conventional Arms Accord." *RFE/RL Online* (April 12, 1999). Internet: http://euro.rferl.org/nca/features/1999/04/ F.RU.990412125005.html. Accessed June 21, 1999.

Farer, Tom. "Collectively Defending Democracy in the Western Hemisphere." In *Beyond Sovereignty: Collectively Defending Democracy in Latin America.* Edited by Tom Farer. Baltimore: Johns Hopkins University Press, 1996.

Fisher, Cathleen S. "The Preconditions of Confidence-Building: Lessons from the European Experience." In *A Handbook of Confidence-Building Measures for Regional Security*. 3rd ed. Edited by Michael Krepon, et al. Washington, D.C.: The Henry L. Stimson Center, 1998.

Fitch, J. Samuel. *The Armed Forces and Democracy in Latin America*. Baltimore: Johns Hopkins University Press, 1998.

Garthoff, Raymond L. *Détente and Confrontation: American-Soviet Relations from Nixon to Reagan*. Washington, D.C.: The Brookings Institution, 1985.

Ghali, Mona. *The Multinational Forces: Non-UN Peacekeeping in the Middle East*. Occasional Paper No. 12. Washington, D.C.: The Henry L. Stimson Center, May 1993.

———. "United Nations Disengagement Observer Force." In *The Evolution of UN Peacekeeping: Case Studies and Comparative Analysis*. Edited by William J. Durch. New York: St. Martin's Press, 1993.

———. "United Nations Emergency Force I: 1956–1967." In *The Evolution of UN Peacekeeping: Case Studies and Comparative Analysis*. Edited by William J. Durch. New York: St. Martin's Press, 1993.

———. "United Nations Emergency Force II: 1973–1979." In *The Evolution of UN Peacekeeping: Case Studies and Comparative Analysis*. Edited by William J. Durch. New York: St. Martin's Press, 1993.

———. "United Nations Interim Force in Lebanon: 1978–Present." In *The Evolution of UN Peacekeeping: Case Studies and Comparative Analysis*. Edited by William J. Durch. New York: St. Martin's Press, 1993.

Greenwood, Ted. "Experience from European and US-Soviet Agreements." In *Mending Fences: Confidence- and Security-Building Measures in South Asia*. Edited by Šumit Ganguly and Ted Greenwood. Boulder, Colo.: Westview Press, 1996.

Hanrahan, Joseph P., and John C. Kuhn III. *On-Site Inspections Under the CFE Treaty*. Washington, D.C.: U.S. Onsite Inspection Agency, 1997.

Henderson, Jeannie. *Reassessing ASEAN*. Adelphi Paper. No. 328. London: IISS, May 1999.

Holsti, Kalevi J. *The State, War, and the State of War*. New York: Cambridge University Press, 1995.

Institute of Contemporary International Problems, Russian Diplomatic Academy, Moscow. *DPRK Report No. 7*. May–June 1997. Distributed online at http://www.nautilus.org/pub/ftp/napsnet/RussiaDPRK/DPRK_7.txt.

International Crisis Group. "ICG Analysis of the 1997 Municipal Election Results." ICG Bosnia Project Press Release, October 14, 1997. Internet: http://www.intl-crisis-group.org/projects/bosnia/reportbhxxpr10.htm. Accessed April 10, 1998.

International Institute for Strategic Studies [IISS]. *Strategic Survey, 1976*. London: IISS, 1977.

———. *Strategic Survey 1990–91*. London: Brassey's for the IISS, 1991.

———. *Strategic Survey, 1994–95*. London: Oxford University Press for the IISS, 1995.

———. *Strategic Survey, 1998–99*. London: Oxford University Press for the IISS, 1999.

Job, Brian. "Report on the Fourth Meeting of the NPWG, November 1998." CSCAP, March 31, 1999. Internet: http://www.cscap.org/pacific.htm. Accessed June 23, 1999.

Joeck, Neil. *Maintaining Nuclear Stability in South Asia*. Adelphi Paper No. 312. New York: Oxford University Press for the IISS, September 1997.

Jones, Bruce D. "Military Intervention in Rwanda's Two Wars: Partisanship and Indifference." In *Civil Wars, Insecurity, and Intervention*. Edited by Barbara F. Walter and Jack Snyder, New York: Columbia University Press, 1999.

Junnolla, Jill R. "Confidence-Building Measures in the Middle East." In *A Handbook of Confidence-Building Measures for Regional Security*. 3rd ed. Edited by Michael Krepon, et al. Washington, D.C.: The Henry L. Stimson Center, 1998.

Khadiagala, Gilbert M. "Confidence-building Measures in Sub-Saharan Africa." In *A Handbook of Confidence-Building Measures for Regional Security*. 3rd ed. Edited by Michael Krepon, et al. Washington, D.C.: The Henry L. Stimson Center, 1998.

Korean Buddhist Sharing Movement. "Famine Witnessed by 472 Korean Refugees Interviewed in China." March 4, 1998. Internet: ftp://ftp.nautilus.org/napsnet/special_reports/Survey_on_North_Korean_Refugees.txt.

Korean Peninsula Energy Development Organization. "KEDO Executive Board Agrees on Cost Sharing for Light Water Reactor Project." KEDO Press Release. November 10, 1998. Internet: http://www.nyu.edu/globalbeat/asia/KEDO111098.html. Accessed May 27, 1999.

Krepon, Michael. "Conflict Avoidance, Confidence-building, and Peacemaking." In *A Handbook of Confidence-building Measures for Regional Security.* 3rd ed. Edited by Michael Krepon et al. Washington, D.C.: The Henry L. Stimson Center, March 1998.

————, and Mishi Faruqee, eds. *Conflict Prevention and Confidence-Building Measures in South Asia: The 1990 Crisis.* Occasional Paper No. 17. Washington, D.C.: The Henry L. Stimson Center, April 1994.

Laurance, Edward J., and Tracy Keith. "An Evaluation of the Third Year of Reporting to the United Nations Register of Conventional Arms." *Nonproliferation Review* (Spring 1996).

Leifer, Michael. *The ASEAN Regional Forum: Extending ASEAN's Model of Regional Security.* Adelphi Paper No. 302. London: Oxford University Press for the IISS, July 1996.

Mamdani, Mahmood. "Why foreign invaders can't help Congo." *Electronic Mail and Guardian* (Johannesburg), November 2, 1998. Internet: http://www.mg.co.za/mg/news/98nov1/2nov-congo.html. Accessed November 15, 1998.

Mandela, Nelson. "New Directions for SADC." Speech at the concluding session of the Southern Africa International Dialogue, Kasane, Botswana, May 6, 1997. Internet: http://www.sadc-online.com/sadc/said/spemand.htm.

Manning, Robert A. "Building Community or Building Conflict? A Typology of Asia Pacific Security Challenges." In *Asia Pacific Confidence and Security Building Measures.* Edited by Ralph A. Cossa. Washington, D.C.: Center for Strategic and International Studies, 1995.

McCausland, Colonel Jeffrey D. "The Conventional Armed Forces in Europe (CFE) Treaty: Threats from the Flank." nd. Internet: http://www.cdsar.af.mil/apj/mccaus.html.

Moon, Chung-in. "Arms Control and Peace on the Korean Peninsula." Paper No. 5. *NAPSNet* Policy Forum Online, June 26, 1997. Available at http://www.nautilus.org/napsnet/fora/5a_Moon.html.

Organization of American States. *Annual Report of the Secretary General, 1997–98.* Washington, D.C., 1998. Introduction. Internet: http://www.cidi.oas.org/SGannualchap-e.html. Accessed June 18, 1999.

Pineo, Paul F., and Lora Lumpe. "Recycled Weapons: American Exports of Surplus Arms, 1990–1995." Study by the Arms Sales Monitoring Project of the Federation of American Scientists. Washington, D.C.: FAS, May 1996. Internet: http:www.fas.org/asmp/library/publications/recycle.html.

Rajmaira, Sheen. "Indo-Pakistani Relations: Reciprocity in Long-Term Perspective." *International Studies Quarterly* 41, no. 3 (September 1997): 557–59.

Ray, James Lee. *Democracy and International Conflict.* Columbia: University of South Carolina Press, 1995.

Rudolph, Matthew C. J. "Confidence-building Measures between Pakistan and India." In *A Handbook of Confidence-building Measures for Regional Security.* 3rd ed. Edited by Michael Krepon et al. Washington, D.C.: The Henry L. Stimson Center, March 1998.

Ruggie, John G. *Winning the Peace.* A Twentieth Century Fund Book. New York: Columbia University Press, 1996.

Sandrock, John. "Prerequisites for Success." In *Mending Fences: Confidence- and Security-Building Measures in South Asia.* Edited by Šumit Ganguly and Ted Greenwood. Boulder, Colo.: Westview Press, 1996.

Stockholm International Peace Research Institute. *SIPRI Yearbook 1994: Armaments, Disarmament, and International Security.* New York: Oxford University Press, 1994.

Tacsan, Joaquín. Director, Center for Peace and Reconciliation, Arias Foundation. "The Failure in San Pedro Sula." *Inforpress Central America* No. 3 (January 1996). Internet: http://www.us.net/cip/dialogue/9601in01.html.

United Nations General Assembly. *General and Complete Disarmament: Transparency in Armaments, United Nations Register of Conventional Arms.* Report of the Secretary-General, A/51/300 (August 20, 1996) and A/51/300/Add.1–5 (July 14, 1997).

———. *General and Complete Disarmament: Transparency in Armaments.* United Nations Register of Conventional Arms, Report of the Secretary-General, A/52/312 (August 15, 1997) and A/52/312/Add.1 (October 17, 1997). Annex.

United Nations Office for the Coordination of Humanitarian Affairs, Integrated Regional Information Network [IRIN]. "Angola: DeBeers says complying with diamond embargo," June 17, 1999. Internet: http://www.reliefweb.int/IRIN/sa/countrystories/angola/19990617a.htm.

United Nations Register of Conventional Arms. "Country Reports." Internet: http://domino.un.org/REGISTER.NSF. Accessed June 21, 1999.

United Nations Security Council. "Final report of the UN International Commission of Inquiry (Rwanda)," UN Document S/1998/1096, November 18, 1998. Annex.

U.S. Arms Control and Disarmament Agency. "The Central American Democratic Security Treaty." Fact sheet, May 20, 1996. Internet: http://www.acda.gov/factshee/secbldg/cademosb.html.

———. "Confidence and Security Building Measures in Europe," Updated to May 20, 1996. Internet: http://www/acda.gov/factshee/secbldg/csbms394.html.

———. "Open Skies Consultative Commission." Fact sheet, as of May 20, 1996. Internet: http://www.acda.gov/factshee/secbldg/opnskicc.html.

———. "Organization of American States (OAS) General Assembly Resolution on Conventional Arms Transparency and Confidence-Building in the Americas." July 30, 1997. Internet: http://www.acda.gov/factshee/secbldg/unoas.htm.

———. "Organization of American States Conference on Confidence- and Security Building Measures." Fact sheet, May 20, 1996. Internet: http://www.acda.gov/factshee/secbldg/orgsbp.htm

U.S. Congress. House International Relations Committee. Subcommittee on the Western Hemisphere. Testimony of Deputy Assistant Secretary of State John Hamilton. 105th Cong., 1st sess., June 25, 1997. Internet: http://www.house.gov/international_relations/wh/wswh6252.html.

U.S. Department of State. Bureau of Inter-American Affairs. "Major Accomplishments of the OAS General Assembly, Lima, Peru, June 1–5, 1997," June 27, 1997. Internet: http://www.state.gov/www/regions/ara/fs_oasga_accomp062797.html.

U.S. General Accounting Office. *Nuclear Nonproliferation: Implementation of the U.S./North Korean Agreed Framework on Nuclear Issues.* GAO/RCED/NSIAD-97-165. June 1997.

U.S. Information Agency. "Background Briefing on Korea Four-Party Talks," April 27, 1999. Internet: http://www.nyu.edu/globalbeat/asia/USIA050399.html. Accessed May 27, 1999.

U.S. Senate. Committee on Foreign Relations. Subcommittee on African Affairs. "Central African Conflict and Its Implications for the Future of U.S. Policy Goals and Strategies," testimony by U.S. assistant secretary of state Susan Rice, June 8, 1999. Posted on the Internet at: http://members.tripod.com/~heritiersjus/rice.htm. Accessed June 21, 1999.

University of California, Institute on Global Conflict and Cooperation. "About Northeast Asia Cooperation Dialogue." Internet: http://www-igcc.ucsd.edu/igcc2/neacd/aboutneacd.html. Accessed June 24, 1999.

———. "NEACD VII Reaches Watershed Agreements." *NEACD Newsletter,* April 29, 1998. Internet: http://neacd.llnl.gov/news/neacd.html. Accessed June 24, 1999.

———. "Virtual Dialogue in Northeast Asia: A Virtual Research Center Supporting Regional Multilateral Peacebuilding." Internet: http://www-igcc.ucsd.edu/igcc2/NEACD/USIP-NEAC.html. Accessed June 24, 1999.

Vaccaro, J. Matthew. "The Politics of Genocide: Peacekeeping and Disaster Relief in Rwanda." In *UN Peacekeeping, American Politics, and the Uncivil Wars of the 1990s.* Edited by William J. Durch. New York: St. Martin's Press, 1996.

Wanandi, Jusuf. "ASEAN's Challenges for its Future." *Pacific Forum CSIS PacNet* No. 3 (January 22, 1999). Internet: http://www.nyu/globalbeat/asia/ Wanandi012399.html. Accessed May 27, 1999.

Wortzel, Col. Larry M. *The ASEAN Regional Forum: Asian Security without an American Umbrella.* Carlisle Barracks, Pa.: U.S. Army War College, Strategic Studies Institute, December 1996. pt. 3. Internet: http://carlisle-www.army.mil/usassi/ssipubs/pubs96/asean/aseanai.html.

Chapter 6

Biddle, Stephen D. "Recent Trends in Armor, Infantry, and Artillery Technologies: Developments and Implications." In *The Diffusion of Advanced Weaponry: Technologies, Regional Implications, and Responses.* Edited by W. Thomas Wander, Eric H. Arnett, and Paul Bracken. Washington, D.C.: American Association for the Advancement of Science, 1994.

Boulding, Kenneth E. *Three Faces of Power.* Newbury Park, Calif.: SAGE Publications, 1990.

Council of Freely Elected Heads of Government. "Statement." In *Final Report: Agenda for the Americas for the 21st Century.* Atlanta: The Carter Center, April 1997.

Curran, Brian P., William J. Durch, and Jolie M. F. Wood. "Alternating Currents: Technology and the New Context for U.S. Foreign Policy." Foreign Policy Project Occasional Paper No. 5. Washington, D.C.: The Henry L. Stimson Center, October 1997.

Fetter, Steve. "Ballistic Missiles and Weapons of Mass Destruction." *International Security* 16 (Summer 1991): 5–42.

Forecast International. "MGM–140 ATACMS." *Forecast International/DMS Market Intelligence Report: Missile Forecast,* January 1996.

Frye, Alton. "Banning Ballistic Missiles." *Foreign Affairs* (Nov./Dec. 1996): 99–112.

Gompert, David C. "Sharing the Burdens of Global Security." Foreign Policy Project Occasional Paper No. 1. Washington, D.C.: Henry L. Stimson Center and Overseas Development Council, August 1996.

Holton, J. Jerome, Lora Lumpe, and Jeremy J. Stone. "Proposal for a Zero Ballistic Missile Regime." In *1993 Science and International Security Anthology.* Washington, D.C.: American Association for the Advancement of Science, 1993.

Honwana, Joao. "Implementing Peace Agreements in Civil Wars: The Case of Mozambique." Paper presented to the authors' conference of the project on Peace Implementation, Stanford University Center for International Security and Cooperation, October 24–26, 1998, Stanford, California.

Huntley, James R. *Pax Democratica: A Strategy for the 21st Century.* Foreword by Lawrence Eagleburger. New York: St. Martin's Press, 1998.

Ikenberry, G. John. "Institutions, Strategic Restraint, and the Persistence of American Postwar Order." *International Security* 23 (Winter 1998–99).

International Institute for Strategic Studies. *The Military Balance, 1997–1998.* Oxford: Oxford University Press, 1997.

Lumpe, Lora. "Arms Control Options for Delivery Vehicles." *Physics and Society* 24 (January 2, 1995).

———. "Zero Ballistic Missiles and the Third World." *Arms Control* 14, no. 1 (April 1994).

Nathan, Laurie. *The Changing of the Guard: Armed Forces and Defence Policy in a Democratic South Africa.* Pretoria: HSRC Publishers, 1994: 2, 85–86.

Nolan, Janne E. *Trappings of Power: Ballistic Missiles in the Third World.* Washington, D.C.: The Brookings Institution, 1991.

Organization for Economic Cooperation and Development. *Convention on Combating Bribery of Foreign Public Officials in International Business Transactions.* Signed in Paris, December 17, 1997. Internet: http://www.unog.ch/frames/disarm/review/conweap.htm.

Pearson, Frederic S. *The Global Spread of Arms: Political Economy of International Security.* Boulder, Colo.: Westview Press, 1994.

Pierre, Andrew J. "Toward an International Regime for Conventional Arms Sales." In *Cascade of Arms: Managing Conventional Weapons Proliferation.* Edited by A. J. Pierre. Washington, D.C.: The Brookings Institution Press, 1997.

Reinecke, Wolfgang. "Cooperative Security and the Political Economy of Nonproliferation." In *Global Engagement.* Edited by Janne E. Nolan. Washington, D.C.: Brookings Institution, 1994.

Ronfeldt, David F. "Tribes, Institutions, Markets, and Networks: A Framework About Societal Evolution." RAND Paper P-7967. Santa Monica, Calif.: RAND Corporation, 1996.

United Nations. International Criminal Court. Internet: http://www.un.org/icc/pressrel/lrom22.htm. July 17, 1998. Accessed November 2, 1998.

U.S. Senate. Committee on Armed Services. *Hearing before the Strategic Subcommittee on Strategic Nuclear Forces and Policy in Review of the Defense Authorization Request for Fiscal Year 2000 and the Future Years Defense Program.* Testimony of Hon. Edward L. Warner, III, Assistant Secretary of Defense for Strategy and Threat Reduction, April 14, 1999. Internet: http://www.clw.org/coalition/warner0599.htm.

University of Toronto. G8 Information Centre. "G7 Statement." June 18, 1999, and "Communique, Köln 1999." June 20, 1999. G8 Summit, Köln, Germany, June 18–20, 1999. Internet: http://www.library.utoronto.ca/g7/99koln. Accessed July 5, 1999.

Zawels, E. A., et al. *Managing Arms in Peace Processes: The Issues.* Geneva: United Nations Institute for Disarmament Research, 1996.

Appendices

Beaver, Paul, ed. "Global Update: Flashpoints and Conflicts." Special report No. 2, *Jane's Defence Weekly,* 1994.

Berman, Robert P. *Soviet Air Power in Transition.* Washington, D.C.: The Brookings Institution, 1978.

Blalock, Hubert M., Jr. *Social Statistics.* 2nd ed. New York: McGraw-Hill Book Co., 1972.

Brzoska, Michael, and Thomas Ohlson. *Arms Transfers to the Third World, 1971–85.* New York: Oxford University Press, 1987.

Cohen, Eliot. "Toward Better Net Assessment: Rethinking the European Conventional Balance." *International Security* 13 (Summer 1988).

Cook, Thomas D., and Donald T. Campbell. *Quasi-Experimentation: Design and Analysis Issues for Field Settings.* Boston: Houghton Mifflin Co., 1979.

Dunnigan, James F., and William Martel. *How to Stop a War: The Lessons of Two Hundred Years of War and Peace.* New York: Doubleday, 1987.

Epstein, Joshua M. "Security Regimes and Arms Race Dynamics." Presentation to the Defense and Arms Control Studies Seminar, Massachusetts Institute of Technology, Cambridge, Mass., December 2, 1992.

———. "The 3:1 Rule, the Adaptive Dynamic Model, and the Future of Security Studies." *International Security* 13 (Spring 1989): 90–127.

———, Kim R. Holmes, John J. Mearsheimer, and Barry R. Posen. "Policy Focus: The European Conventional Balance." *International Security* 12 (Spring 1988): 152–202.

Etcheson, Craig. *Arms Race Theory: Strategy and Structure of Behavior.* New York: Greenwood Press, 1989.

Fox, John. *Applied Regression Analysis, Linear Models, and Related Methods.* Thousand Oaks, Calif.: SAGE Publications, 1997.

Gochman, Charles S., and Zeev Maoz. "Militarized Interstate Disputes, 1816–76." *Journal of Conflict Resolution* 28 (1984): 585–615.

Goldstein, Joshua. "A Conflict-Cooperation Scale for WEIS Events Data." *Journal of Conflict Resolution* 36 (June 1992): 369–385.

Grant, Neil. *Chronicle of 20th Century Conflict.* New York: Smithmark Publishers, Inc., 1993.

Grimmett, Richard F. *Conventional Arms Transfers to Developing Nations, 1989–96.* CRS Report 97–778F. Washington, D.C.: Congressional Reference Service, 1997.

———. *Conventional Arms Transfers to the Third World, 1986–93.* CRS report 94 612F. Washington, D.C.: Congressional Reference Service, 1994.

Gurr, Ted Robert. *Why Minorities Rebel: A Global View of Ethnopolitical Conflicts.* Washington, D.C.: U.S. Institute of Peace, 1993.

Heise, David R. *Causal Analysis.* New York: John Wiley & Sons, 1975.

Hilton, Gordon. *Intermediate Politometrics.* New York: Columbia University Press, 1976.

Mearsheimer, John J. "Assessing the Conventional Balance: The 3:1 Rule and Its Critics." *International Security* 13 (Spring 1989): 54–89.

Richelson, Jeffrey T., Lewis W. Snider, and Abraham R. Wagner. *Arms Transfer Control Criteria: Quantitative Measures and Analytical Approach.* Report TR-6001/78 for the U.S. Arms Control and Disarmament Agency. Marina del Ray, Calif.: Analytical Assessments Corporation, July 1978.

Saperstein, Alvin M. "Alliance Building versus Independent Action: A Non-linear Modeling Approach to Comparative International Stability." *Journal of Conflict Resolution* 36 (September 1992): 518–45.

SAS Institute. *JMP Statistics and Graphics Guide,* vol. 3.1. Cary, N.C.: SAS Institute, Inc., 1995.

Sherwin, Ronald G., and Edward J. Laurance. "Arms Transfers and Military Capability: Measuring and Evaluating Conventional Arms Transfers." *International Studies Quarterly* 23 (September 1979): 360–89.

Shi, Yuhang. "What to Gain from Technical Sophistication?" *PS: Political Science and Politics* (September 1995): 505–506.

Snider, Lewis W. *Arabesque: Untangling the Patterns of Supply of Conventional Arms to Israel and the Arab States.* Denver Monograph Series in World Affairs, vol. 15, book one. Denver: University of Denver, 1977.

Stockholm International Peace Research Institute. *SIPRI Yearbook 1996, Armaments, Disarmament and International Security.* Oxford: Oxford University Press, 1996.

U.S. Arms Control and Disarmament Agency. *World Military Expenditures and Arms Transfers 1997.* Washington, D.C.: ACDA, 1998. Internet: http://www.acda.gov/wmeat97/wmeat97.htm. Accessed January 2, 1999.

U.S. Army Concepts Analysis Agency, War Gaming Directorate. *Weapon Effectiveness Indices/Weighted Unit Values III (WEI/WUV III).* CAA-SR-79-12. Bethesda, Md.: CAA, November 1979. Declassified December 31, 1986.

U.S. Congressional Budget Office. *A Look at Tomorrow's Tactical Air Forces.* Washington, D.C.: U.S. GPO, January 1997.

U.S. Department of Defense. Office of the Undersecretary of Defense (Comptroller). *National Defense Budget Estimates for FY 1999.* March 1998.

U.S. Navy. Office of Naval Intelligence. *Worldwide Challenges to Naval Strike Warfare.* Washington, D.C.: ONI, February 1997.

Wayman, Frank Whelon, and J. David Singer. "Evolution and Directions for Improvement in the Correlates of War Project Methodologies." In *Measuring the Correlates of War.* Edited by J. David Singer and Paul F Diehl. Ann Arbor: University of Michigan Press, 1990.

Wolfson, Murray, Anil Puri, and Mario Martelli. "The Non-linear Dynamics of International Conflict." *Journal of Conflict Resolution* 36 (March 1992): 119–49.

Index

Printed in the United States
By Bookmasters